AF352881

Recent Progress in Urbanisation Dynamics Research

Recent Progress in Urbanisation Dynamics Research

Editors

Iwona Cieślak
Andrzej Biłozor
Luca Salvati

MDPI • Basel • Beijing • Wuhan • Barcelona • Belgrade • Manchester • Tokyo • Cluj • Tianjin

Editors

Iwona Cieślak
Department of Socio-Economic
Geography
University of Warmia and
Mazury in Olsztyn
Olsztyn
Poland

Andrzej Biłozor
Department of Socio-Economic
Geography
University of Warmia and
Mazury in Olsztyn
Olsztyn
Poland

Luca Salvati
Department of Economics and
Law
University of Macerata
Macerata
Italy

Editorial Office
MDPI
St. Alban-Anlage 66
4052 Basel, Switzerland

This is a reprint of articles from the Special Issue published online in the open access journal *Land* (ISSN 2073-445X) (available at: www.mdpi.com/journal/land/special_issues/Urbanisation_ Dynamics).

For citation purposes, cite each article independently as indicated on the article page online and as indicated below:

LastName, A.A.; LastName, B.B.; LastName, C.C. Article Title. *Journal Name* **Year**, *Volume Number*, Page Range.

ISBN 978-3-0365-3023-9 (Hbk)
ISBN 978-3-0365-3022-2 (PDF)

Contents

About the Editors

Iwona Cieślak

Iwona Cieślak is a professor at the Department of Socio-Economic Geography at the Institute of Spatial Management and Geography, Faculty of Geoengineering, University of Warmia and Mazury in Olsztyn (Poland). Her research interests are mainly focused on the fundamental aspects of spatial analysis; GIS; and urbanism. These activities concern the development of methods for investigating the phenomena of city planning and sustainable development as well as evaluation of spatial conflicts. She is the author of more than 70 scientific papers on the subject. She leads and participates in scientific projects. She has also been an academic teacher for many years, teaching in this field.

Andrzej Biłozor

Andrzej Biłozor is a professor at the Department of Socio-Economic Geography at the Institute of Spatial Management and Geography, Faculty of Geoengineering, University of Warmia and Mazury in Olsztyn (Poland). His research interests are mainly focused on spatial planning; urban analysis; optimization of urban space; spatiotemporal analysis; fuzzy set theory; changes in land use and land cover. These activities concern the development of methods for investigating the phenomena of city planning and sustainable development, evaluation of spatial conflicts and optimization of spatial processes. He is the author of the 76 scientific papers on the subject. He leads and participates in scientific projects. He has also been an academic teacher for many years, teaching in this field.

Luca Salvati

Luca Salvati, M.Sc. Statistics and Economics, Sapienza University of Rome, PhD at the Faculty of Economics of the same university, is a researcher (Class B, holding a full professor habilitation in economic statistics) at the Department of Economics and Law of the University of Macerata. Since 2001, he has been staff researcher at the National Council for Research in Agriculture and the Analysis of Agricultural Economics (CREA). From 2006 to 2010, he was permanent researcher at the National Institute of Statistics (ISTAT). He has carried out continuous research periods abroad (2012-2014) at various public universities and research institutions in France (Toulouse), Spain (Barcelona) and Greece (Athens). He is an expert in the topics of official statistics, spatial statistics, Geographic Information Systems and Remote sensing applied to socioeconomic and environmental issues. He has conducted studies on topics of economic statistics, urban economics, regional demography and sustainable development, using exploratory multivariate statistics and geographic information systems for decision support, more recently developing studies on the monetary evaluation of natural resources. He has held courses in Economic Statistics (University of Rome Tor Vergata), Basic Mathematics (University of Rome La Sapienza), Multivariate Statistics (University of Roma Tre), Regional Economics (University of Camerino, University of Eastern Piedmont), and Strategic Evaluation of environmental impact (University of Roma Tre). He is a member of the teaching staff of the doctorate of for economics and finance, curriculum of Economic Statistics (University of Rome La Sapienza). He has supervised master's and doctoral theses, both in Italian universities and abroad. He has published more than 30 printed books and over 500 scientific publications.

Preface to "Recent Progress in Urbanisation Dynamics Research"

This book is devoted to in depth analysis of past, present and future urbanization processes all over the world. New methods and assessment techniques were also investigated extensively. The development of science and technology has provided many new tools for the observation of urbanization processes and the formulation of conclusions about this phenomenon. Recent Progress in Urbanization Dynamics Research contains a broad range of papers presenting the complexity of the urbanization process, and varied approaches focusing on different aspects of urban development in relation to land development. This issue of *Land* is dedicated to the dynamics of urbanization processes, rapid changes in urbanization and scattered development of urban areas, zoning, sustainable development, urban sprawl, spatial conflicts, the real estate market, transport accessibility, spatial analyses involving GIS tools, and map-making methods.

Iwona Cieślak, Andrzej Biłozor, Luca Salvati
Editors

 *land*

Editorial

Land as a Basis for Recent Progress in the Study of Urbanization Dynamics

Iwona Cieślak [1,*], Andrzej Biłozor [1] and Luca Salvati [2]

[1] Department of Socio-Economic Geography, Faculty of Geoengineering, University of Warmia and Mazury in Olsztyn, Prawocheńskiego 15, 10-720 Olsztyn, Poland; abilozor@uwm.edu.pl

[2] Department of Economics and Law, University of Macerata, Via Armaroli 43, 62100 Macerata, Italy; luca.salvati@crea.gov.it

* Correspondence: isidor@uwm.edu.pl

Citation: Cieślak, I.; Biłozor, A.; Salvati, L. Land as a Basis for Recent Progress in the Study of Urbanization Dynamics. *Land* **2022**, *11*, 118. https://doi.org/10.3390/land11010118

Received: 4 January 2022
Accepted: 10 January 2022
Published: 12 January 2022

Publisher's Note: MDPI stays neutral with regard to jurisdictional claims in published maps and institutional affiliations.

Urbanization is one of the most dynamic processes occurring on the Earth. The urban population continues to increase due to the benefits of the urban lifestyle and the economic and social aspects of urbanization. This process became especially intensified in the second half of the 20th century, when the number of urban dwellers rose from 751 million to 4.2 billion in 2018. According to international forecasts, by 2050, the world's urban population will grow by another 2.5 billion, i.e., by 68%. However, the rapid rate of urbanization leads to serious environmental, spatial and socio-economic problems such as soil degradation, loss of urban ecosystem services, urban heat islands and air pollution. Health problems, urban poverty, rising crime and overcrowding are also becoming very acute. However, it should be noted that urbanization is perceived positively in many dimensions. Planned and sustainable urban development is the basis of a properly functioning economy, promoting a rise in living standards through higher quality of education and improved access to healthcare, culture and art. Researchers thus devote much attention to this process and monitor it constantly. Until recently, statistical data, including population or investment growth in administrative units, were the primary source of information for studying urbanization processes in the global or local dimension. This process could not be followed with a frequency suited to the rate of its progress based on data on spatial transformation. However, enormous databases containing a broad spectrum of spatial information, as well as new and rapid tools for processing spatial data have been available to researchers for more than 20 years, not only locally, but also on the international and global scale. The scope and quality of research on urbanization processes based on land data are also increasing. According to estimates, in 1990–2000, the average annual increase in built-up land was approximately 3.6% in developing countries and only 2.9% in industrialized countries. The most dynamic changes in land use towards urbanization have been observed in East Asia, including the Pacific region, and South-East Asia, where the increase in urbanized land reached 7.2% and 6.4%, respectively. In Europe, the annual increase in urban area does not exceed 2% in the most rapidly developing areas and is close to zero in rural areas.

Land use (LU) studies have become easier thanks to land cover (LC) observations. Land cover analyses are becoming increasingly advanced and provide knowledge on all dimensions of urbanization at various scales of reference. Most of the world's urban areas have experienced significant changes in land cover over the years. Land data, generally referred to as Spatial Information System (SIS) data, are increasingly used in research. The SIS provides information on LU/LC and currently covers various areas of interest, including administrative boundaries, transport and hydrographic networks, terrain, settlement and anthropogenic structures. These data are expanded to include environmental and social dimensions that describe not only the changes taking place in the environment, but also in the quality of life. SIS databases are being created at all levels of detail. Various types of national spatial information systems provide access to key data. In this case, spatial data refer to location (coordinates in the adopted reference system), the geometric properties and spatial relationships of objects that are identified in relation to the Earth and can be used in the analyses of urbanization processes. Spatial information can be presented cartographically and analyzed to identify urbanization patterns. Based on this reasoning, the Earth is one of the primary sources of information on urbanization processes.

Urbanization causes deep changes in space and gradually modifies the land use structure. There is an enormous need for monitoring urban space as well as changes that occur in areas directly subjected to urbanization pressure. The development of science and technology has provided many new tools for the observation of urbanization processes and the formulation of conclusions about this phenomenon. *Recent Progress in Urbanization Dynamics Research* contains a broad range of papers presenting the complexity of the urbanization process, and varied approaches focusing on different aspects of urban development in relation to land development emphasize the need for further research on urbanization. Different chapters are dedicated to the dynamics of urbanization processes, rapid changes in urbanization and scattered development of urban areas, zoning, sustainable development, urban sprawl, spatial conflicts, the real estate market, transport accessibility, spatial analyses involving GIS tools and map-making methods. Specifically, the authors focus on:

- The latest research results and the most interesting methods and information sources used in studies of urbanization dynamics. Numerous examples and potential uses of spatial data for the analyses of urbanization processes are also presented [1].
- A geospatial analysis of urban development patterns aiming to identify the characteristic features of urban expansion with the use of the Geographic Information System (GIS) and remote detection techniques (Landsat images) [2].
- Monitoring and modeling urban development patterns and trends with the use of the Urban Sprawl Matrix and CA-Markov Model, analyzing changes in the urban landscape and predicting LULC changes [3].
- Spatial and temporal characterization of urban expansion with the use of a concentric-ring and grid-based analysis. These approaches were combined to describe urban expansion processes in five large Latin American cities in 2000–2014 [4].
- An analysis of temporal and spatial patterns and urban expansion factors in the Texas Triangle Megaregion with the use of land cover data and the imperviousness of the National Land Cover Database transport data for 2001–2016, transport data from the Texas Department of Transportation (TxDOT) and auxiliary socio-demographic data [5].
- Urban Innovation Efficiency Improvement in the Guangdong–Hong Kong–Macao Greater Bay Area from the Perspective of Innovation Chains, presenting a three-step "knowledge innovation–scientific research innovation–product innovation" model that proposes effective solutions and suggestions regarding the promotion and optimization of innovations based on cooperation in the Greater Bay Area, in view of various factors, industrial structure and urban agglomeration innovation networks [6].
- Analysis of susceptibility to land-use conflict aimed at developing a procedure for evaluating the risk of conflict in land management based on the specific attributes of land with the use of databases, GIS tools and statistical data-processing methods [7].
- Analysis and evaluation of the spatial structure of Cittaslow cities based on examples of several regions in central Italy and north-eastern Poland, including an assessment of the urban layout, architectural features and the composition of urban and architectural factors that are largely responsible for the perception of multidimensional spaces [8].
- Spatial differences in land-use change indicators per capita in Rome in central Italy (in 1949, 1974, 1999, 2008 and 2016) with the aim of quantifying the discrepancies between urban expansion and population growth. The empirical results of this study encourage a discussion on the (presumed) unsustainability of the current urban expansion compared to former settlement structures, which can be attributed to land fragmentation, loss of relict habitats and traditional croplands on the outskirts of the city [9].

- Spatial processes accompanying the loss and expansion of forests in the Rome metropolitan area with the use of diachronic maps from 1936-2018, representing different socioeconomic conditions [10].
- Promoting the high-quality development of areas subjected to urbanization pressure, improving the effectiveness of green innovations in urban agglomerations, analyzing the impact of network structure characteristics (such as network scale and network structure hole) on green innovations in urban agglomerations with the use of the unexpected output SBM model to measure green innovation efficiency in eight prefecture-level cities in the Great Changsha-Zhuzhou-Xiangtan City Group, and to analyze the contributing factors using the panel Tobit model [11].
- Improved methodology for selecting the optimum solutions for sustainable spatial and traffic planning by combining quantifiable results of traffic microsimulation with the multi-criteria optimization method [12].
- Transport accessibility in the suburban zone and its impact on the local real estate market, in view of commuting time and demographic changes. The spatial differentiation in population distribution was analyzed with the use of the Gini index and geostatistical interpolation techniques [13].
- The role and significance of a pedestrian freeway overpass in Trabzon. Safety issues and the motivation for using the overpass among respondents from different age groups were described in detail [14].

All of the discussed topics relate to urbanization processes that are observed on a daily basis as well as the methods and techniques for monitoring these phenomena. The relationship between urbanization and space seems obvious, but it is also interesting and multi-layered. Scientific and technological advancement has generated numerous tools for observing urbanization processes, conducting analyses and formulating conclusions about urbanization. *Recent Progress in Urbanization Dynamics Research* presents the latest research trends by relying on the extensive body of knowledge relating to urban geography. Special emphasis was placed on the global effects of urbanization, and this multidisciplinary phenomenon was analyzed with the use of satellite and photogrammetric observation techniques and GIS tools.

Author Contributions: Conceptualization, I.C. and A.B.; writing—original draft preparation, I.C. and A.B.; writing—review and editing, I.C., A.B. and L.S. All authors have read and agreed to the published version of the manuscript.

Acknowledgments: We thank all the reviewers for their feedback on earlier versions of the manuscripts in this Special Issue.

Conflicts of Interest: The authors declare no conflict of interest.

References

1. Biłozor, A.; Cieślak, I. Review of Experience in Recent Studies on the Dynamics of Land Urbanisation. *Land* **2021**, *10*, 1117. [CrossRef]
2. Seevarethnam, M.; Rusli, N.; Ling, G.H.T.; Said, I. A Geo-Spatial Analysis for Characterising Urban Sprawl Patterns in the Batticaloa Municipal Council, Sri Lanka. *Land* **2021**, *10*, 636. [CrossRef]
3. Baqa, M.F.; Chen, F.; Lu, L.; Qureshi, S.; Tariq, A.; Wang, S.; Jing, L.; Hamza, S.; Li, Q. Monitoring and Modeling the Patterns and Trends of Urban Growth Using Urban Sprawl Matrix and CA-Markov Model: A Case Study of Karachi, Pakistan. *Land* **2021**, *10*, 700. [CrossRef]
4. Wu, S.; Sumari, N.S.; Dong, T.; Xu, G.; Liu, Y. Characterizing Urban Expansion Combining Concentric-Ring and Grid-Based Analysis for Latin American Cities. *Land* **2021**, *10*, 444. [CrossRef]
5. Guo, J.; Zhang, M. Exploring the Patterns and Drivers of Urban Expansion in the Texas Triangle Megaregion. *Land* **2021**, *10*, 1244. [CrossRef]
6. Ye, W.; Hu, Y.; Chen, L. Urban Innovation Efficiency Improvement in the Guangdong–Hong Kong–Macao Greater Bay Area from the Perspective of Innovation Chains. *Land* **2021**, *10*, 1164. [CrossRef]
7. Cieślak, I.; Biłozor, A. An Analysis of an Area's Vulnerability to the Emergence of Land-Use Conflicts. *Land* **2021**, *10*, 1173. [CrossRef]

8. Zagroba, M.; Pawlewicz, K.; Senetra, A. Analysis and Evaluation of the Spatial Structure of Cittaslow Towns on the Example of Selected Regions in Central Italy and North-Eastern Poland. *Land* **2021**, *10*, 780. [CrossRef]
9. Bianchini, L.; Egidi, G.; Alhuseen, A.; Sateriano, A.; Cividino, S.; Clemente, M.; Imbrenda, V. Toward a Dualistic Growth? Population Increase and Land-Use Change in Rome, Italy. *Land* **2021**, *10*, 749. [CrossRef]
10. Bianchini, L.; Salvia, R.; Quaranta, G.; Egidi, G.; Salvati, L.; Marucci, A. Forest Transition and Metropolitan Transformations in Developed Countries: Interpreting Apparent and Latent Dynamics with Local Regression Models. *Land* **2022**, *11*, 12. [CrossRef]
11. Wang, L.; Ye, W.; Chen, L. Research on Green Innovation of the Great Changsha-Zhuzhou-Xiangtan City Group Based on Network. *Land* **2021**, *10*, 1198. [CrossRef]
12. Ištoka Otković, I.; Karleuša, B.; Deluka-Tibljaš, A.; Šurdonja, S.; Marušić, M. Combining Traffic Microsimulation Modeling and Multi-Criteria Analysis for Sustainable Spatial-Traffic Planning. *Land* **2021**, *10*, 666. [CrossRef]
13. Szczepańska, A. Transport Accessibility in a Suburban Zone and Its Influence on the Local Real Estate Market: A Case Study of the Olsztyn Functional Urban Area (Poland). *Land* **2021**, *10*, 465. [CrossRef]
14. Hełdak, M.; Kurt Konakoglu, S.S.; Kurdoglu, B.C.; Goksal, H.; Przybyła, B.; Kazak, J.K. The Role and Importance of a Footbridge Suspended over a Highway in the Opinion of Its Users—Trabzon (Turkey). *Land* **2021**, *10*, 340. [CrossRef]

Review

Review of Experience in Recent Studies on the Dynamics of Land Urbanisation

Andrzej Biłozor and Iwona Cieślak *

Department of Socio-Economic Geography, Faculty of Geoengineering, University of Warmia and Mazury in Olsztyn, Prawocheńskiego 15, 10-720 Olsztyn, Poland; abilozor@uwm.edu.pl
* Correspondence: isidor@uwm.edu.pl

Abstract: Urbanisation rapidly accelerated in the 20th century. Along with the increasing dynamics of this phenomenon, the desire to know its origins and its course as well as to anticipate its effects is also growing. Investigations into the mechanisms governing urbanisation have become the subject of numerous studies and research projects. In addition, there has been a rapid increase in the number of tools and methods used to track and measure this phenomenon. However, new methods are still being sought to identify changes in space caused by urbanisation. Some of the indicators of urbanisation processes taking place include quantitative, qualitative and structural changes in land use, occurring at a certain time and place. These processes, related to human activity at a given time and in a given area, are determined by spatial diffusion, usually spreading from the city center towards the peripheral zones. Changes in land use involve the transition from less intensive to more intensive forms of land use. The constant effort to acquire new land for development, the search for alternative solutions for the location of investments and the need to determine the correct direction of development generates the need to constantly apply newer methods in the study of the dynamics of urbanisation processes. This paper presents an overview of recent studies and the most interesting—in the authors' opinion—methods used in research into the dynamics of urbanisation processes. The main objective of the authors was to produce a compendium to guide the reader through the wide range of topics and to provide inspiration for their own research.

Keywords: urbanisation; land use; dynamics of urbanisation; methods for mapping

check for updates

Citation: Biłozor, A.; Cieślak, I. Review of Experience in Recent Studies on the Dynamics of Land Urbanisation. *Land* **2021**, *10*, 1117. https://doi.org/10.3390/land10111117

Academic Editor: Dagmar Haase

Received: 15 September 2021
Accepted: 18 October 2021
Published: 21 October 2021

Publisher's Note: MDPI stays neutral with regard to jurisdictional claims in published maps and institutional affiliations.

1. Introduction

Urbanisation is one of the most visible manifestations of global change occurring for anthropogenic reasons [1]. It can be defined as a process of population concentration, mainly in urban areas, which also determines the growth of the urban population and its share in the population of a given area. Urbanisation is described as a cultural and civilizational process reflected in the development of cities, the increase in their number, the expansion of their area, the progressive concentration of the population in their immediate vicinity, the expansion of non-agricultural sources of income, the acceptance and assimilation of urban standards, customs, etc. It is a complex and diversified process, taking place with differing intensity and speed and with different effects in various countries and regions of the world [2].

Urbanisation is both a process and a state [3]. As a process, it involves the changes in human activities and socio-economic behavior described above, etc. As a state, i.e., the result of a process, urbanisation most often refers to the number and size of urban settlement units, the proportion of the urban population in relation to the total population of the country or region (urbanisation coefficient), the concentration of the population in large cities or its dispersion, and the area occupied by city buildings [4]. These two dimensions are interrelated and are dependent on each other. It should also be noted that urbanisation is a global phenomenon, which is irreversible and inherent within human development. It is associated with the scientific and technological revolution, with the concentration of productive forces and forms of social relations, with the spread of the urban style of life, with changes in social relations and ties, with the transformation of the rural population into diversified non-agricultural groups and with the modernization of the entire settlement network [5,6].

Nowadays, urbanisation is approached as a multidimensional complex of economic, social, demographic, cultural and spatial events and phenomena leading to the growth of cities and an increase in the share of the urban population, the population concentration in spatially expanding urban areas, the concentration of economic and administrative activities generating an increase in the importance of cities, shaping specific cultural patterns of urban lifestyle and specific arrangements of landscape and architecture [7].

The study of urbanisation processes can be based on each of these dimensions. The demographic dimension of urbanisation is related to the increase in the number of city residents, caused both by rural population migration to cities and by natural growth in cities and the creation of new settlement units. With regard to rural areas, demographic urbanisation, especially in the suburban zones of highly developed countries, is related to the population outflow from cities to the countryside. The social dimension of urbanisation is expressed in the spread and deepening of the "urban lifestyle", i.e., in attitudes, skills, behavioral patterns and even personality traits that are characteristic of metropolitan communities, both in the city area and in the countryside [8–10]. The economic dimension of urbanisation involves changes in occupational structure and employment. The occupational diversification of the population is becoming more pronounced in the city. The number of people employed in various services is growing at the expense of agriculture and crafts. The proportion of employment in traditional agricultural jobs is changing in rural areas in favor of non-farm jobs. All of the above dimensions of urbanisation are linked by spatial changes in the appearance and organization of the surrounding areas. Therefore, the spatial dimension of this phenomenon is one of the most interesting topics for research into the dynamics of urbanisation, since it provides a basis for inferring the causes and effects of the aforementioned aspects. The spatial dimension of urbanisation applies to both urban and rural areas. Spatial urbanisation is understood as the expansion of the urban landscape [11–15] and includes an increase in the urban investment area as well as the saturation of villages with urban-like infrastructure and buildings. In rural areas, spatial changes are associated with transformations in land use. The area of land used for farming is decreasing in favor of other, non-agricultural forms of use. Changes are taking place in the physiognomy of the buildings and the village morphology, as well as in the modernization of the technical infrastructure.

Current research primarily focuses on demographic and spatial urbanisation. It should be stressed that these are essentially just objectively separated manifestations of the same phenomenon. However, no spatial effects of urbanisation exist without demographic changes. The anthropogenic background of this phenomenon makes the demographic aspect an overriding factor in understanding the causes of spatial change. In practice, demographic urbanisation is measured and expressed, first of all, in the form of census-based statistics, the purpose of which is (at least in principle) is to distinguish the inhabitants of cities and towns from those living in rural areas [16]. The demographic dimension of urbanisation is related to the increase in the proportion of the urban population. Nevertheless, quantitative changes are also accompanied by qualitative ones. Changes are taking place in the gender and age structures of the population, the level of population growth, the balance of migration, the occupational structure of the population, the size of the family and its socio-professional status, and the size of the household, etc. [17] Accordingly, in recent studies, demographic urbanisation is examined in a much broader context, taking into account all the transformations occurring in demographic structures and processes, not only in quantitative but also in qualitative terms [4,18,19].

Spatial urbanisation is most often reflected in the rapid growth of cities, which has exerted pressure on land, local resources and especially on rural areas in their immediate vicinity [20,21]. The concentration of human activity in specific locations and regions favors the development of large cities and various forms of urban settlement [22]. Urban growth and transformation are driven by the forces of attraction between specific locations. This process of spatial diffusion results from interactions between multiple factors and contributes to the development of new spatial patterns [23]. This is primarily due to the specific topography of the city surroundings, the existing road network, the land ownership structure and the pace of its development. These changes are catalyzed by the development of transport networks, as well as the advantages arising from the spatial agglomeration of resources, including human resources. Urbanisation is viewed as a process of organic growth, starting in the city center and leading to the spontaneous formation of 'mini' urban areas, especially throughout the transport network [24]. The majority of these processes are very dynamic and cause rapid changes in the structure and organization of land use [23]. The process of urbanisation, particularly spatial urbanisation, requires the application of continuously evolving research methods and techniques, which render it possible to grasp its essence and pace. The aim of this paper is to present the most interesting, according to the authors, methods used in research on the dynamics of the spatial urbanisation processes. Sources of information employed in the analyses of urbanisation processes are described and a number of examples of the use of spatial data in analyses of urbanisation processes are presented. The solutions described apply to multiple countries and continents and the issue is a common one. Studies addressing this subject usually test new methods for a universal understanding of the urbanisation phenomenon. The added value of this paper is in the fact that it describes ready-to-use solutions in a single publication, in the form of data sources and their use in analysing urbanisation processes. This should facilitate the selection of an appropriate method and its application in research on the dynamics of urbanisation processes for many researchers. The paper presents an overview of recent research and the most interesting, according to the authors, methods used in research on the dynamics of urbanisation processes. The main aim of the article was to produce a compendium that would guide the reader through the wide range of topics and provide inspiration for their own research. The first chapter defines the concept of urbanisation. Chapter 2 presents demographic changes and the spatial dimension of urbanisation from a global perspective. Chapter 3 describes recent trends in research on urbanisation processes. Sources of information used in the analyses of urbanisation processes are described in detail in Section 3.1. Section 3.2 presents examples of applying spatial data in analyses of urbanisation processes. Section 3.3 examines the possibilities of using spatial data in the study of urbanisation processes. Discussion, conclusions and further directions for research are presented in Chapter 4.

2. Demographic Changes and the Spatial Dimension of Urbanisation from a Global Perspective

Demographic processes are the driving force behind urbanisation. According to data published in World Urbanisation Prospects—The 2018 Revision [1]—approximately 55% of the world's population lived in towns and cities in 2013. It is projected that this value will reach 68% by 2050. This is due to changing lifestyles and a gradual reduction in agricultural employment in favor of the endogenous functions of cities [25]. According to the data published by the United Nations, two-thirds of the world's population will live in urban areas (Figure 1).

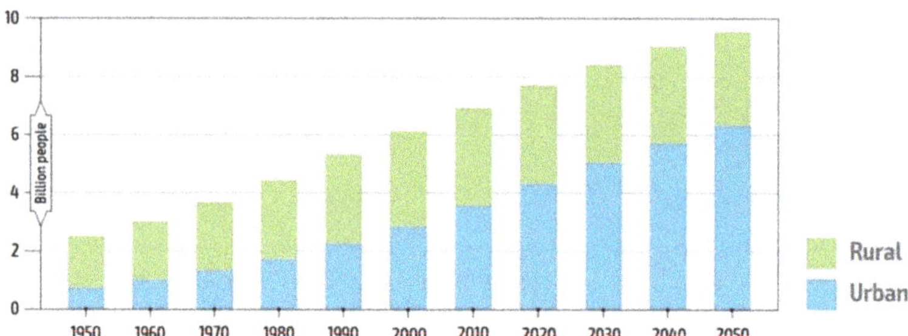

Figure 1. Global urban and rural populations: historical and projected. Source: [1].

In the second half of the 20th century, urbanisation rapidly accelerated. Dynamic growth of cities was observed in the more developed regions of the world, such as Europe and later North America. Yet, in recent decades, it has been Asia that has been urbanising at a tremendous rate—it now has more than half of the world's 40 megacities (with more than 10 million inhabitants). Africa is also gaining urbanisation momentum and now has three megacities in Cairo, Lagos and Kinshasa [26]. People migrate in search of a higher standard of living and the benefits of an urban lifestyle [27–29]. This process is particularly visible in Asia and Africa (90% urban population growth), where the highest rates of population growth are also found [30,31]. The migration of people to towns and cities was particularly high after 1950. From the level of 751 million, the urban population grew to 4.2 billion in 2018. It is predicted that by 2050 this number could rise by another 2.5 billion [32] (Figure 2).

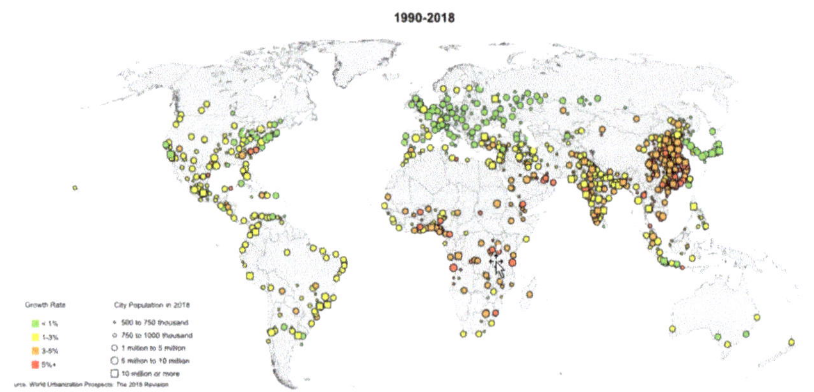

Figure 2. Growth rates of urban agglomerations by size class. Source: https://population.un.org/wup/Maps/ (accessed on 2 August 2021).

Today, the most urbanised areas include North America (82% of the population live in urban areas), Latin America and the Caribbean (81%), Europe (71%) and Oceania (68%). Although these percentages are much lower in Asia (50%) and Africa (43%) in view of the relatively lower numbers of cities and towns, these are the areas with the largest cities (over 10 million) [1]. It is expected that by 2030, the number of cities with over 10 million inhabitants will rise to 43, while in 1970, there were only three such locations. The scale of these changes is shown in Figure 3.

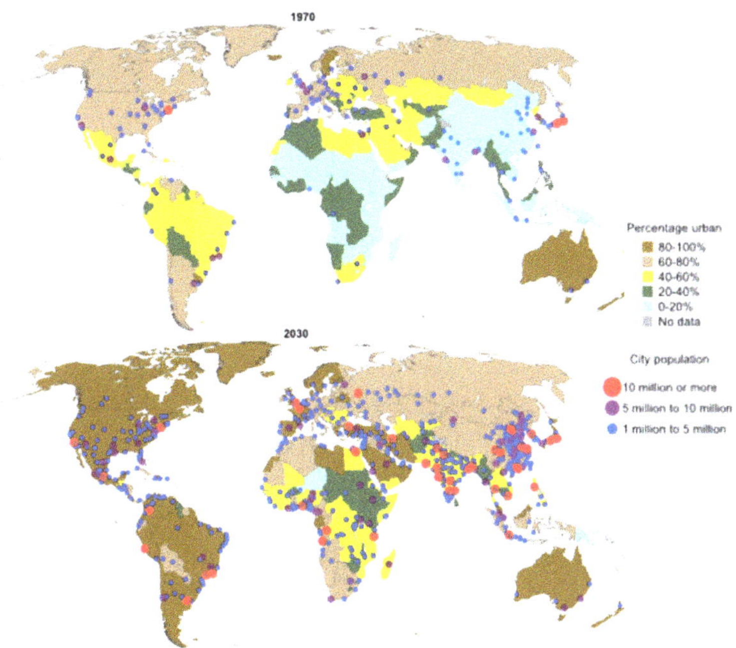

Figure 3. The projected growth of urban population and cities in 2030 compared to 1970. Source: https://population.un.org/wup/Maps/ (accessed on 2 August 2021).

Urban areas are the focal point of economic activity in most areas of the globe. Cities provide driving forces for development, bringing economies of scale, developing markets, creating jobs and encouraging new forms of business. As economies progress from basic activities—agriculture, fishing and mining—to industrial production and then services, the role of cities in the global economy expands with each transition. At present, about 55% of the world's population (4.2 billion people) live in cities, and by 2050, almost 7 out of 10 people worldwide will live in urban areas. According to the figures provided by the World Bank, more than 80% of global GDP is generated in cities. Major urban areas, particularly in developed countries, are economic giants [33]. The 600 largest urban centres, with a fifth of the world's population, generate 60 per cent of global GDP. The 380 largest cities in the top 600 in terms of GDP accounted for 50 per cent of global GDP [34].

Research studies conducted on the phenomenon of urbanisation are usually based on an analysis of statistical data concerning demographic and investment growth in administrative units. They are very significant as they clearly show the scale of the phenomenon. However, these studies often lack a spatial reference indicating the direction and the territorial extent of the changes.

The development of towns and cities is no longer controlled. The main reason behind this is the scale of economic and social change. The increased population mobility, the development of technology and the exchange of information have also resulted in city boundaries becoming indistinct [35]. The constantly growing urban population needs more space and the rate of growth results in a low level of space urbanisation processes [20,36]. This poses a serious threat to the spatial order and ultimately reduces the quality of this space and, consequently, the quality of people's lives [37]. Housing land, which accounts for over 70% of land use in most urban areas, determines the form and density of cities, provides employment and contributes to their growth. Housing policy and the provision of adequate infrastructure to residents for their safety and health is one of the most important challenges for these cities. In addition, it should be remembered that the shortage of space causes price increases in the property markets, which leads to an increase in the number of residents living in extremely poor conditions or slums [22,38]. It is estimated that 881 million urban dwellers live in slums, a number that has increased globally by 28% in the last 24 years. Although the share of urban populations living in slums has declined over the past two decades, the number of slum dwellers continues to rise [24,39].

Forecasts show that the amount of urban land on Earth by 2100 may range from around 1.1 million to 3.6 million km^2 depending on the approach and the application of sustainable development principles—Figure 4.

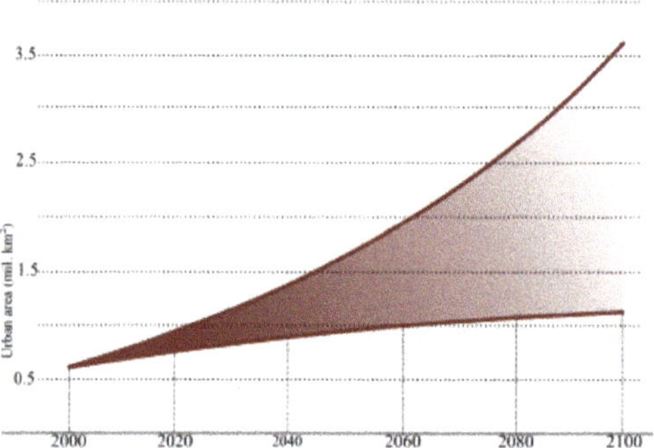

Figure 4. Projected growth of urban areas between 2000 and 2100. Source: https://population.un.org/wup/Maps/ (accessed on 2 August 2021).

Transformations of land use triggered by new human demand affects many aspects of the environment at spatial and temporal scales, including freshwater quality and availability, extreme precipitation and flooding, biodiversity and habitat loss, and global warming [40]. Although cities around the world are the driving forces for economic value creation and income generation, fulfilling an essential role in many aspects of society, the spatial changes they induce and do not control can pose a huge threat to human beings.

Uncontrolled urban growth is clearly visible in the surroundings of metropolitan cities and often results from insufficient space that can be allocated for investment purposes within the city to satisfy the needs of an economically and demographically expanding metropolis. The spatial dimension of urbanisation applies to both urban and rural areas, which is reflected in the amount of space used in a characteristic urban manner. Rural areas, in particular, are experiencing a constant transformation, which is caused by the interest in these areas of urban dwellers. These processes are usually very dynamic and produce rapid changes, based mainly on the principle of reorganisation of the structure of space, and manifest themselves as advanced suburbanisation and exurbanisation [35]. In the literature, this phenomenon has become known as 'urban sprawl', denoting the process of spreading and the enlargement of big cities [41,42]. This phenomenon is seen as a process resulting from socio-economic changes due to spontaneous and disorderly urban expansion [43]. For this reason, it is extremely important to monitor the direction, the level and the pace of spatial urbanisation. Urban sprawl is often measured by the population density gradient (as the percentage decrease in density over increasing distance from the city centre) [44] or by the spatial size of the urban area [45]. The analysis of these values shows a disproportionate increase in urban space to the increase in urban population [46].

Suburbanisation is of particular importance in a period of rapid urban growth. Areas under direct urbanisation pressure are described in the literature as the urban–rural transition zone [47,48], the urban–rural continuum [49–53] or the suburban zone [54–57]. Studies of these areas focus not only on the progressive integration of new land into the urban sphere of influence, but also on changing land use, the extent of infrastructure, access to services and markets, and exposure to urban production processes and environmental pollution [58]. Changes in land use in areas are referred to as a rural–urban fringe [59–64], in so-called "green belts" surrounding the city [65–67] and areas referred to as "urban villages" [68] are also explored. Areas under urbanisation pressure are also investigated in terms of investment attractiveness [69,70] or real estate market analyses [71–75].

3. Methodology of Proceeding and the Determination of Recent Trends in Research on Urbanisation Processes

In order to describe the investigated area of knowledge, based on an extensive literature review, a number of methods and techniques used to study the dynamics of urbanisation processes were examined. The variety of attributes and the large range of data make the process of analysing spaces under urbanisation pressure both complex and time-consuming. The key to the choice of the spatial data presented and the methods applied in the research on the dynamics of urbanisation processes was their availability, the difficulty of classifying certain forms of development, the time of implementation, costs and their spatial and temporal continuity. The main focus of the paper is on the use of photogrammetric and remote sensing sources to study urbanisation changes. Among these, sources were separated according to the type of information and their spatial scope. The examined sources demonstrate land cover changes, which can be divided into four classes: land-use change analyses, environmental change analyses and demographic and economic change analyses. The spatial extent of the data gathered is highly diverse. Its scope can be classified as local, regional, national or global. Taking into account these two aspects, the most interesting examples of applying particular databases created with photogrammetric and remote sensing sources for the purpose of analysing the dynamics of urbanisation processes were identified by a literature review. The presented research results are based on data from 1990–2020. The classification of spatial data sources and the selection procedure of application cases are shown in Figure 5.

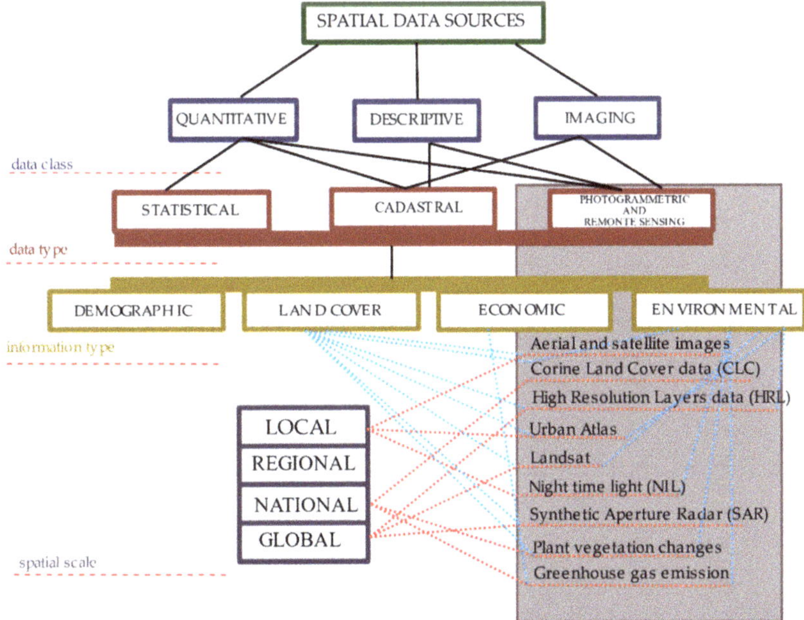

Figure 5. Classification of spatial data sources.

3.1. Sources of Information Used in Analyses of Urbanisation Processes

Urbanisation produces profound changes in space and gradually modifies the structure of land use. There is a great need to monitor urban space, as well as the changes that are taking place in areas subject to the direct pressure of urbanisation. The development of science and technology provides many new tools to observe urbanisation processes and to analyse and draw conclusions regarding this phenomenon. Sources of information data used in analyses of urbanisation processes can be generally divided into statistical, cadastral, photogrammetric and remote sensing.

Statistical data used in analyses of urbanisation processes take into account changes occurring over time in land-use forms, settlement unit boundaries, changes in the territorial division, in the number of towns and cities, as well as in the area, population and structure. Statistical data collected and processed, e.g., by the European Statistical Office (Eurostat) and governmental statistical offices (Statistics Poland), also refer to internal and external population migration processes.

Cadastral data used in an analyses of urbanisation processes are gathered in public registers and cartographic studies. The data in the registers include, among others, information on persons, their rights, legal transactions, obligations, goods, etc. For example, in Poland there are about 280 public registers serving different purposes, e.g., the registration of the actual state of real estate is assigned to the land and building register, while the registration of the legal status of real estate belongs to the scope of the land and mortgage register [76]. Descriptive and spatial attributes of real estate used in an analyses of urbanisation processes include the number of parcels, numerical description of boundaries, surface areas, land use contours and soil classes, designation of the land and mortgage register or sets of documents defining the legal status of the real estate, street names and ordinal numbers of buildings, names of physiographic objects, GESUT (geodetic records of public utilities) data, providing information on ground, aboveground and underground technical infrastructure. Data from cartographic studies also contain information on the condition, distribution and interrelationships of various types of terrain objects and phenomena, their qualitative and quantitative characteristics, as well as their names and descriptions. The analysis of urbanisation processes with the use of digital or analogue cartographic works, such as a topographic map, takes into account surface elements of the landscape, such as residential estates, industrial and agricultural facilities, water and related facilities, vegetation, crops, land, boundaries, landforms, etc. In general, cadastral data can be divided into qualitative (describing the type, equipment, buildings, utilities), quantitative (area, volume) and situational information (location, access).

Photogrammetric and remote sensing data seem to be particularly important in the analysis of urbanisation processes. Photogrammetry (terrestrial and aerial) is used to develop and update various types of maps, orthophotomaps, digital elevation models and spatial information systems. Remote sensing data (data on phenomena and objects acquired without direct contact with them), which are produced at various altitudes and by various sensors, can be divided into active (radars-SAR, scanners-LiDAR) and passive data (cameras or thermal scanners-placed on the board of satellites, aircraft and drones). The sensors provide images in various ranges of electromagnetic radiation. The resulting databases, complemented by spatial databases, which are already common today, provide enormous research material and make it possible to study the phenomenon of urbanisation with greater intensity and in a more up-to-date context.

Photogrammetric and remote sensing (RS) data and GIS tools provide accurate information on land use and land cover changes. Currently, the main sources of data are the Copernicus (program supervised by ESA) and Landsat (program supervised by NASA) satellites. The data acquired and shared from the Copernicus (in particular Sentinel-2 and Sentinel-1) and Landsat satellites provide a high level of detail. The use of a variety of sensors, including optical and radar, facilitate monitoring land cover and support crisis management. Additionally, the photogrammetry and remote-sensing products obtained through the Copernicus and Landsat programs, generating high-resolution images of land cover [77], make it possible to monitor urbanisation processes [78–80]. Monitoring land cover with high-resolution imagery, especially in urban areas, is a key task and has applications in many fields, such as land development planning, urban planning and architecture, ecology and environmental protection, etc. These data are also of fundamental importance for understanding spatial urbanisation processes. With the availability of urbanisation data, photogrammetry and remote sensing techniques are becoming increasingly popular for monitoring changes in land use processes. The Copernicus program services, coordinated by the European Environment Agency, provide information on land cover and land use on a European, national and local scale. The pan-European component includes, among others, the uniform land cover databases (CLC1990, CLC2000, CLC2006, CLC2012 and CLC2018) and High Resolution Layers (HRL).

The CORINE Land Cover database includes all land cover forms occurring on the European continent, leaving no unclassified areas (clc.gios.gov.pl, access date: 1 February 2018). The database contains information on both land cover and land use. In the CLC, land cover forms are hierarchically divided into three levels. The database distinguishes five basic forms of land cover: anthropogenic land (built-up areas, used for housing, services, industry or mines, and municipal green spaces), agricultural land (arable land, permanent crops, meadows and pastures, and wooded and shrubby land used for agriculture), forests and ecosystems (land covered with or partially devoid of forest vegetation), wetlands (inland marshes, peat bogs, salt marshes and mudflats) and water areas (inland waters and sea waters). These forms are further subdivided into levels two and three, which specify the form of land use within the group [81]. The CLC database is a tool for conducting complex spatial analyses based on diverse land-use types. One of the greatest advantages of the CLC dataset is that it is regularly updated, which makes this resource particularly useful for analysing the rate of changes in land use and developing forecasts. HRLs contain detailed information on characteristic land cover forms: impervious areas, wooded areas, areas with grass cover, water bodies and wetlands. These layers are complementary to the Corine Land Cover databases.

The local component also offers Urban Atlas, Riparian Zones and Natura 2000. The Urban Atlas data represents the functional zones of urban areas and contains detailed data on land cover (land use) compiled for the most populous European cities (for most cities with more than 50,000 inhabitants). The classification recognises 17 urban classes (including five classes of mixed-density development) and ten other classes associated with other land cover forms. The Urban Atlas project currently comprises nine products, including land cover data from 2006, 2012, 2018 and Urban Atlas Change 2006–2012 and 2012–2018. Riparian Zones data concern land cover and land use in areas along rivers, i.e., riparian areas. The primary objective is the need to monitor biodiversity at the European level, including as part of the improvement of 'green' and 'blue' infrastructure in the European Union. Natura 2000 data relate to grassland-rich areas and their assessment for conservation efficiency. Coastal Zones data are used to monitor the dynamics of land-use change in coastal zones.

The above-mentioned types of sources of information on urbanisation processes are merged to create Spatial Information Systems (SIP). The data necessary for analyses of urbanisation processes can be found in various databases of this system, such as Euro-Geographics. EuroGeographics facilitates access to official, comparable and verifiable geospatial data from European national authorities responsible for maps, cadaster and land and mortgage registers. EuroGlobalMap is a geodatabase at a 1:1,000,000 accuracy level covering 45 European countries and territories. The thematic scope of the data includes administrative boundaries, transport and hydrographic networks, landforms, settlements and anthropogenic objects. EuroGlobalMap provides an ideal background for a wide range of activities, from network planning, monitoring and analysis to the presentation of environmental policies.

Access to key data is also provided by all types of national spatial information systems. Spatial information refers here to information on the location (coordinates in an assumed reference system), geometric properties, and spatial relationships of objects that are identified in relation to the Earth and that can be used in an analyses of urbansation processes. In Poland, for example, the National Geographic Information System (KSIG) consists of standardised reference databases containing information on objects located on and below the Earth's surface, together with their location, situated in the territory of Poland, as well as procedures and techniques for the systematic collection, updating, processing and disseminating of the data. The most important components of KSIG include vector map level 2 (VMAP2), topographic database (TBD), topographic and base maps, general geographic database (BDO), photogrammetric database, orthophotomaps and the land and building register (EGIB). The most important feature distinguishing the spatial information systems in operation, which is the consequence of the presence of spatial information contained in them, is the possibility of its cartographic presentation and of an analysis allowing answers to be obtained regarding the real world modelled by the system [82].

3.2. Examples of Using Spatial Data in Analyses of Urbanisation Processes

The examples presented below reflect the latest and, in the authors' opinion, the most interesting trends in the research on the dynamics of spatial urbanisation processes.

3.2.1. The Analysis of Urbanisation Processes Using Aerial and Satellite Images

Aerial and satellite imagery is very often used in spatial analyses. The result of processing these images is an orthophotomap, which, unlike a traditional map, presents the state of land cover and land use in the most realistic way. The use of publicly available orthophotomaps enables the identification and location of the urban investment boundary, the degree and rate of urbanisation of the area (Figure 6).

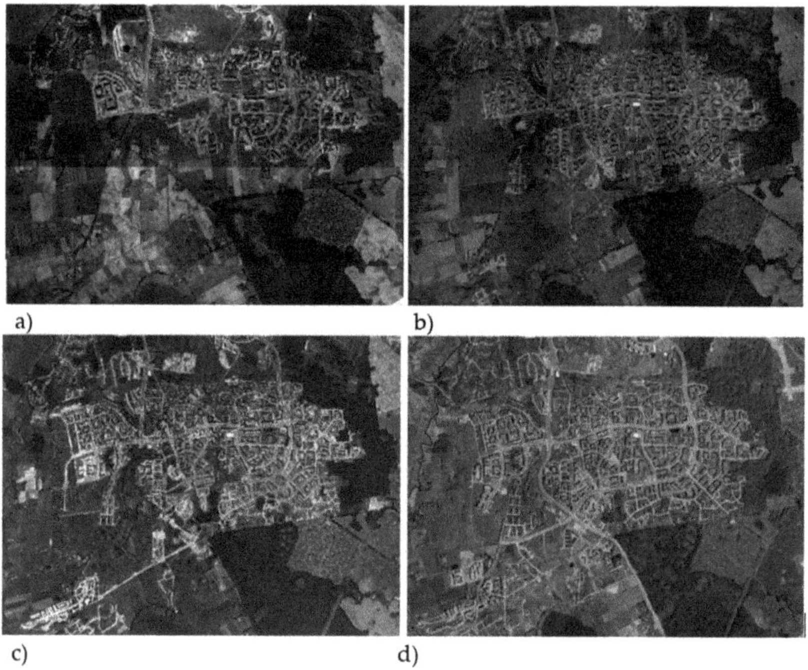

Figure 6. An orthophotomap of the southern part of Olsztyn: (**a**)—1995, (**b**)—2005, (**c**)—2009, (**d**)—2017. Source: https://msipmo.olsztyn.eu/imap/ (accessed on 20 July 2021).

The expert method of visual interpretation of the orthophotomap (photointerpretation) and additional data—e.g., from the land and building register and from the field visits, make it possible to identify current and historical land-use forms (Figure 7).

Figure 7. Inventory of the existing development—an example, 2017. Source: [20].

At present, orthophotomaps are produced with a field resolution of 10 cm, which allows the precise identification of objects. Orthophotomaps are successfully used as separate layers in geographic systems and form the background for spatial planning and the creation of thematic studies.

3.2.2. The Analysis of Urbanisation Processes Using Corine Land Cover Data (CLC) Data

The European Earth observation program known as Copernicus Land Monitoring provides two datasets on land cover characteristics in Corine Land Cover data (CLC) and High Resolution Layers (HRL). The Corine Land Cover initiated in 1985, updated in 2018, provides information for the whole area of the European Union—39 countries. Mainly, the identification of land cover is mapped by a visual interpretation of high-resolution satellite images. The CLC dataset is divided into 44 classes, which describe five categories: artificial surfaces, agricultural areas, forest and seminatural areas, open spaces with little or no vegetation and wetlands and water bodies [23,48,83,84].

A land cover model based on Corine Land Cover data (CLC) is presented in a study conducted by Biłozor A. et al. (Identification and Location of a Transitional Zone between an Urban and a Rural Area Using Fuzzy Set Theory, CLC, and HRL Data). The result of the aggregation of individual CLC data classes defining the range of land uses for the selected area is shown in Figure 8.

The CORINE Land Cover (CLC) databases compiled for the period 1990–2000–2006–2012–2018 provide another reliable source of information about ongoing urbanisation processes. One of the uncontrolled urbanisation phenomena is suburbanisation connected with the emergence of urbanised areas far beyond the city limits in the form of often chaotically located buildings. This is common practice but is undesirable due to the quality of these areas, which tend to be poorly equipped with technical infrastructure and significantly increase the cost of commuting to work and of basic services. An example of using the CLC database, Geographical Information System (GIS) tools and the over-urbanisation (OU) indicator described in the publication: The Use of the CORINE Land Cover (CLC) Database for Analyzing Urban Sprawl [84] is presented in Figure 9.

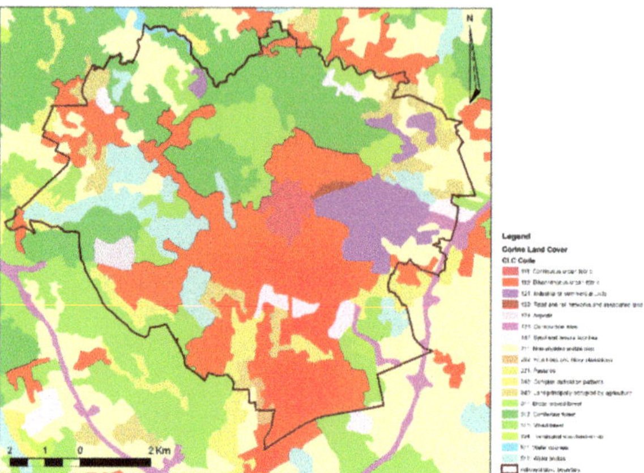

Figure 8. Land cover model in the area under investigation based on the CLC data. Source: [48].

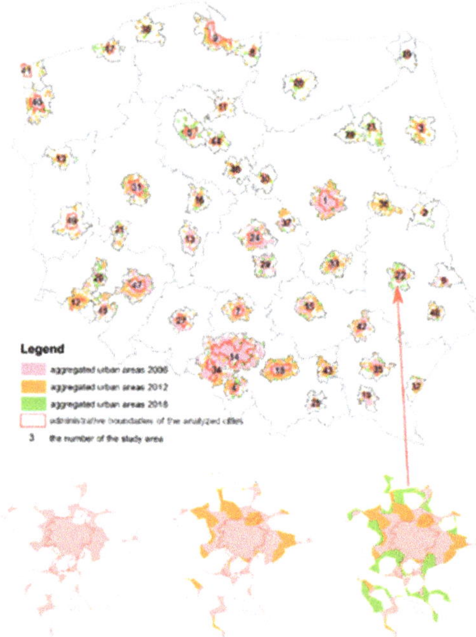

Figure 9. Aggregated urban areas. Source: [84].

3.2.3. The Analysis of Urbanisation Processes Using High Resolution Layers Data (HRL)

Another type of data-High Resolution Layers (HRL) provides more detailed information about land cover than CLC. This dataset was first produced in 2012 from satellite imagery through a combination of automatic processing and interactive rule-based classification. Currently, the main source of data are satellites created in the Sentinel project, in particular, Sentinel-2 and Sentinel-1, which allows the use of different sensors, including optical and radar. In this dataset, which is presented in spatial resolution of 20 × 20 m for 39 countries in the EU, five themes can be identified which correspond with the main categories in CLC. The level of sealed soil (imperviousness degree 1–100%) which is produced using a semi-automated classification based on calibrated NDVI, captures the different types of specific land cover. Those five products capture the spatial distribution of Imperviousness, Forest, Natural Grassland, Wetlands, Permanent Water-bodies, Wetness and Water and Small Woody Features [85–87]. The land cover model based on HRL Imperviousness data is presented in Biłozor A., Czyża Sz., Bajerowski T.-Identification and Location of a Transitional Zone between an Urban and a Rural Area Using Fuzzy Set Theory, CLC, and HRL Data [48] (Figure 10). The conducted analyses included raster reclassification followed by the polygonization of urbanised areas. The measures taken enabled the indication of the boundary of the urban area, which, according to the adopted assumptions, determined areas with imperviousness at a level of 30%.

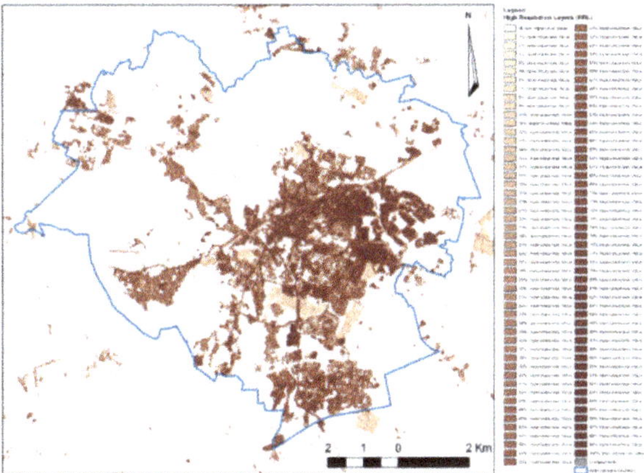

Figure 10. Land cover model in the area under investigation based on HRL Imperviousness data Source: [48].

HRL databases are also a reliable source of information about ongoing urbanisation processes. In their research, Liua X. et al. (High-resolution multi-temporal mapping of global urban land using Landsat images based on the Google Earth Engine Platform) used the Normalized Urban Areas Composite Index (NUACI) method and utilised the Google Earth Engine to facilitate the global urban land classifications from an extensive number of Landsat images [87]. High-resolution multi-period mapping (with a 5-year interval) of global urban areas using Landsat imagery based on the Google Earth Engine platform is shown in Figure 11.

3.2.4. The Analysis of Urbanisation Processes Using Urban Atlas Data

The Urban Atlas contains pan-European comparable data on land cover and land use comprising a range of functional urban areas. The presented example describes changes to the urban landscape identified and assessed using the Urban Atlas data. The aim of the research conducted by Pazúr R. et al. (Changes of urbanised landscape identified and assessed by the Urban Atlas data: Case study of Prague and Bratislava [88]) was to document, examine and compare changes in land use/cover (LUCC) of LUZ (Large Urban Zones) in Bratislava and Prague in 2006–2012 using the UA data, and to demonstrate how these changes are recorded in official statistics—Figure 12.

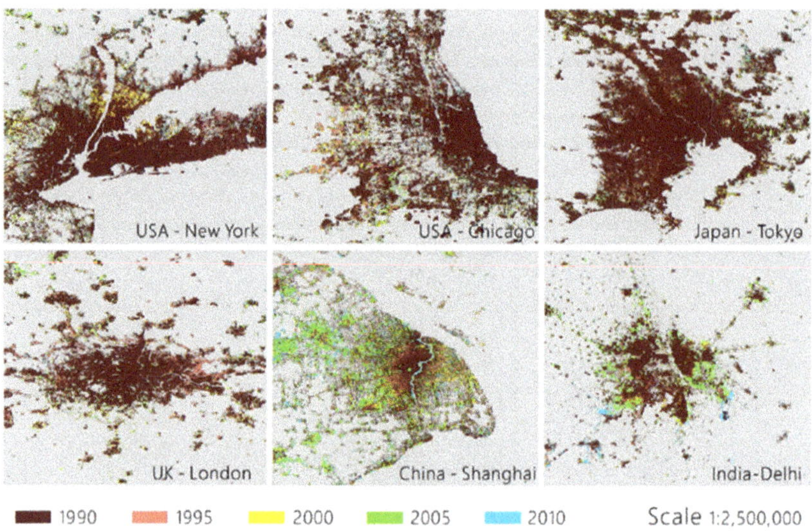

Figure 11. Urban land expansion from 1990 to 2010 in the representative cities. Source: [87].

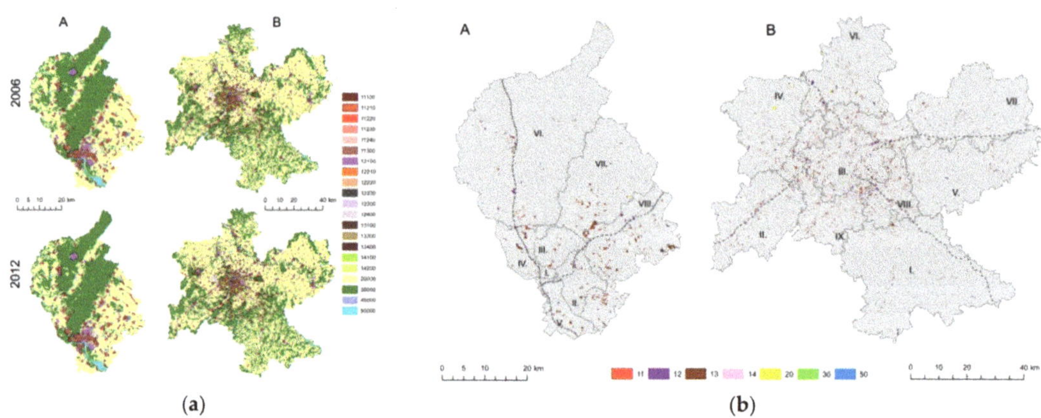

Figure 12. (**a**) LUC in (A) FUA Bratislava and B) FUA Prague in the years 2006 and 2012 according to the UA data; (**b**) LUC gains in (A) FUA Bratislava and (B) FUA Prague in the period 2006–2012 according to the UA data. Source: [88].

3.2.5. Monitoring Changes in Land Use with Landsat Time Series

The presented example shows the mapping of urban development patterns in Ouagadougou, Burkina Faso, using machine learning regression modelling with Landsat two-season time series [89]—Figure 13. The study was designed to quantify land cover for the Ouagadougou metropolitan area from 2002 to 2013 using Landsat-TM/ETM + /OLI time series.

3.2.6. The Use of Nighttime Light (NTL) Data in Analyses of Urbanisation Processes

Nighttime Light (NTL) data are increasingly often used in an analyses of urbanisation processes because of their strong relationship with human activities. Systems such as the Visible Infrared Imaging Radiometer Suite (VIIRS) Day/Night Band (DNB) onboard the Joint Polar Satellite System (JPSS) are designed with very high sensitivity to detect even the faintest light in the visible range. This makes them ideal for observing the lighting on the ground. An example of the use of these data is presented in The Annual Cycling of Nighttime Lights in India [90]—Figure 14. By classifying the ACF (autocorrelation function) profiles for each pixel location, interesting spatial patterns were revealed in different regions.

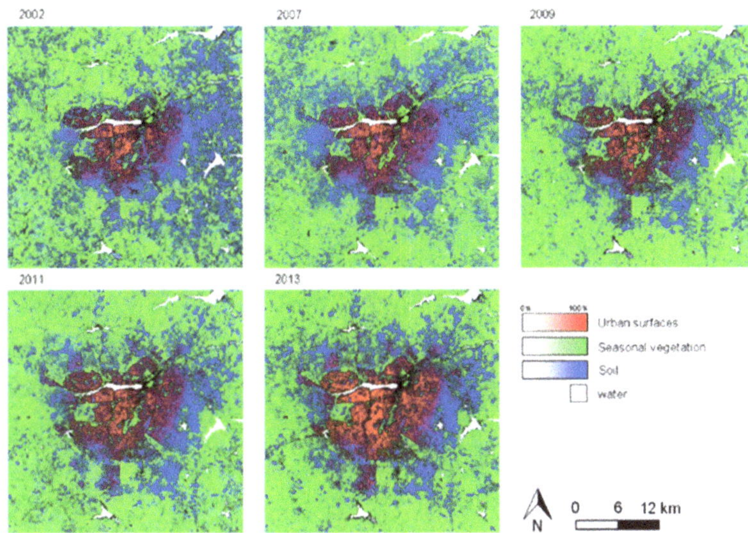

Figure 13. Fraction development of urban surfaces, seasonal vegetation and soil from 2002 to 2013. Source: [89].

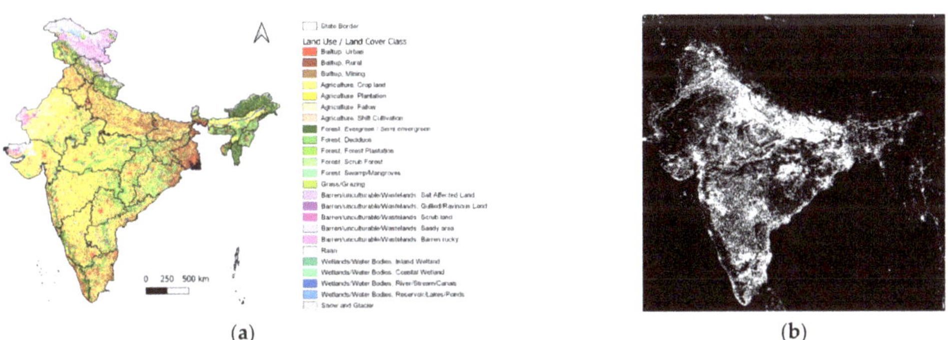

(**a**)   (**b**)

Figure 14. (**a**). Land use/land cover map for India. (**b**). Binary background mask from the 2019 annual VIIRS Nighttime Light (VNL) product. Source: [90].

Up-to-date and accurate information on the dynamics of urban expansion is essential to reveal the relationship between this phenomenon and the ecosystem in order to optimise land-use patterns and to promote efficient urban development. A publication entitled "Extracting the dynamics of urban expansion in China using DMSP-OLS nighttime light data from 1992 to 2008" [91] describes a method for systematically correcting multi-year stable night light (NSL) data from multiple satellites and rapidly extracting urban expansion dynamics based on the corrected data—Figure 15.

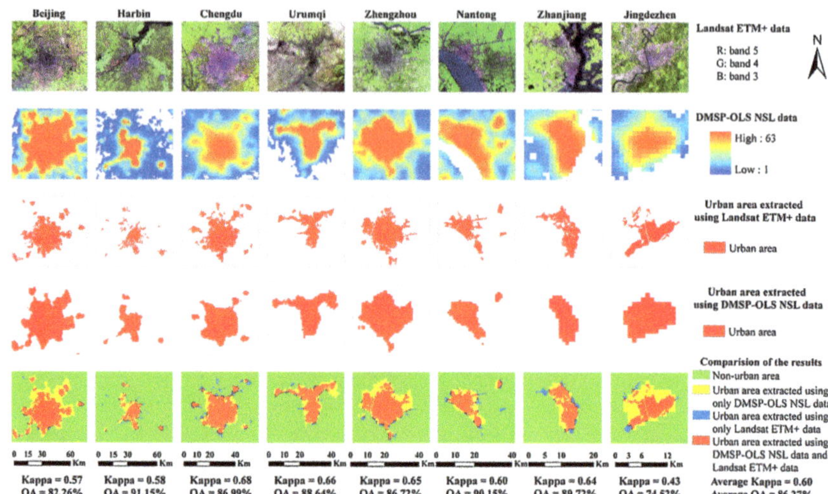

Figure 15. Accuracy assessment of selected urban areas in China in 2008 extracted using NSL data. Source: [91].

The dynamics of city expansion in China from 1992 to 2008 were extrapolated with an overall average accuracy of 82.74% and a mean Kappa value of 0.40—Figure 16.

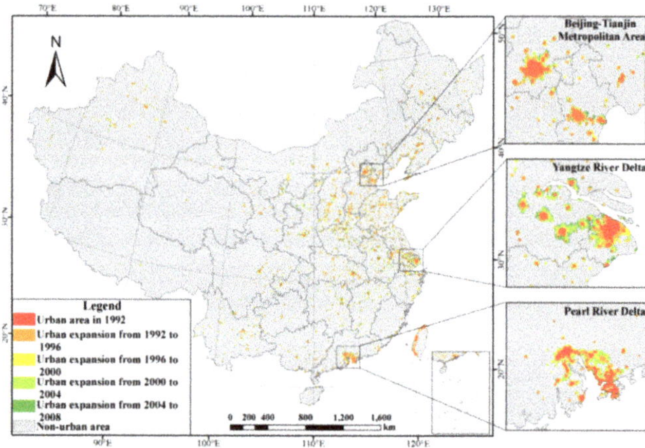

Figure 16. The dynamics of urban expansion in China from 1992 to 2008. Source: [91].

3.2.7. The Use of Greenhouse Gas Emission Data in Analyses of Urbanisation Processes

The research conducted by Christopher Jones and Daniel M. Kammen in the USA revealed that the spatial distribution of household carbon footprints identifies a process of suburbanisation by analysing the effects of greenhouse gas emissions linked to urban density [92]. The econometric models of demand for energy, transportation, food, goods and services were developed using national household surveys and were used to determine the average household carbon footprint (HCF) for U.S. zip codes, cities, counties and metropolitan areas. A lower HCF was observed in city centres and a higher carbon footprint in the peripheries (~50 tCO2e), ranging from ~25 to > 80 tCO2e in the 50 largest metropolitan areas. Population density demonstrates a weak, though positive, correlation with HCF until a density threshold is reached, after which the range, mean and standard deviation of HCF decrease. Population density contributes to the relatively low HCF in the urban centres of large metropolitan areas. In contrast, more extensive suburbanisation in these regions contributes to an overall net increase in HCF—Figure 17.

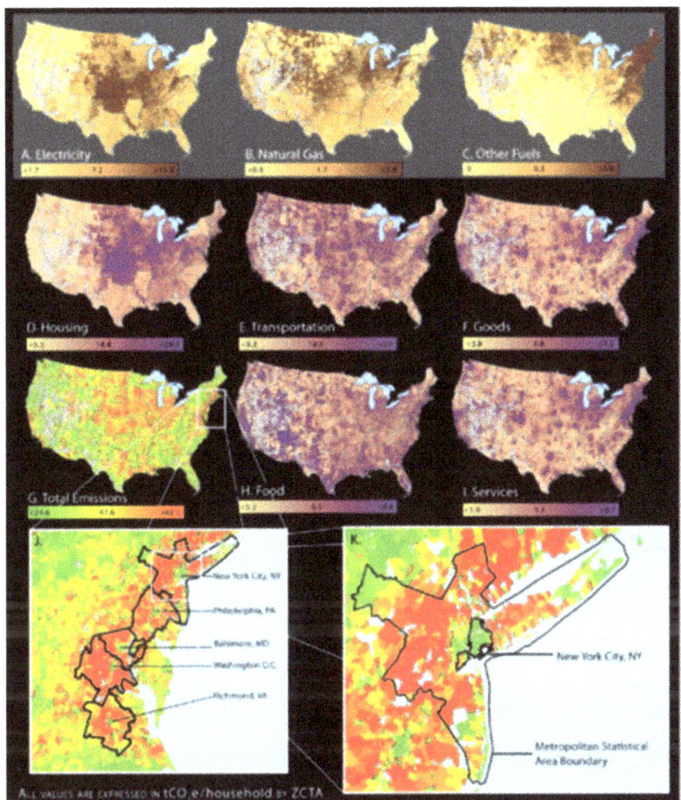

Figure 17. HCF from (**A**) electricity, (**B**) natural gas, (**C**) fuel oil and other fuels, (**D**) housing = A + B + C + water, waste, and home construction, (**E**) transportation, (**F**) goods, (**H**) food, (**I**) services, and (**G**) total = D + E + F + H + I. Transportation includes motor vehicle fuel, lifecycle emissions from fuel, motor vehicle manufacturing, air travel direct and indirect emissions, and public transit. Scales below each map show gradients of 30 colours, with labels for the upper value of lowest of quantile, median value and lowest value of highest quantile, in metric tons CO2e per household, for zip code tabulation areas (ZCTAs). East Coast metropolitan statistical areas (**J**), with a larger map of New York metropolitan area (**K**, outer line) and New York City (**K**, inner line), highlight the consistent pattern of relatively low GHG urban core cities and high GHG suburbs. Source: [92].

3.2.8. The Use of Data on Plant Vegetation Changes in Analyses of Urbanisation Processes

Urbanisation destroys and divides large amounts of natural habitats, with serious consequences for ecosystems, which is particularly evident in developing countries. The term 'urban landscape' refers to a region where specific landscape elements within an urban area prevail, including buildings, roads, infrastructure and green spaces. A number of studies have been carried out to monitor the dynamics of urban landscape changes and the ecological impact of urbanisation. Sufficient evidence has been found to demonstrate that changes in the spatial arrangement of urban landscapes (e.g., distribution, composition and configuration) lead to mass plant die-offs [93], a decrease in biodiversity [94], a reduction in atmospheric humidity [95] and a change in ground surface temperature [96,97], and an increase in fine particulate pollution [98]. Therefore, the change in the urban landscape pattern is one of the most striking aspects of urbanisation, with a negative impact on ecosystems. The presented example describes the response of vegetation to a change in the urban landscape spatial pattern in the Yangtze River Delta, China [99]. Figure 18 depicts the spatial patterns of vegetation cover in 2004, 2008, and 2013 in the Yangtze River Delta Urban Agglomeration (YRDUA). The remote sensing datasets used in this study include the Defense Meteorological Satellite Program's Operational Linescan System (DMSP/OLS) and the normalised difference vegetation index (NDVI) obtained from a Moderate Resolution Imaging Spectroradiometer (MODIS) dataset. The vegetation level with a *fc* (fractional vegetation cover) value higher than 0.75 extended over a large part of the study area, which was mostly dominated by trees, shrubs or green plants. Scattered vegetation in the northern and central parts of the YRDUA experienced the most notable reduction. It should be noted that since these two regions mainly consist of arable land and building land, the decline in *fc* can be attributed to land use transformation caused by intensive human activities, and vice versa, as the *fc* values in the southern part of the YRDUA, i.e., most of Zhejiang Province, have basically remained unchanged. Until 2013, the *fc* value for the entire area under investigation presented a clear spatial disproportion between the north and the south.

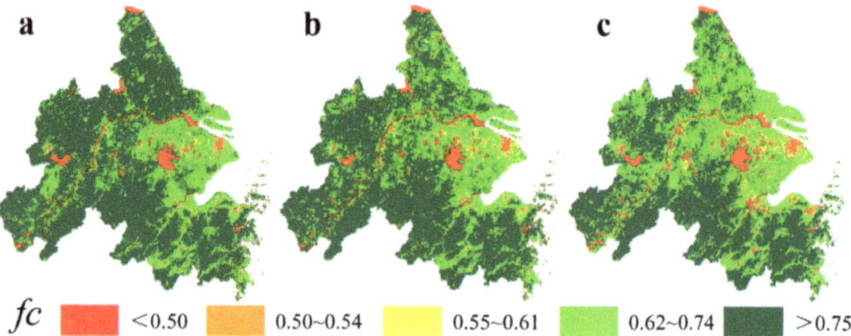

Figure 18. Spatial patterns of vegetation cover for (**a**) 2004, (**b**) 2008, and (**c**) 2013. Source: [99].

The spatial pattern of the vegetation change trend is shown in Figure 19.

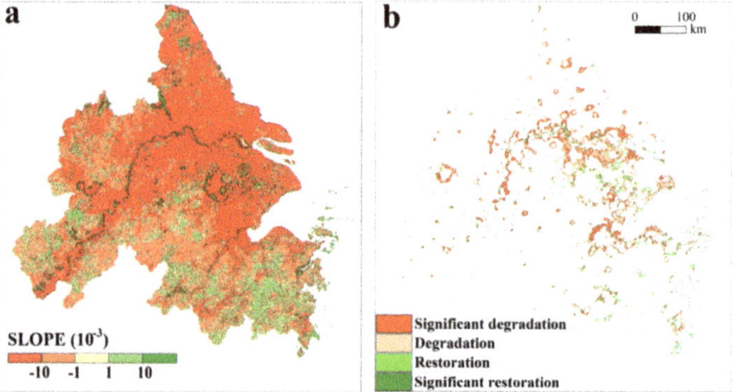

Figure 19. Trends of vegetation change in (**a**) YRDUA and (**b**) urban sprawl area in 2004–2013. Source: [99].

3.2.9. The Use of Synthetic Aperture Radar (SAR) in the Analysis of Urbanisation Processes

Synthetic Aperture Radar (SAR) for high-resolution images of stationary objects is used to create images of the land surface, of the Earth and other planets using remote sensing techniques. Remote sensing satellites, especially the German TerraSAR-X radar system, independent of weather and time of day, and which takes up to three days to revisit a particular area, have been successfully providing short-term data on urbanisation processes for many years. The presented example investigates the potential and limitations of TerraSAR-X in the context of automated, object-based detection of human settlements. The approach presented by Thiel M. et al. (Object-oriented detection of urban areas from TERRASAR-X DATA) [100] for urban detection using TerraSAR-X data, achieves an overall accuracy of around 95% and 89%—an indication of the high potential of TerraSAR-X data to identify developed areas—Figure 20.

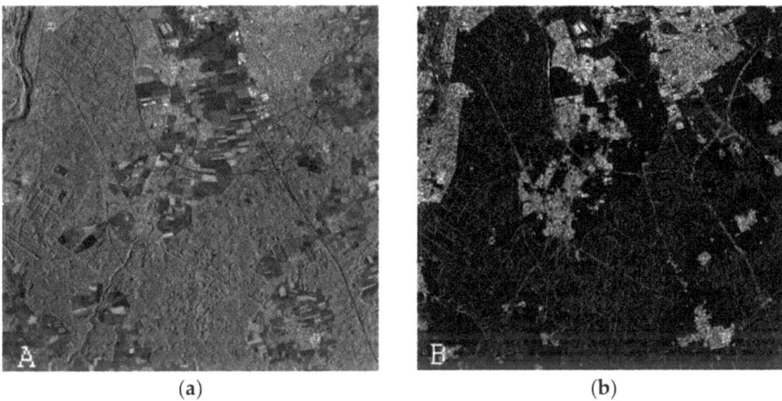

(a)   (b)

Figure 20. Filtered TerraSAR-X intensity layer (**a**) and calculated speckle divergence layer (**b**) of a subset of the region of Munich. The built-up areas are characterised by high values of speckle divergence. Source: [100].

3.3. Possibilities of Using Spatial Data in the Research on Urbanisation Processes

The development of a relatively simple measure of urban sprawl based on the observation of changes in actual land cover using GIS and a tool supporting rational land management provides a substantive basis for city planning. An example is the research described in a study entitled "An analysis of urbanisation dynamics with the use of the fuzzy set theory—A case study of the city of Olsztyn", enabling the application of fuzzy set theory as a tool supporting rational space management and provides a substantive basis for urban spatial planning [20]. The established degrees of membership determine which forms of space use have a so-called "more city-like" character than the others, i.e., have more characteristics of urban space. The results according to the assumptions of the fuzzy set theory are determined in the interval [0, 1]. The determined degrees of membership in urban-type uses and the data from the inventory of the existing state of development (interpretation of satellite images) in 2005, 2010 and 2018 enable the development of a fuzzy city model—Figure 21.

Boundaries of urbanised areas determined by the developed method based on the fuzzy set theory for 2005, 2010 and 2018 make it possible to determine the degree of urbanisation of the area in the interval [0, 1] and the dynamics of changes in urbanisation processes in the years of 2005–2010–2018 (Figure 22).

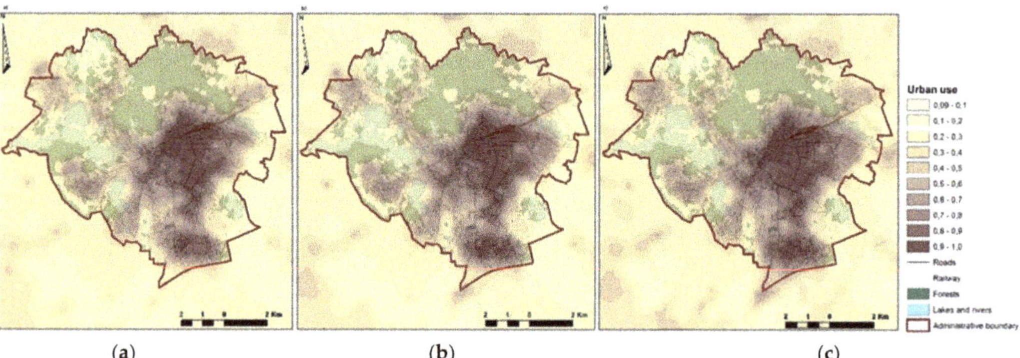

Figure 21. Urban use in the study area in the degree of membership interval of [0.00–1.00]: (**a**)—2005; (**b**)—2010; (**c**)—2018. Source: [20].

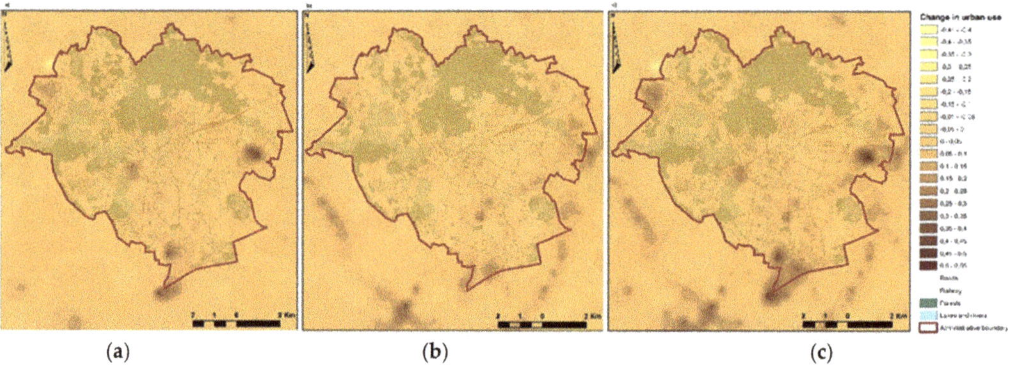

Figure 22. Changes to land use in the years: (**a**): 2005–2010; (**b**): 2010–2018; (**c**): 2005–2018. Source: [20].

The spatial development of cities and the related population growth has become a subject of broadly understood spatial management research. Planners and geographers are replacing previous models of cities that described the process of their development with new models that express how uncoordinated local decisions affect their global growth [101]. Urban compactness studies can identify cities which are expanding to make optimal use of suburban areas and those where suburban sprawl is chaotic. An example of research into the compactness of urban systems is the establishment of a compactness index based on the spatial length of the boundaries of the area covered with a particular form of development [84]. Studies of this type can apply GIS-type data; in this case, the Corine Land Cover Data CLC databases. Such data, for example, was used to examine the compactness of the urbanised areas of 49 district towns in Poland. For the 49 urban areas, the zone of the neighbouring municipalities was separated as the area most exposed to suburbanisation, and subsequently, using the CLC databases according to the developed research methodology, the areas related to urban use were separated for the three moments of time 2006, 2012 and 2018—Figure 23.

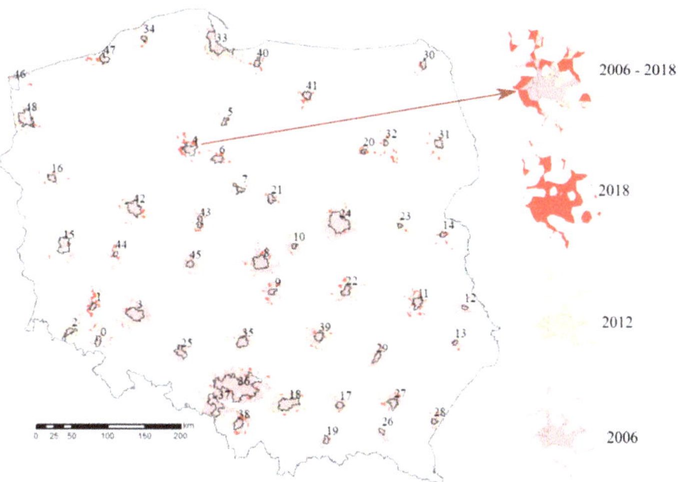

Figure 23. The areas related to urban use separated for the three moments of time 2006, 2012 and 2018. Source: own elaboration.

Based on the data concerning the area of these zones, the area compactness index AC was calculated, which is equal to the ratio of the circumference of a circle with the surface area equal to that of the urbanised areas to the length of the boundaries of these zones. The index ranges from 0 to 1. The values closer to zero indicate lower compactness. The value of 1 is theoretical and would represent the area of the city with an urbanised area forming a dense circular surface.

This indicator increases for most cities in 2018 as compared to previous years, indicating that areas that were heavily dispersed between 2006 and 2012 are becoming more compact—Figure 24. Unfortunately, for all towns, the index for the years 2006, 2012 and 2018 included in the analysis is very low (apart from 4 examples, it does not exceed the value of 0.4, and for more than a half, it is below 0.2), which indicates a strong dispersion of development in the vicinity of the towns under analysis. The examples provided in the discussion illustrate how spatial data can be used to assess the level and pace of urbanisation. As spatial databases become more accurate, complete and accessible, with images taken and made available at a better resolution, more refined and novel methods for mapping and analysing urbanisation processes will emerge.

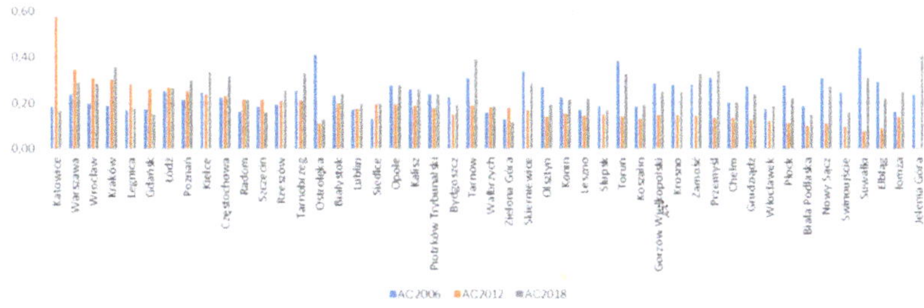

Figure 24. The AC index in the analysed cities in Poland. Source: own elaboration.

4. Discussion and Conclusions

The dynamics of urbanisation processes have been the subject of multiple studies differing in both scale and aspect. Some interesting publications on this subject include works which are well known to researchers in this field, such as [102], but also some other very interesting items [103,104]. These include some studies that use land cover information, such as [105,106]. In the majority of these papers, the authors identify specific methods for environmentally, economically or demographically oriented research. Contrary to other manuscripts of this kind, this paper shows the wide possibilities of using spatial data to study urbanisation processes at different scales and with different underlying meaning.

The analytical methods and databases discussed above differ in their availability, ease of classification of selected land uses, labour intensity, cost, spatial and temporal continuity. Most frequently, they also require a specialised knowledge and dedicated algorithms to process the collected data. Automatic classification can result in errors in identifying and separating different land-use types when a large number of similar and overlapping land cover types are processed, such as single-family homes and multi-family housing. In view of the variety of colours, structures and textures in urban areas, spatial conditions in the vicinity of the identified areas should also be taken into account. The interpretation of aerial and satellite photographs alone is time-consuming and requires a team of several researchers. However, from a longer perspective, such analyses are less time- and cost-intensive and can be carried out on a large number of cities to generate sufficient data to draw conclusions concerning the degree and nature of urbanisation.

Urbanisation is a global phenomenon, irreversible and inherent within human development. The inevitability of these processes stimulated by developing cities requires systematic research into the characteristics and parameters that form them and the various processes of transformation taking place. Urban development significantly affects areas located at the interface between urban and rural uses, resulting in the permeation and overlapping of different land-use forms in the fringe zone. Spatial development of cities manifested by an increased demand for new land exerts a significant impact on the surrounding grassland and agricultural areas. The demand for new non-urbanised building land, much cheaper than property in cities, is triggering a response from municipal governments, environmentalists, farmers and the non-agricultural rural population. Identifying the specific nature of urbanisation processes occurring in and around cities can minimise the uncontrolled and unplanned creation of zones of chaotic development, degrading natural resources and could limit uncontrolled and unplanned city growth. This is also of fundamental importance in the context of sustainable urban development. New methods of urban management, environmental protection, uncontrolled suburbanisation, acquisition of new urban land (agricultural as well as post-industrial) transport, social participation, etc., are new challenges for sustainable urban areas and practices. Planning for sustainable urban development requires a knowledge of the individual elements of the city system and its surroundings and the relationships between them, the adoption of certain assumptions and goals leading to sustainability, and the adaptation of these assumptions to local conditions. This is particularly important from a spatial perspective [107–114].

Rapid urban growth and the resulting challenges require precise techniques for identifying and locating urbanised areas and for their mapping in order to represent complex land cover characteristics in adequate detail. Geospatial data provide a rich source of knowledge and a reference for the development of new tools for identifying and monitoring urbanisation processes. This applies in particular to areas subject to urbanisation pressures which cannot be clearly classified as urban or rural. These discrepancies are reflected in the existing land use and land cover types. For this reason, the degree and rate of urbanisation can be most effectively analysed by monitoring land cover changes. Virtually unlimited access to land cover data provided by current and historical orthophotomaps and databases, as well as developing GIS techniques, enable the rapid processing of spatial information. Research on spatial urbanisation consumes much less time and concerns a large number of cities, which may provide new sources for drawing conclusions about the scale and nature of this phenomenon.

The possibilities and constraints related to decision-making regarding the spatial development of the city are primarily attributable to the multidimensional nature of the spatial development space and its probabilistic and fuzzy nature. The variety of spatial features (attributes) and the wide range of data result in the process of investigating and planning a suburban area, which is complex and lengthy and, consequently, subject to high risk. Studies of urbanisation processes must take into account the nature of the explored component, characteristics, as well as the frequency, pace and magnitude of these changes. The analyses presented in this paper provide an attempt to further define current trends, methods, techniques and databases used in the analysis of the ongoing urbanisation processes.

Author Contributions: Conceptualisation, A.B. and I.C.; methodology, A.B. and I.C.; software, A.B. and I.C.; validation, A.B. and I.C.; formal analysis A.B. and I.C.; investigation A.B. and I.C.; resources, A.B. and I.C.; data curation, A.B. and I.C.; writing—original draft preparation, A.B.; writing—review and editing, I.C.; visualisation, A.B. and I.C.; supervision, A.B. and I.C.; project administration, A.B. and I.C.; funding acquisition, A.B. and I.C. All authors have read and agreed to the published version of the manuscript.

Funding: This research received no external funding.

Institutional Review Board Statement: Not applicable.

Informed Consent Statement: Not applicable.

Data Availability Statement: Not applicable.

Conflicts of Interest: The authors declare no conflict of interest.

References

1. United Nations. *World Urbanization Prospects—Population Division*; United Nations: New York, NY, USA, 2019.
2. Parysek, J. Duże Miasta Europy i ich rola w Procesie Urbanizacji, Rozwoju Społeczno-Gospodarczego i Europejskiej Integracji u Schyłku XX Wieku. *Przegląd Geogr. Pol. Akad. Nauk.* **1995**, *67*, 3–4.
3. Müller, N.L. Brazil. In *Essays on World Urbanization*; Ronald, J.E., Ed.; George Philip and Son Limited: London, UK, 1975; pp. 212–222.
4. Szymańska, D.; Biegańsk, J. *The Phenomenon of Urbanizationand Processes Connected with It*; Studia Miejskie; University of Opole Publishing House: Opole, Poland, 2011; Volume 4, pp. 13–38. (In Polish)
5. Szymańska, D. Zjawisko Urbanizacji i jej Konsekwencje. In *Badania Środowiska*; Turło, J.E., Ed.; Wyd. Naukowe UMK: Toruń, Poland, 1995; pp. 71–79.
6. Szymańska, D. *Urbanizacja na Świecie*; Wyd. Naukowe PWN: Warszawa, Poland, 2007; p. 392.
7. Węcławowicz, G. *Geografia Społeczna Miast. Zróżnicowania Społeczno-Przestrzenne*; Wyd. Naukowe PWN SA: Warszawa, Poland, 2003; p. 163.
8. Ziółkowski, J. Miejsce i rola procesu urbanizacji w prZeobrażeniach społecznych w polsce ludowej. In *Procesy Urbanizacyjne w Powojennej Polsce*; Nowakowski, S., Ed.; PWN: Warsaw, Poland, 1967.
9. Nowak, S. *Obiektywne i Psychologiczne Parametry Przemian Struktury Społecznej*; Materiały III Ogólnopolskiego Zjazdu Socjologicznego, z: Warsaw, Poland, 1967; p. 2.
10. Gałęski, B. *Innowacje a Społeczność Wiejska*; Wyd. Książka i Wiedza: Warszawa, Poland, 1971.
11. Smiraglia, D.; Salvati, L.; Egidi, G.; Salvia, R.; Giménez-Morera, A.; Halbac-Cotoara-Zamfir, R. Toward a New Urban Cycle? A Closer Look to Sprawl, Demographic Transitions and the Environment in Europe. *Land* **2021**, *10*, 127. [CrossRef]
12. Divigalpitiya, P.; Nurul Handayani, K. Measuring the Urban Expansion Process of Yogyakarta City in Indonesia Urban expansion process and spatial and temporal characteristics of growing cities. *Int. Rev. Spat. Plan. Sustain. Dev.* **2015**, *3*, 18–32. [CrossRef]
13. Meng, L.; Sun, Y.; Zhao, S. Comparing the spatial and temporal dynamics of urban expansion in Guangzhou and Shenzhen from 1975 to 2015: A case study of pioneer cities in China's rapid urbanization. *Land Use Policy* **2020**, *97*, 104753. [CrossRef]
14. Antrop, M. Landscape change and the urbanization process in Europe. *Landsc. Urban Plan.* **2004**, *67*, 9–26. [CrossRef]
15. Ramachandra, T.V.; Bharath, H. Aithal Durgappa, D. Sanna. Insights to urban dynamics through landscape spatial pattern analysis. *Int. J. Appl. Earth Obs. Geoinf.* **2012**, *18*, 329–343. [CrossRef]
16. Smailes, A.E. *The Definition and Measurement of Urbanization, [w:] Essays on World Urbanization*; Jones, R., Ed.; George Philip and Son Ltd.: London, UK, 1975; pp. 1–19.
17. Szymańska, D. *Geografia Osadnictwa*; Wyd. Naukowe PWN: Warszawa, Poland, 2009.
18. You, H. Quantifying the coordinated degree of urbanizationin Shanghai, China. *Qual. Quant.* **2016**, *50*, 1273–1283. [CrossRef]
19. Sato, Y.; Yamamoto, K. Population concentration, urbanization, and demographic transition. *J. Urban Econ.* **2005**, *58*, 45–61. [CrossRef]
20. Biłozor, A.; Cieślak, I.; Czyza, S. An analysis of urbanisation dynamics with the use of the fuzzy set theory—A case study of the city of Olsztyn. *Remote Sens.* **2020**, *12*, 1784. [CrossRef]

21. Wojtkun, G. Do Cities Really Expand? *Czas. Tech. Archit.* **2010**, *107*, 311–317.
22. Mulder, C.H. Population and housing: A two-sided relationship. *Demogr. Res.* **2006**, *15*, 401–412. [CrossRef]
23. Cieślak, I.; Szuniewicz, K.; Pawlewicz, K.; Czyza, S. Land Use Changes Monitoring with CORINE Land Cover Data. *IOP Conf. Ser. Mater. Sci. Eng.* **2017**, *245*, 052049. [CrossRef]
24. UN-Habitat, Urbanization and Development: Emerging Futures. 2016. Available online: https://unhabitat.org/world-cities-report (accessed on 15 July 2021).
25. Montgomery, M.R.; Stren, R.; Cohen, B.; Reed, H.E. *National Research Council. Cities Transformed: Demographic Change and Its Implications in the Developing World*; The National Academies Press: Washington, DC, USA, 2003. [CrossRef]
26. Pictet. The Role of Cities in the Global Economy. Available online: https://www.group.pictet/wealth-management/role-cities-global-economy (accessed on 2 September 2021).
27. Weinstein, E. Migration for the benefit of all: Towards a new paradigm for economic immigration. *Int. Labour Rev.* **2002**, *141*, 225. [CrossRef]
28. Kruszyna, M.; Śleszyński, P.; Rychlewski, J. Dependencies between Demographic Urbanization and the Agglomeration Road Traffic Volumes: Evidence from Poland. *Land* **2021**, *10*, 47. [CrossRef]
29. Schneider, C.; Achilles, B.; Merbitz, H. Urbanity and Urbanization: An Interdisciplinary Review Combining Cultural and Physical Approaches. *Land* **2014**, *3*, 105–130. [CrossRef]
30. Hou, B.; Nazroo, J.; Banks, J.; Marshall, A. Are cities good for health? A study of the impacts of planned urbanization in China. *Int. J. Epidemiol.* **2019**, *48*, 1083–1090. [CrossRef]
31. Güneralp, B.S.; Lwasa, H.; Masundire, S.; Parnell, S.; Seto, K.C. Urbanization in Africa: Challenges and opportunities for conservation. *Environ. Res. Lett.* **2018**, *13*, 015002. [CrossRef]
32. United Nations. *The World's Cities 2018-Data Booklet*; Book Series: Statistical Papers-United Nations (Ser. A), Population and Vital Statistics Report; United Nations: New York, NY, USA, 2018. [CrossRef]
33. The World Bank. Urban Development. Available online: https://www.worldbank.org/en/topic/urbandevelopment/overview (accessed on 2 September 2021).
34. Dobbs, R.; Smit, S.; Remes, J.; Manyika, J.; Roxburgh, C.; Restrepo, A. Urban World: Mapping the Economic Power of Cities. McKinsey Global Institute. March 2011. McKinsey & Company 2011. Available online: https://www.mckinsey.com/~{}/media/McKinsey/Featured%20Insights/Urbanization/Urban%20world/MGI_urban_world_mapping_economic_power_of_cities_full_report.ashx (accessed on 15 July 2021).
35. Cieślak, I. Identification of areas exposed to land use conflict with the use of multiple-criteria decision-making methods. *Land Use Policy* **2019**, *89*, 104225. [CrossRef]
36. Cieślak, I.; Biłozor, A.; Źróbek-Sokolnik, A.; Zagroba, M. The use of geographic databases for analyzing changes in land cover—A case study of the Region of Warmia and Mazury in Poland. *ISPRS Int. J. Geo-Inf.* **2020**, *9*, 358. [CrossRef]
37. Wojtkun, G. Współczesne osadnictwo osiedlowe-aspekt metodyczny. *Sr. Mieszk.* **2009**, *3*, 159–161.
38. Mikolai, J.; Kulu, H.; Mulder, C.H. Family life transitions, residential relocations, and housing in the life course: Current research and opportunities for future work: Introduction to the Special Collection on "Separation, Divorce, and Residential Mobility in a Comparative Perspective. *Demogr. Res.* **2020**, *43*, 35–58. [CrossRef]
39. Caprotti, F.; Cowley, R.; Datta, A.; Broto, V.C.; Gao, E.; Georgeson, L.; Herrick, C.; Odendaal, N.; Joss, S. The New Urban Agenda: Key opportunities and challenges for policy and practice. *Urban Res. Pract.* **2017**, *10*, 367–378. [CrossRef]
40. Gao, J.; O'Neill, B.C. Mapping global urban land for the 21st century with data-driven simulations and Shared Socioeconomic Pathways. *Nat. Commun.* **2020**, *11*, 2302. [CrossRef] [PubMed]
41. Czerny, M. *Globalizacja i Rozwój. Wybrane Zagadnienia Geografii Gospodarczej Świata*; Wydawnictwo Naukowe PWN: Warsaw, Poland, 2005.
42. Parysek, M.; Parysek, J.J.; Mierzejewska, L.; Jażdżewska, I. *Między Dezurabnizacją a Reurbanizacją: Nowe Oblicze Urbanizacji w Pol-sce, [w:] Współczesne Procesy Urbanizacji i Ich Skutki*; Wyd. UŁ: Łódź, Poland, 2005.
43. Degórska, B. *Spatial Urbanization of Rural Areas in the Warsaw Metropolitan Area. Ecological and Landscape Context*; Instytut Geografii I Przestrzennego Zagospodarowania im. Stanisława Leszczyckiego, Polska Akademia Nauk: Warsaw, Poland, 2017. (In Polish)
44. Mils, E. *Studiesin fhe Structureofthe Urban Economy*; Johns Hopkins Press: Baltimore, MD, USA, 1972.
45. O'sullivan, A. *Urban Economics*; McGraw-Hill/Irwin: New York, NY, USA, 2007; pp. 225–226.
46. Geshkov, M. Urban sprawl in Eastern Europe: The Sofia City example. *Econ. Altern.* **2015**, *2*, 101–116.
47. Biłozor, A. Urban land use changes forecasting. In Proceedings of the 9th International Conference on Environmental Engineering. Section: Sustainable Urban Development, Vilnius, Lithuania, 22–23 May 2014; ISBN 978-609-457-640-9.
48. Biłozor, A.; Czyża, S.; Bajerowski, T. Identification and Location of a Transitional Zone between an Urban and a Rural Area Using Fuzzy Set Theory, CLC, and HRL Data. *Sustainability* **2019**, *11*, 7014. [CrossRef]
49. Simon, D. Urban Environments: Issues on the Peri-Urban Fringe. *Annu. Rev. Environ. Resour.* **2008**, *33*, 167–185. [CrossRef]
50. Siemiński, J. Rural-urban continuum and some of its infrastructural problems. *Infrastruct. Ecol. Rural. Areas* **2010**, *2*, 215–228.
51. Sobotka, S. Transformation of the historical spatial arrangements of the villages in the suburban area of Olsztyn, with particular emphasis on Brąswałd, Dorotowo and Jonkowo. *Acta Sci. Pol. Adm. Locorum* **2014**, *13*, 39–68. (In Polish)

52. Szmytkie, R. *Methods of Analysis of the Morphology and Physiomic of Sedate Units*; Scientific Hearings of the Institute of Geography and Regional Development of the University of Wrocław no. 35; Institute of Geography and Regional Development of the University of Wrocław: Wrocław, Poland, 2014; Available online: http://www.geogr.uni.wroc.pl/data/files/publikacje-rozprawy-naukowe-igrr/rozprawy_35.pdf (accessed on 15 July 2021). (In Polish)

53. Konecka-Szydłowska, B. The smallest cities in Poland in terms of the concept of urban-rural continuums. *Reg. Dev. Reg. Policy* **2018**, *41*, 151–165. (In Polish)

54. Degórska, B. Spatial urbanization of rural areas in the metropolitan area of Warsaw. Ecological and landscape context. In *Institute of Geography and Spatial Development im. Stanisława Leszczyckiego, Polish Academy of Sciences, Geographic Work, no 262*; Institute of Geography and Spatial Development: Warsaw, Poland, 2017. (In Polish)

55. Ravetz, J.; Warhurst, P. Manchester: Re-inventing the local–global in the peri-urban city-region. In *Peri-Urban Futures: Scenarios and Models for Land Use Change in Europe*; Nilsson, K., Pauleit, S., Bell, S., Aalbers, C., Sick Nielsen, T., Eds.; Springer: Berlin/Heidelberg, Germany, 2013. [CrossRef]

56. Loibl, W.; Piorr, A.; Ravetz, J. Concepts and methods. In *Peri-Urbanisation in Europe: Towards a European Policy to Sustain Urban-Rural Futures*; Piorr, A., Ravetz, J., Tosics, I., Eds.; University of Copenhagen, Academic Books Life Sciences: Copenhagen, Denmark, 2011; pp. 24–29.

57. Labbé, D. *Facing the Urban Transition in Hanoi: Recent Urban Planning Issues and Initiatives*; Institut National de la Reherche scientifique Centre-Urbanisation Culture Société: Québec City, QC, Canada, 2010; Available online: http://espace.inrs.ca/id/eprint/4986/1/HanoiUrbanization.pdf (accessed on 15 July 2021).

58. Acevedo, W.; Masuoka, P. Time-series animation techniques for visualizing urban growth. *Comput. Geosci.* **1997**, *23*, 423–435. [CrossRef]

59. Csatári, B.; Farkas, J.Z.; Lennert, J. Land Use Changes in the Rural-Urban Fringe of Kecskemét after the Economic Transition. *J. Settl. Spat. Plan.* **2013**, *4*, 153–159.

60. Nabielek, K.; Kronberger-Nabielek, P.; Hamers, D. The rural-urban fringe in the Netherlands: Recent developments and future challenges. *SPOOL* **2013**, *1*, 1–18. [CrossRef]

61. Simon, D.; McGregor, D.; Nsiah-Gyabaah, K. The changing urban–rural interface of African cities: Definitional issues and an application to Kumasi, Ghana. *Environ. Urban.* **2004**, *16*, 235–248.

62. Tacoli, C. The links between rural and urban development. *Environ. Urban.* **2003**, *15*, 3–12. [CrossRef]

63. Hoffmann, E.; Jose, M.; Nölke, N.; Möckel, T. Construction and Use of a Simple Index of Urbanisation in the Rural–Urban Interface of Bangalore, India. *Sustainability* **2017**, *9*, 2146. [CrossRef]

64. Gallent, N. The Rural–Urban fringe: A new priority for planning policy? *Plan. Pract. Res.* **2006**, *21*, 383–393. [CrossRef]

65. Hao, P.; Geertman, S.; Hooimeijer, P.; Sliuzas, R. The land-use diversity in urban villages in Shenzhen. *Environ. Plan.* **2012**, *44*, 2742–2764. [CrossRef]

66. Gallent, N.; Shaw, D. Spatial planning, area action plans and the rural-urban fringe. *J. Environ. Plan. Manag.* **2001**, *50*, 617–638. [CrossRef]

67. Gant, R.; Robinson, G.; Shahab, F. Land-use change in the 'edgelands': Policies and pressures in London's rural–urban fringe. *Land Use Policy* **2011**, *28*, 266–279. [CrossRef]

68. Datta, R. Territorial Integration: An approach to address urbanising villages in the planning for Delhi metropolitan area, India. In Proceedings of the Territorial Integration of Urbanising Villages 40th ISoCaRP Congress, Geneva, Switzerland, 22 September 2004.

69. Biłozor, A.; Renigier-Biłozor, M.; Cellmer, R. Assessment Procedure of Suburban Land Attractiveness and Usability for Housing. In Proceedings of the Baltic Geodetic Congress, Olsztyn, Poland, 21–23 June 2018. [CrossRef]

70. Renigier-Bilozor, M.; Bilozor, A. Optimization of the Variables Selection in the Process of Real Estate Markets Rating. *Oeconomia Copernic.* **2015**, *6*, 139–157. [CrossRef]

71. Renigier-Biłozor, M.; Biłozor, A.; Wiśniewski, R. 2017. Real estate markets rating engineering as the condition of urban areas assessment. *Land Use Policy* **2017**, *61*, 511–525. [CrossRef]

72. Ready, R.; Abdalla, C. *GIS Analysis of Land Use on the Rural-Urban Fringe: The Impact of Land Use and Potential Local Disamenities on Residential Property Values and on the Location of Residential Development in Berks County, Pennsylvania*; Final Report; Staff Paper 364; Pennsylvania State University: State College, PA, USA, 2003.

73. Renigier-Bilozor, M.; Wisniewski, R.; Bilozor, A. Rating attributes toolkit for the residential property market. *Int. J. Strateg. Property Manag.* **2017**, *21*, 307–317. [CrossRef]

74. Renigier-Bilozor, M. Modern classification system of real estate markets. *Geodetski Vestnik* **2017**, *61*, 441–460. [CrossRef]

75. Renigier-Biłozor, M.; Walacik, M.; Źróbek, S.; d'Amato, M. Forced sale discount on property market—How to assess it? *Land Use Policy* **2018**, *78*, 104–115. [CrossRef]

76. Konieczna, J.; Trystuła, A. Źródłą danych geoprzestrzennych na potrzeby waloryzacji. In *Współczesna Waloryzacja Przestrzeni Zurbanizowanej*; Iwona, C., Ed.; Wydawnictwo UWM w Olsztynie: Olsztyn, Poland, 2012.

77. Lefebvre, A.; Sannier, C.; Corpetti, T. Monitoring Urban Areas with Sentinel-2A Data: Application to the Update of the Copernicus High Resolution Layer Imperviousness Degree. *Remote Sens.* **2016**, *8*, 606. [CrossRef]

78. Zhang, M.; Chen, F.; Tian, B.; Liang, D. Multi-temporal SAR image classification of coastal plain wetlands using a new feature selection method and random forests. *Remote Sens. Lett.* **2019**, *10*, 312–321. [CrossRef]

79. Renigier-Biłozor, M.; Biłozor, A. The use of geoinformation in the process of optimization the use of land. In Proceedings of the 9th International Conference on "Environmental Engineering", Vilnius, Lithuania, 22–23 May 2014. Section: Sustainable Urban Development Vilnius, Lithuania, eISSN 2029-7092/eISBN 978-609-457-640-9.

80. Lisini, G.; Salentinig, A.; Du, P.; Gamba, P. SAR-Based Urban Extents Extraction: From ENVISAT to Sentinel-1. *IEEE J. Sel. Top. Appl. Earth Obs. Remote Sens.* **2018**, *11*, 2683–2691. [CrossRef]

81. Bielecka, E.; Ciołkosz, A. *CLC-2006-Database of Land Cover in Poland*; PPK t. 41; 2009; Volume 3, pp. 227–236.

82. Kantak, T. Spatial Information Systems (Auditory). Self-Study Support Materials for 3rd Year Part-Time Students of Navigation Major 2013/2014. Available online: https://gisplay.pl/gis/definicje-gis.html (accessed on 16 July 2021).

83. Heymann, Y.; Steenmans, C.; Croisille, G.; Bossard, M. *Corine Land Cover*; Technical Guide 1994; Office for Official Publications of European Communities: Luxembourg, 1994.

84. Cieślak, I.; Biłozor, A.; Szuniewicz, K. The Use of the CORINE Land Cover (CLC) Database for Analyzing Urban Sprawl. *Remote Sens.* **2020**, *12*, 282. [CrossRef]

85. Kuc, G.; Chormański, J. Sentinel-2 imagery for mapping and monitoring imperviousness in urban areas. *Int. Arch. Photogramm. Remote Sens. Spat. Inf. Sci.* **2019**, *42*, 43–47. [CrossRef]

86. Furberg, D.; Ban, Y.; Nascetti, A. Monitoring of Urbanization and Analysis of Environmental Impact in Stockholm with Sentinel-2A and SPOT-5 Multispectral Data. *Remote Sens.* **2019**, *11*, 2408. [CrossRef]

87. Liu, X.; Hu, G.; Chen, Y.; Li, X.; Xu, X.; Shaoying, L.; Pei, F.; Wang, S. 2018. High-resolution multi-temporal mapping of global urban land using Landsat images based on the Google Earth Engine Platform. *Remote Sens. Environ.* **2018**, *209*, 227–239. [CrossRef]

88. Pazúr, R.; Feranec, J.; Štych, P.; Kopecká, M.; Holman, L. 2017. Changes of urbanised landscape identified and assessed by the UrbanAtlas data: Case study of Prague and Bratislava. *Land Use Policy* **2017**, *61*, 135–146. [CrossRef]

89. Schug, F.; Okujeni, A.; Hauer, J.; Hostert, P.; Nielsen, J.; van der Linden, S. Mapping patterns of urban development in Ouagadougou, Burkina Faso, using machine learning regression modeling with bi-seasonal Landsat time series. *Remote Sens. Environ.* **2018**, *210*, 217–228. [CrossRef]

90. Hsu, F.; Zhizhin, M.; Ghosh, T.; Elvidge, C.; Taneja, J. The Annual Cycling of Nighttime Lights in India. *Remote Sens.* **2021**, *13*, 1199. [CrossRef]

91. Liua, Z.; Hea, C.; Zhangc, O.; Huangd, O.; Yang Yanga, Y. 2012. Extracting the dynamics of urban expansion in China using DMSP-OLS nighttime light data from 1992 to 2008. Landsc. *Urban Plan.* **2012**, *106*, 62–72. [CrossRef]

92. Jones, C.; Kammen, D.M. 2014. Spatial Distribution of U.S. Household Carbon Footprints Reveals Suburbanization Undermines Greenhouse Gas Benefits of Urban Population Density. *Environ. Sci. Technol.* **2014**, *48*, 895–902. [CrossRef]

93. Hahs, A.K.; Mcdonnell, M.J.; Mccarthy, M.A.; Vesk, P.A.; Corlett, R.T.; Norton, B.A.; Clemants, S.E.; Duncan, R.P.; Thompson, K.; Schwartz, M.W. A global synthesis of plant extinction rates in urban areas. *Ecol. Lett.* **2010**, *12*, 1165–1173. [CrossRef]

94. Ives, C.D.; Lentini, P.E.; Threlfall, C.G.; Ikin, K.; Shanahan, D.F.; Garrard, G.E.; Bekessy, S.A.; Fuller, R.A.; Mumaw, L.; Rayner, L.; et al. Cities are hotspots for threatened species. *Glob. Ecol. Biogeogr.* **2016**, *25*, 117–126. [CrossRef]

95. Luo, M.; Lau, N.C. Urban Expansion and Drying Climate in an Urban Agglomeration of East China. *Geophys. Res. Lett.* **2019**, *46*, 6868–6877. [CrossRef]

96. Fan, H.; Yu, Z.; Yang, G.; Liu, T.Y.; Liu, T.Y.; Hung, C.H.; Vejre, H. How to cool hot-humid (Asian) cities with urban trees? An optimal landscape size perspective. *Agr. For. Meteorol.* **2019**, *265*, 338–348. [CrossRef]

97. Jenerette, G.D.; Harlan, S.L.; Brazel, A.; Jones, N.; Larsen, L.; Stefanov, W.L. Regional relationships between surface temperature, vegetation, and human settlement in a rapidly urbanizing ecosystem. *Landsc. Ecol.* **2007**, *22*, 353–365. [CrossRef]

98. Tu, M.; Liu, Z.; He, C.; Fang, Z.; Lu, W. The relationships between urban landscape patterns and fine particulate pollution in China: A multiscale investigation using a geographically weighted regression model. *J. Clean Prod.* **2019**, *237*, 117744. [CrossRef]

99. Cao, Y.; Wang, Y.; Li, G.; Fang, X. 2019. Vegetation Response to Urban Landscape Spatial Pattern Change in the Yangtze River Delta, China. *Sustainability* **2020**, *12*, 68. [CrossRef]

100. Thiel, M.; Esch, T.; Schenk, A. Object-oriented detection of urban areas from terrasar-x data. 2008. The International Archives of the Photogrammetry. *Remote Sens. Spat. Inf. Sci.* **2008**, *XXXVII Pt B8*, 23–26.

101. Batty, M. New ways of looking at cities. *Nature* **1995**, *377*, 574. [CrossRef]

102. Antrop, M. Changing patterns in the urbanized countryside of Western Europe. *Landsc. Ecol.* **2000**, *15*, 257–270. [CrossRef]

103. Zhu, Y. Changing urbanization processes and in situ rural-urban transformation: Reflections on China's settlement definitions. In *New forms of Urbanization*; Routledge: London, UK, 2017; pp. 207–228.

104. Das, M.; Das, A. Dynamics of Urbanization and its impact on Urban Ecosystem Services (UESs): A study of a medium size town of West Bengal, Eastern India. *J. Urban Manag.* **2019**, *8*, 420–434. [CrossRef]

105. Wang, J.; Lin, Y.; Glendinning, A.; Xu, Y. Land-use changes and land policies evolution in China's urbanization processes. *Land Use Policy* **2018**, *75*, 375–387. [CrossRef]

106. Petrişor, A.I.; Ianoş, I.; Tălângă, C. Land cover and use changes focused on the urbanization pro-cesses in Romania. *Environ. Eng. Manag. J.* **2010**, *9*, 765–771. [CrossRef]

107. Morano, P.; Guarini, M.R.; Tajani, F.; Anelli, D. Sustainable redevelopment: The cost-revenue analysis to support the urban planning decisions. In *International Conference on Computational Science and Its Applications*; Springer: Cham, Switzerland, 2020; pp. 968–980.

108. BenDor, T.K.; Metcalf, S.S.; Paich, M. The dynamics of brownfield redevelopment. *Sustainability* **2011**, *3*, 914–936. [CrossRef]

109. BenDor, T.; Metcalf, S. Conceptual Modeling and Dynamic Simulation of Brownfield Redevelopment. In Proceedings of the 23rd International System Dynamics Conference, Boston, MA, USA, 17–21 July 2005.
110. Attardi, R.; Cerreta, M.; Sannicandro, V.; Torre, C.M. Non-compensatory composite indicators for the evaluation of urban planning policy: The land-use policy efficiency index (LUPEI). *Eur. J. Oper. Res.* **2018**, *264*, 491–507. [CrossRef]
111. Cerreta, M.; Panaro, S.; Poli, G. A spatial decision support system for multifunctional landscape assessment: A transformative resilience perspective for vulnerable inland areas. *Sustainability* **2021**, *13*, 2748. [CrossRef]
112. Burkhard, B.; Kroll, F.; Müller, F.; Windhorst, W. Landscapes' capacities to provide ecosystem services—A concept for land-cover based assessments. *Landsc. Online* **2009**, *15*, 1–22. [CrossRef]
113. Sutton, P.C.; Costanza, R. Global estimates of market and non-market values derived from nighttime satellite imagery, land cover, and ecosystem service valuation. *Ecol. Econ.* **2002**, *41*, 509–527. [CrossRef]
114. Mierzejewska, L. Sustainable urban development—Selected understandings, concepts and models. Problems of Urban Development. *Probl. Rozw. Miast.* **2015**, *3*, 5–11. (In Polish)

Article

A Geo-Spatial Analysis for Characterising Urban Sprawl Patterns in the Batticaloa Municipal Council, Sri Lanka

Mathanraj Seevarethnam [1,*], Noradila Rusli [2], Gabriel Hoh Teck Ling [1] and Ismail Said [3]

1 Department of Urban and Regional Planning, Faculty of Built Environment and Surveying, Universiti Teknologi Malaysia, Skudai, Johor Bahru 81310, Malaysia; gabriel.ling@utm.my
2 Centre for Innovative Planning and Development (CIPD), Department of Urban and Regional Planning, Faculty of Built Environment and Surveying, Universiti Teknologi Malaysia, Skudai, Johor Bahru 81310, Malaysia; noradila@utm.my
3 Department of Landscape Architecture, Faculty of Built Environment and Surveying, Universiti Teknologi Malaysia, Skudai, Johor Bahru 81310, Malaysia; b-ismail@utm.my
* Correspondence: mathanrajs@esn.ac.lk

Abstract: Urban sprawl related to rapid urbanisation in developed and developing nations affects sustainable land use. In Sri Lanka, urban areas have mostly expanded in a rather spontaneous, unplanned manner (based on the current settlers' subjective movement) rather than conforming to the local government's development plan. This growth inevitably leads to uncontrolled urban sprawl in many Sri Lankan cities, including Batticaloa. So far, Sri Lanka's planners or researchers have not yet tackled the sprawling developments in this city. Understanding the different forms and patterns of urban sprawl is the key to address sprawling growth. This study aims to identify the characteristics of urban sprawl in the Batticaloa municipal council using Geographic Information System (GIS) and remote sensing technology. Landsat satellite images for the years 2000, 2010, and 2020 as well as 2002, 2011, and 2019 population data were used and analysed using ArcGIS' maximum likelihood classification tool and the density function, respectively, to delineate the characteristics of urban sprawl. The results revealed that low-density development, leapfrog development, commercial ribbon development, and scattered growth are the influencing characteristics of urban sprawl in the Batticaloa municipality. These characteristics were found mainly in the urban edge of the city and have led to urban sprawl. The finding provides knowledge into recognising the characteristics of urban sprawl with empirical evidence. It affords a clear direction for future studies of urban sprawl in rapidly growing cities that are numerous in Sri Lanka, and the identified characteristics of urban sprawl can be useful in minimising future sprawl. This result can be a tool for future urban planning and management in the Batticaloa municipality.

Keywords: urban sprawl; land use; urbanisation; leapfrog development; scattered development

Citation: Seevarethnam, M.; Rusli, N.; Ling, G.H.T.; Said, I. A Geo-Spatial Analysis for Characterising Urban Sprawl Patterns in the Batticaloa Municipal Council, Sri Lanka. *Land* **2021**, *10*, 636. https://doi.org/10.3390/land10060636

Academic Editors: Luca Salvati, Iwona Cieślak and Andrzej Biłozor

Received: 15 May 2021
Accepted: 9 June 2021
Published: 15 June 2021

Publisher's Note: MDPI stays neutral with regard to jurisdictional claims in published maps and institutional affiliations.

1. Introduction

Urbanisation is a reflection of the human activities affecting the land that has been threatened by the enormous pressure from population growth [1]. Rapid urban growth is generally related to and driven by the concentration of population in an area [2]. According to the United Nations' world population prospects in 2019, there will be an increase in the next 30 years of two billion people, from the current world population of 7.7 billion to 9.7 billion in 2050. Further, this increase will grow to almost 400 cities in the early 21st century, which includes around 70% in developing countries [3], including Sri Lanka. Currently, the urban population of Sri Lanka is almost 25% of the total population, which is expected to increase by 65% by 2030, which will cause cities to grow physically and numerically, creating urban sprawl issues in the future [4].

Urban sprawl has attracted much attention among policymakers and researchers in developed [5,6] and developing [4,7–9] countries worldwide. Most arguments for urban sprawl are not based on strong empirical evidence but rather on speculation and assumptions [10]. Many researchers explain this concept with the urban environment of the research area. So far, there is no consensus on the definition of urban sprawl. Further, urban sprawl is a socioeconomic phenomenon that has gradually become a critical issue in many urban areas [11], including Sri Lankan cities. Built-up area, which is often considered a parameter for measuring urban sprawl [2], especially settlement density or size, can only afford an overall measure of the urban form [12]. Increasing urban sprawl results in household preferences, the locational choice for commercial investment, often being agricultural, and building construction in the cities [13,14]. Vacant lands, which are primarily transformed into housing on a daily basis, are increasing sprawling growth and property costs in the urban periphery and surrounding areas.

As a developing country in the world, Sri Lanka has been studied to define urban sprawl [4] and its impact [15] particularly in Colombo city and the spatiotemporal patterns of urban sprawl in Kandy city [8]. Besides these, the Batticaloa municipality area is highly accompanied by residential and commercial development. According to the Sri Lanka National Physical Planning Policy and Plan 2010–2030, Batticaloa is a rapidly developing city in Sri Lanka that is expected to become a metro city by 2030. Considering the existing situation in the Batticaloa municipality, many ongoing urban development projects have been carried out since the end of the Civil War in 2009. However, these developments were not well designed by planning experts. It led to less effective growth and changes in the urban area. Some studies have examined the urban land use changes in this area [16,17], and these studies provide an insight into why this area should be necessary for studying urban sprawl patterns. The built-up development has been growing rapidly in this area in recent decades [16], creating an irregular pattern or sprawling growth established with empirical evidence.

Furthermore, the characteristics of urban sprawl have been studied in various cities around the world (see Table 1), such as India, Malaysia, China, Romania, and the United States. However, these studies have not identified all the characteristics of urban sprawl in a single city. Several characteristics were only identified in a particular city in a developed country or a developing country. Moreover, since Sri Lanka is a developing country, it is difficult to identify influencing characteristics of urban sprawl based on the experience of previous literature from developing countries. Furthermore, cities in Sri Lanka never studied the characteristics of urban sprawl before, which also makes it difficult to understand the pattern of urban sprawl in the Batticaloa city. More precisely, there is still a lack of knowledge, especially about the characteristics (forms and patterns) of urban sprawl via the analysis of the different parts of the city, which is essential to tackle the sprawl effectively. The built-up patterns are the key parameter to identify the different characteristics such as low-density development, leapfrog development, commercial ribbon development and scattered development to establish the urban sprawl development. So far, the planners or academics have not yet addressed the sprawling development in this city in Sri Lanka. If this growth continues in this city, it will affect its sustainability when it becomes a metro city in 2030. In the end, this study can answer which characteristics influence the Batticaloa municipality through geospatial analysis.

Thus, this study can involve finding the different characteristics of urban sprawl in the Batticaloa municipality through spatial patterns. This finding can minimise the sprawling growth in the future and develop a sustainable city in Sri Lanka. The influencing characteristics in this city can be identified from the experiences of previous studies of urban sprawl in different cities in the world (see Table 1). Therefore, this study aims to identify the characteristics of urban sprawl patterns in the Batticaloa municipal council using Geographic Information System (GIS) and remote sensing technology. The findings can provide knowledge about the characteristics of urban sprawl to understand the sprawling patterns in other cities in Sri Lanka that have not been addressed so far. Apart from the empirical and methodological contributions, the findings of this study, in line with

Sustainable Development Goal 11 and the New Urban Agenda, offer useful insights and measures to control sprawl.

Table 1. Summary of characteristics for urban sprawl based on previous studies.

Authors	Low Density	Leapfrog Development	Commercial Strip or Ribbon Development	Scattered Development	Auto Dependent or Car Dependent	Uncontrolled Growth	Uncoordinated Growth	Unplanned Growth
Hamad [1]						X	X	X
Grigorescu et al. [5]	X	X						X
Lv et al. [7]	X			X	X			
Yue et al. [9]	X	X						
Paul et al. [12]	X	X	X	X				
Farooq & Ahmad [13]	X	X	X					
Prakasa, Soemardiono, & Defiana [18]	X	X	X	X				
Galster et al. [19]	X	X						
Sudhira & Ramachandra [20]						X	X	X
Ottensmann [21]	X	X	X	X				
Shirkhanloo [22]	X			X	X			
Nikolov [23]	X					X		X
Sinha [24]	X					X	X	X
Bhatta et al. [25]				X				
Osman, Nawawi, & Abdullah [26]	X	X	X					
Johnson [27]	X	X						
Pichler-Milanović [28]	X							

2. Definition and Characteristics of Urban Sprawl

Although urban sprawl is still considered an elusive concept, it has been used around the world for almost eighty years [29]. It was first realised through the transformation of agricultural and forestry areas into industrial, residential, and commercial development in the United States in the late 1950s. The term "urban sprawl" appeared in printed documents in 1960 [30] and is used in various fields, such as urban studies, remote sensing, and geography [18].

Urban sprawl has been defined by its characteristics identified in a particular urban area. Urban sprawl is the encroachment of non-urban lands to urban lands that occurred beyond the built-up area with a leapfrog pattern, unorganised pattern, low-density pattern, and unordered development [9]. Some patterns, such as ribbon development, low-density development, and leapfrog growth, were identified in Aligarh city, India, called urban sprawl [13]. However, urban sprawl was defined by eight metrics for land use, such as concentration, clustering, proximity, mixed-use, nuclearity, density, clustering, and centrality [19]. Sprawl is the irregular urban form with different scales of land use that consist of common and institutional facilities [31].

In contrast, similar patterns, such as uncoordinated, uncontrolled, and unplanned growth, were found along highways, called urban sprawl [20,32]. In addition, urban sprawl is unplanned discontinuous growth categorised by low-density growth in urban boundaries [5]. It is a typical pattern with scattered growth [11,21,22]. Urban sprawl is car-dependent, and low-density development has several negative impacts [22]. Low-density patterns and stripe development along major highways were identified in Colombo metropolitan area, Sri Lanka [4]. Sprawl development was identified in three directions along three main roads in Kandy city, Sri Lanka [8], which is almost the similar pattern of Colombo city.

Thus, the definition of urban sprawl varies among researchers who define it based on the characteristics of their urban area [5,7,19,22–24]. Although specialists and researchers still have problems defining the term "urban sprawl" [15,33]. However, many researchers accepted Ewing's (1997) definition is more suitable for recognising the urban sprawling, which stated an urban land use existence, characterised by scattered development, low-density, leapfrog development, and commercial strip development [12,34]. According to Table 1, more than 17 researchers have used more than one of these characteristics to explain urban sprawl.

Based on the review (see Table 1), urban sprawl characteristics, which are low density, leapfrog development, commercial strip or ribbon development, scattered and dispersed development, auto-dependent or car-dependent, uncontrolled growth, uncoordinated growth, and unplanned growth, were mainly identified in the developed and developing countries. Some characteristics, such as low density, leapfrog development, and scattered development (see Figure 1), were found in many countries [12,18,21], for example, Indonesia and India. As a developing country in the world, Sri Lanka has many development plans implemented due to the rapid urban growth in recent decades, which is causing urban sprawl.

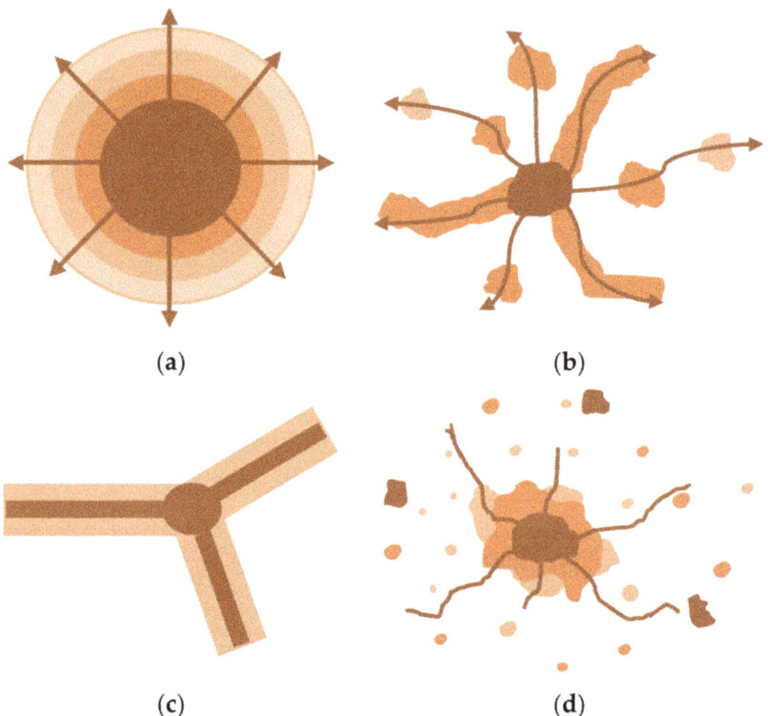

Figure 1. (a) Low-density development [35], (b) Leapfrog development [35], (c) Commercial ribbon development [35] and (d) Scattered development [36].

3. Materials and Methods

3.1. Description of Study Area

The study area, Batticaloa Municipal Council, is the local authority in the Batticaloa district located in the eastern part of Sri Lanka. The average elevation of Batticaloa is 8.523 m above Mean Sea Level. The total population in this area is 93,306 people. The propagation of ethnic population is as a "sandwich pattern" with Tamil (91%), Muslim (5%), Sinhalese (0.14%), and others (3.86%) [37,38].

The total land area of the Batticaloa Municipality is 4311.87 hectares, which is allocated for different purposes of utilisation such as residential, agricultural lands, commercial, wetlands, water bodies, scrub forests, and others (refer to Table 2). Five (5) land parcels separate this city from the inland water bodies. These land parcels are connected to the bridges for transportation. As a clustered city, the land area uses for multiple purposes and these links with various sectors, such as fishing, agriculture, small industries, and commercial. Each land parcel has different property uses and has cluster development in each sector, such as commercial, recreational, and residential. The most dominant land use of this area is residential (1170.24 hectares), and the next is agricultural lands (935.6 hectares). One of the low proportions of the land-use class is commercial (23.5 hectares), which compares to other major land uses. The natural lands, including wetlands (82.5 hectares), water bodies (58 hectares), and scrub forests (185.71 hectares), are also a certain portion in this area [37].

Table 2. Major land use categories in Batticaloa Municipality.

Land Use Type	Area (Hectares)	Area (%)
Residential	1170.24	27.1
Agricultural lands	935.6	21.7
Commercial	23.5	0.6
Wetlands	82.5	1.9
Water bodies	58	1.4
scrub forests	185.71	4.3

Considering the infrastructure facilities in the Batticaloa municipal area, it is undeniable that ongoing development projects will improve their current conditions. However, these development activities are not planned by urban planners and relevant development officials. Arbitrary development occurs highly in this area which affects the pattern of sustainable land use. Since 1990, the rapid growth of the population has caused several changes in the built pattern that were not assessed by the authorities for sustainable development. Permanent and temporary migration to Batticaloa municipality increased from other parts of the district and the Eastern province due to the effects of the Civil War because this area is the major urban centre in the Eastern Province, Sri Lanka, with all amenities. In addition, the proposed development plan in Batticaloa municipality for 2030 contained many rules and regulations on land use, especially built-up development, which is not considered much further in current development. Thus, Batticaloa municipality began to face the urban sprawl development in the core city and the periphery. As rapidly developing cities in Sri Lanka, Batticaloa received more concern from planners as the city was expected to become a metro city by 2030. Therefore, this area has been chosen to study the urban sprawl in Sri Lanka, which is more significant to understand the characteristics of sprawling. Figure 2 shows the Batticaloa Municipal Council area, the study area in this research.

3.2. Data Collection—Source of Data

Understanding the characteristics of urban sprawl requires the pattern of land use, especially the built-up changes in the area. Google Earth Pro and ArcGIS 10.6.1 applications were utilised to produce various layers, such as built-up maps, density maps using satellite images, boundary map for Batticaloa municipal council, and Grama Niladhari division map for Batticaloa Municipality.

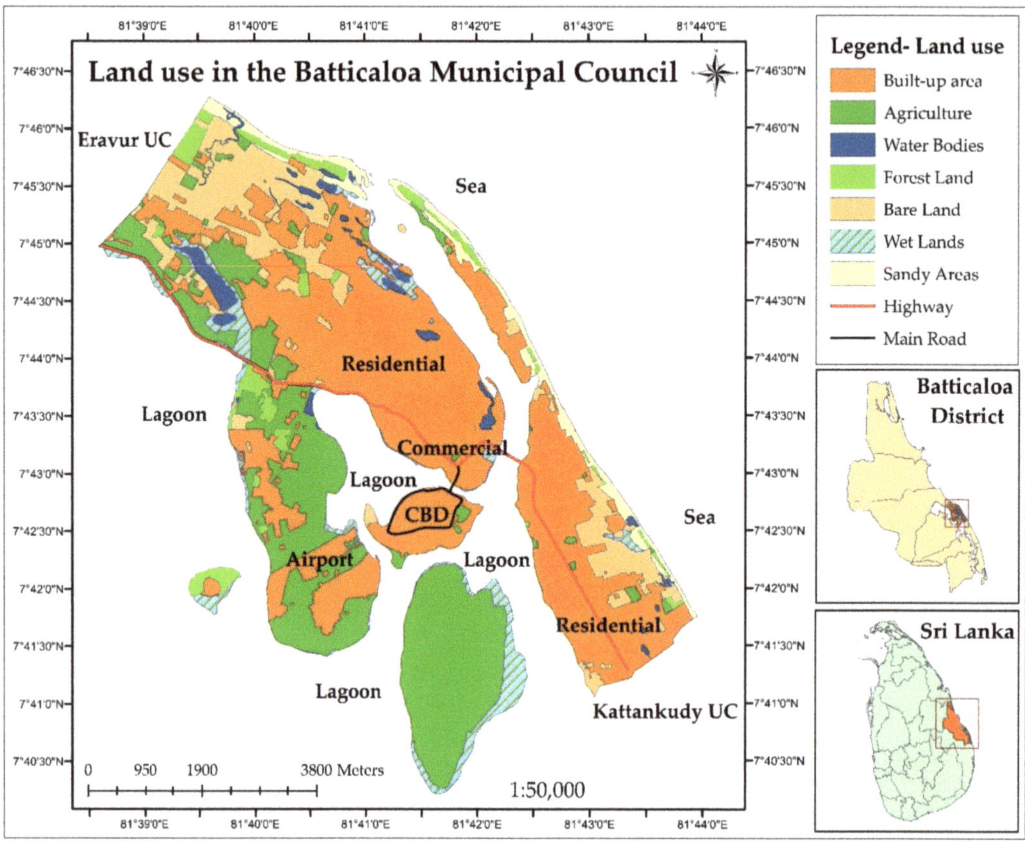

Figure 2. The Study Area—Land use pattern in the Batticaloa Municipal Council—2020. Source: Modified from the Batticaloa Municipal Council Profile, 2021.

There are two (2) types of data, namely, remote sensing data (satellite images) and demographic data, used to generate the maps. Satellite images were downloaded from the Earth Explorer, United States Geological Survey [39]. Details of this data presented in Table 3 with all the information. These time-series images for 2000, 2010, and 2020 used to produce the land use maps to identify the sprawling and extract the built-up area. Meanwhile, the demographic data were collected from the Department of Census and Statistics, Sri Lanka. This data was utilised to produce the population density map in order to identify the density changes. Then, census data for the years 2001, 2012, and 2019, which are the most recent years with satellite images, were compared with the built-up images to understand the influence of population growth on urban sprawl.

Table 3. Details of satellite imagery data.

Type of Satellite	Image ID	Acquisition Date	Resolution
Landsat ETM+ (2000)	LE07_L1TP_140055_20000928_20170209_01_T1	28-SEP-00	30 m
Landsat TM (2010)	LT05_L1TP_140055_20100924_20161212_01_T1	24-SEP-10	30 m
Landsat 8 (2020)	LC08_L1TP_140055_20200303_20200314_01_T1	03-MAR-20	30 m, Pan-15 m

Source: Earth Explorer, 2021.

3.3. Data Processing and Analysis

The downloaded satellite imageries were geo-referenced in World Geodetic System 84 (WGS84) and then projected to the Kandawala local coordinate system. Filters, brightness, and contrast tools were used to improve the quality of the satellite images. The Batticaloa Municipality's boundary was digitised as a shapefile using the current map to demarcate the study area. Based on this boundary, three (3) satellite images, which are from the years 2000, 2010, and 2020, were clipped to delineate the satellite images based on Batticaloa Municipality's boundary. Then, these images were classified into six (6) classes, namely, built-up, agriculture, forest, water bodies, vacant land, and others, according to the training land samples using Supervised Maximum Likelihood classification in ArcGIS. Then, the classified images were validated using the accuracy assessment technique. About 85% of overall accuracy is usually considered enough in the map data [25]. The overall accuracy of land use can be obtained by the following Equation (1) [40]:

$$OA = (1/N) \sum_{i=1}^{r} n_{ii} \tag{1}$$

where OA is overall accuracy, n_{ii} is correctly classified pixels' number, N is pixels' total number, and r is rows' number.

Ground truth data was considered in comparing the classified Landsat images. Overall, 127 training samples for ground truth were obtained as random points at specific locations using the grid layout in Google Earth Pro and using known coordinate points. Each point has valid land-use class values, which are:

(1) Built-up;
(2) Agriculture;
(3) Water bodies;
(4) Scrubland;
(5) Mangroves;
(6) Vacant land;
(7) Others, including the playground, transportation, park, and public land, for the classified and ground truth fields.

The confusion matrix was formulated to find the accuracy and obtain individual accuracy between the classified classes and the reference data, such as coordinate points collected from the field. The classification accuracy was determined to obtain the level of precision. The final output of this process was a land use map of Batticaloa Municipality from the years 2000, 2010, and 2020.

Then, density mapping was used, which is a method to show the location of points or lines which can be concentrated in a given area. Such maps often use interpolation methods to estimate a given surface where the concentration of a given function can be. The population density is 1372 persons per km^2 in Batticaloa [37], which is rapidly increasing in recent decades due to rapid urbanisation. Therefore, the population density map was produced using the density analysis tool in ArcGIS. The population for each Grama Niladhari divisions was utilised to produce the density map. The spatial boundary data for Grama Niladhari was developed using the base map of the Batticaloa Municipality. The population data for each Grama Niladhari division were added to the spatial file to show the population's spatial distribution. Based on the standard calculation of the population density, the number of people divided by land area is calculated. The Equation (2) for population density is as follows:

$$PD = (Tn/a) \tag{2}$$

where PD is population density, Tn is the total population in a particular area, and a is the land area in hectares.

It is categorised by suitable data clustering methods, which is Jenks's natural breaks classification, designed to determine the best arrangement of values. It is used to classify the data into five (5) categories, such as very high, high, moderate, low, and very low. This density distribution helps to understand the changes clearly as to which area is high and low density. Figure 3 shows the complete methodological framework for identifying urban sprawl characteristics.

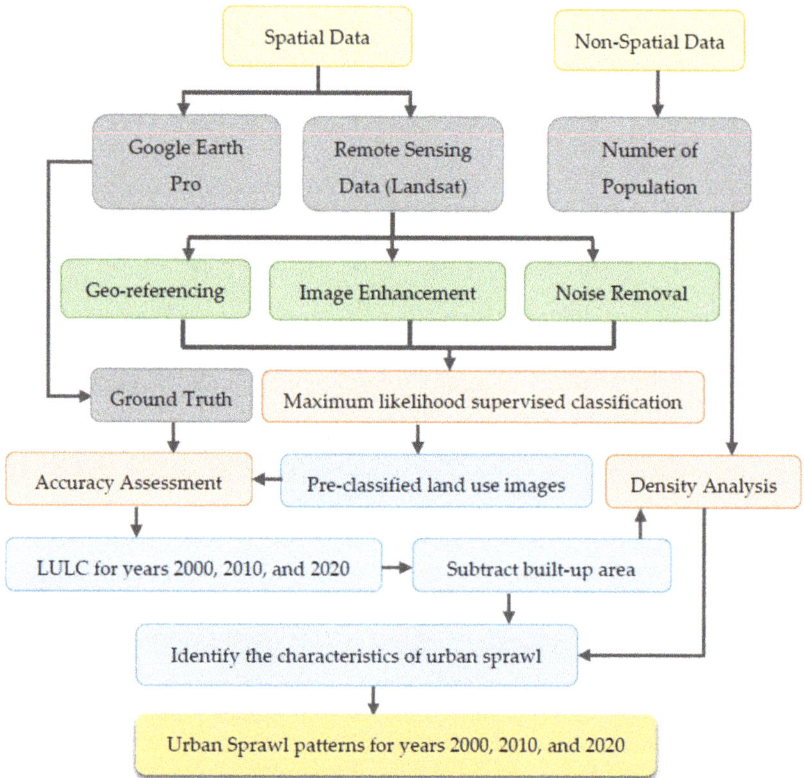

Figure 3. The methodological framework for identifying urban sprawl characteristics.

Further, the built-up density map was also produced to establish the low-density development. The buildings of the study area were digitised using Google Earth Pro. Each building's features were converted into points using the feature to point tool in ArcGIS. Then, buildings for each Grama Niladhari divisions were clipped, and density analysis for buildings was conducted to generate the built-up density for the Grama Niladhari division in Batticaloa municipality. The density was categorised by Jenks's natural breaks method into five (5) classes as very high, high, moderate, low, and very low. The density changes in the study area were identified by using these maps. Based on the results also, we can clearly understand which area has more sprawl.

4. Results and Discussion

The dynamics of the built-up area, known as a typical process of urban sprawl, are of particular importance for understanding spatial patterns for sprawl development. The built-up spatial patterns identified the characteristics of the urban sprawl in the Batticaloa Municipal Council.

4.1. Built-up Patterns

The built-up patterns are presented in the maps of years 2000, 2010, and 2020 (refer to Figure 4) to understand the sprawling characteristics in the Batticaloa municipal council. The built-up area has an extent of 1162 hectares in 2000, which increased to 1439 hectares in 2010. It increased to around 1557 hectares in 2020, showing the increases in the city's built-up pattern (refer to Table 4). Based on this, a rapid increase in built-up growth was identified during the selected periods.

Table 4. The extent of built-up area in Batticaloa Municipality.

Category	2000	2010	2020
Built-up area	1162 Hectares	1439 Hectares	1557 Hectares

The accuracy of the built-up pattern is identified by the classified land use maps of the study area. The analysis revealed that the producer accuracy and the user accuracy varied during the selected periods. Producer accuracy refers to the accuracy of the map with how often real features on the ground are displayed correctly. User accuracy refers to the map user accuracy, which indicates how often the land use class on the map is actually present on the ground. User accuracy shows the reliability of the map. According to the calculation, the producer accuracy for the built-up area in 2000 is 89.43%, while the user accuracy is 96.72%; in 2010 the producer accuracy is 94.6%, while the user accuracy is 99.72%; and in 2020, the producer accuracy is 87.71%, while user accuracy is 93.29%, which is an excellent precision for analysis. The overall accuracy of around 85% is considered enough to prove the precision of the map data [25].

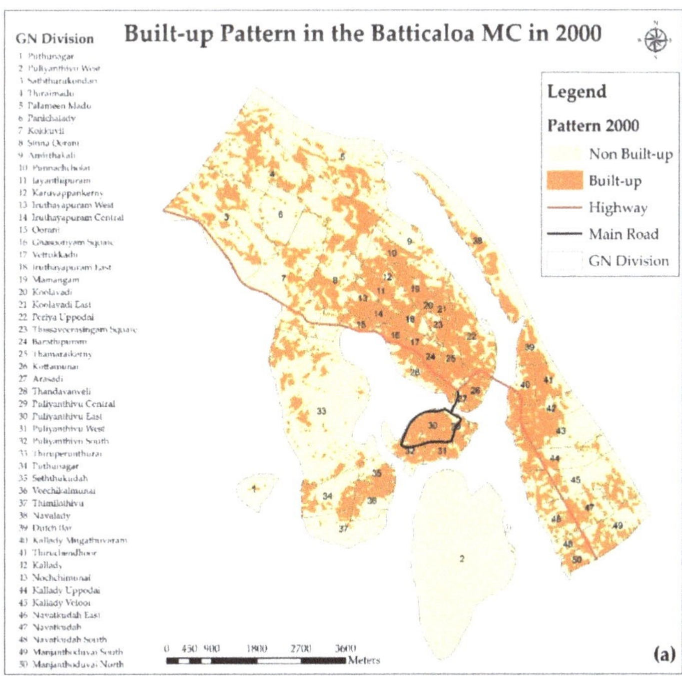

Figure 4. *Cont.*

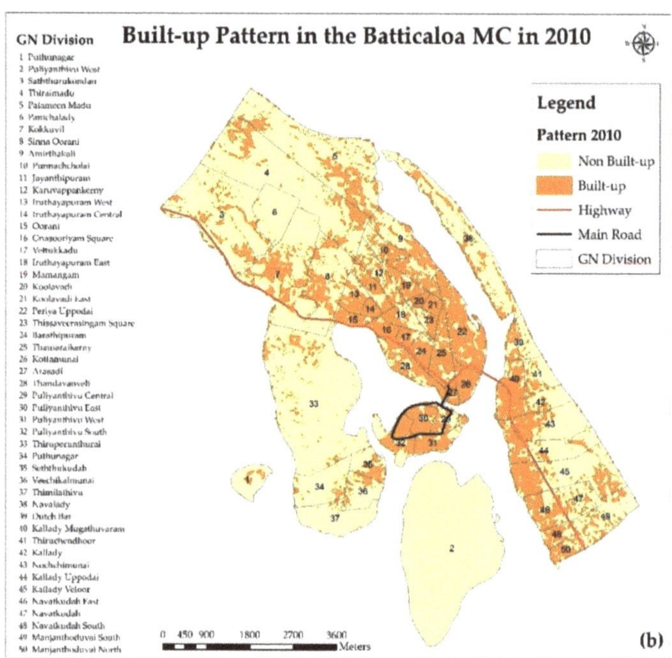

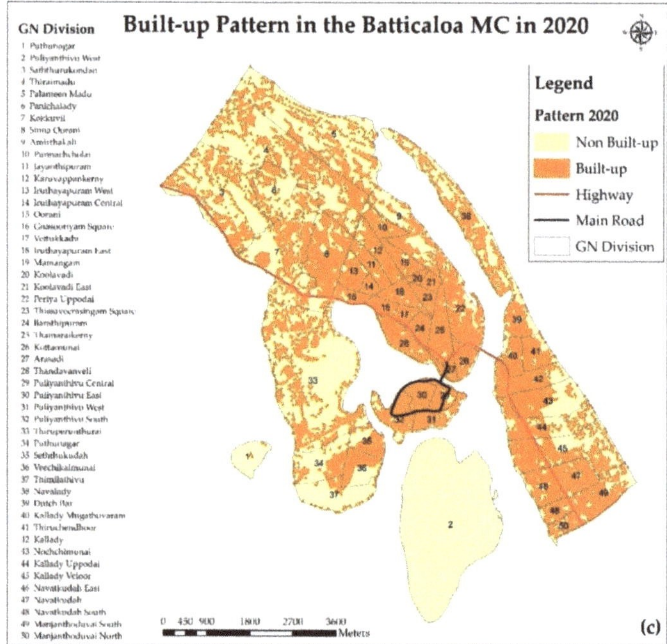

Figure 4. Built-up patterns in the Batticaloa Municipal Council in (**a**) 2000, (**b**) 2010, and (**c**) 2020. Note: The Grama Niladhari Division (GN) is a subdivision of the Divisional Secretariat in Sri Lanka. A total of 14,022 Grama Niladhari divisions are in charge of 331 Divisional Secretariat divisions in Sri Lanka; of these, 48 of Grama Niladhari divisions are in the Batticaloa municipality.

The Batticaloa municipality area consists of 48 Grama Niladhari divisions. Thiraimadu is one of the divisions that emerged with housing developments after the tsunami disaster. This area is also developing as an administrative zone in the city that encourages people to construct housing, which is one reason for rapid built-up development in this area. In addition, Puliyanthivu Island is the city centre, with densely developed commercial and residential buildings. However, the lower land value in the Navalady and Thiraimadu areas attracts people with low and middle incomes to buy the land and build a house. The main reason for the lower land value is that this area is often affected by disasters, especially flood. Besides this, high land value has been identified in the Puliyanthivu, Oorani, and Thandavanveli areas because these areas are close to the city centre, the highway, and several infrastructure facilities. One of the best policies is Land Value Capture and Taxation, which is beneficial for affordable housing in this city with a lower land value. This system is in place to increase revenue and fix up the downtown buildings in the city. This income can be used to develop housing for people with low incomes. However, the tax system is already implemented in the city, which is not strictly followed annually. Although everyone is aware of the property tax in the city, they sometimes forget to pay the annual renewal tax. The municipality does not remember and observe these activities regularly, which leads to the illegal land formation as well as sprawling development in the city.

Several groups of people own the total land area in the Batticaloa municipality. This land has been distributed around 73.1% to the inhabitants, 9.1% to the government, 4.7% to Batticaloa municipality, and 13.1% of obscure details. One of the principal regulations is that people cannot construct any buildings in the Batticaloa municipality area without obtaining an approved development permit. However, some people carry out illegal construction development, which represents around 13.1% of the total area of Batticaloa municipality, and those do not contain explicit information about the property. These owners build houses or other buildings on other people's land without getting the proper approval from the municipality. These activities increase the most illegal construction in the city. These developments triggered the formation of scattered and leapfrog development in the city, which are the main reasons for the sprawling growth in this city. Therefore, the property documentation system should be appropriately maintained by the municipality. A survey for property owners should be conducted at the turn of the year to identify illegal land and minimise sprawling growth. This survey can also inspire people to pay taxes without fail. In addition, the municipality should gently remind people before the annual tax period ends. For this, a smart application should be developed for the municipality to identify the pending cases quickly. These practices can control land occupancy and the maintenance of more than one piece of land of a person in the city.

In addition, land ownership problems related to communities in the Batticaloa municipality were identified in the Nochchimunai and Sinna Oorani Grama Niladhari divisions. At the same time, the Puliyanthivu South, Kallady uppodai, Kokkuvil, Punnaicholai, Karuvappankeny Amirthakali, and Mamangam areas identified landowners' issues associated with low-income people. These types of problems caused to form vacant land and more subdivided lands in the city. In addition, illegally divided lands were identified mainly for sale in the Saththurukondan and Uppukarachai area, where the land value is relatively high today. These activities mainly lead to sprawling growth in the city. Therefore, the municipality must circulate a pre-approval method to divide the land in the city. People should inform the municipality about the subdivision of the land, and then the municipal official should visit the specific area to observe the land. After that, the municipal guidelines must be adapted to that particular area, and the entire previous land document must be verified to confirm the land entitlement for subdivision and sale. This practice can find illegal subdividing of land and sale, minimise future land problems within communities, and reduce the amount of vacant land in the city.

Figure 5 shows the built-up changes between the years 2000 and 2010, and between 2010 and 2020 to understand the expansion. This comparison showed a gradual increase in built-up changes in the study area. The built-up growth increased 227 hectares between

2000 and 2010, and 118 hectares between 2010 and 2020 (see Table 5). These gradual changes in the built-up area have illustrated the growth of sprawling in the city. The rapid growth was registered between the years 2000 and 2010. The Batticaloa area was one of the severely affected areas by the civil war in Sri Lanka. Thus, most people from the other parts (rural areas) of the Batticaloa district, such as Porativu, Vellaveli, Mandur, Thikkodai, and Vaharai, migrated to the Batticaloa city for survival, including security, education, and livelihood. Furthermore, the living standards, access to more facilities, and admired city life are the reasons for the migration of people to the city. The movement to the city led to the demand for land and housing for the people. The desire for their own property has created a haphazard development in the city and the urban fringe of the Batticaloa area. Thus, different characteristics were identified in the different areas of Batticaloa city. However, many urban sprawl characteristics exist in the world's cities, but the study area is occupied with some characteristics.

Table 5. The changes of built-up area in Batticaloa Municipality.

Category	Timespan	Expansion
Built-up Area	2000–2010	277 Hectares
	2010–2020	118 Hectares

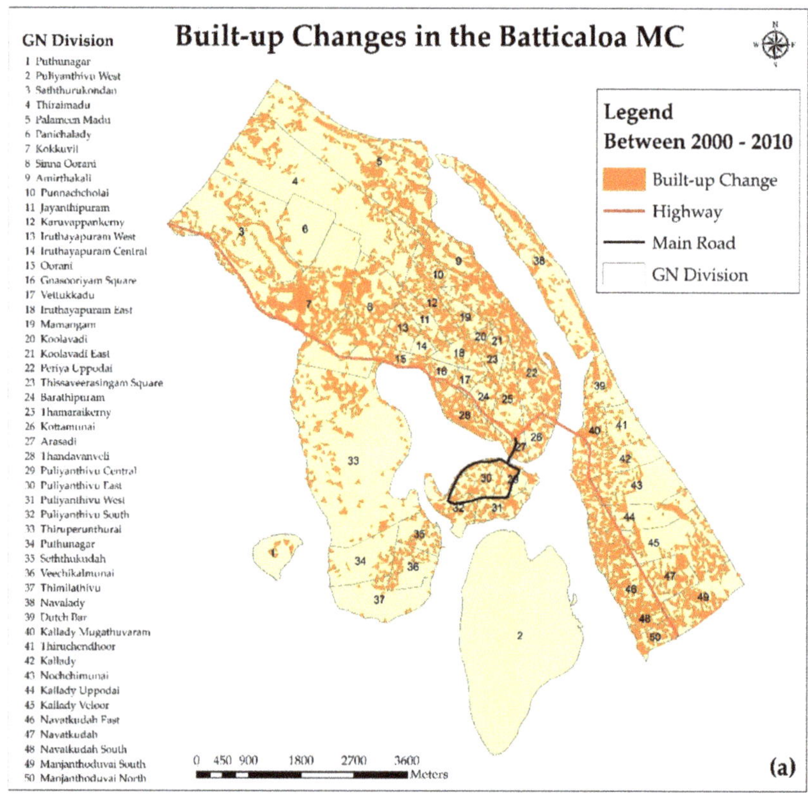

Figure 5. *Cont.*

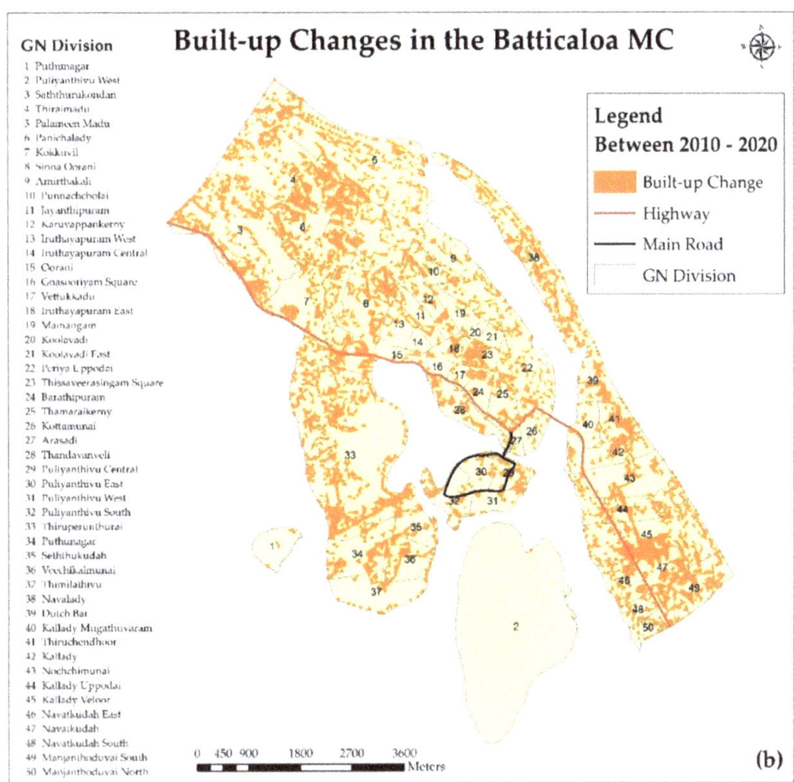

Figure 5. Built-up changes in the Batticaloa Municipal Council (**a**) between 2000 and 2010, and (**b**) between 2010 and 2020.

4.2. Characteristics of Urban Sprawl

The built-up area was extracted from the land use map to identify the urban sprawling characteristics. The spatial and temporal built-up patterns reveal that sprawling characteristics identified as low density, leapfrog development, scattered growth, and commercial ribbon development influenced the irregular urban development pattern. Most of these characteristics are identified in the city limits, and some are in the core city, which affects the city's sustainable growth.

4.2.1. Low-Density Development

Low-density development is one of the main phenomena of urban sprawl generally risky to the urban environment. The primary units for identifying the urban sprawl, including density, are buildings, especially housing units of a particular area [19]. The residential developments are mainly identified in the marginal low-density areas in the city. Low-density development is a piecemeal extension of the built-up area, which consumes much land in the urban fringe. It is the most generally indicated characteristic of urban sprawl in many pieces of literature [41]. Residential housing mostly consumes the vast land, which was vacant land previously, leading to the low density. The rise of land and property value in the city cannot afford a vast population; however, this value is meagre in the urban fringe. Thus, the sprawl areas are occupied mostly by the low-income people for their permanent residence. They are attracted by these vast, spacious living areas to build an affordable house [24], also experienced by the Batticaloa area. The housing preference of the lower class and some middle-class people pushed them to settle in these low land value areas. The people who migrate from the village areas admired the city limits, which is more spacious and affordable for their own housing. In addition, a single dwelling unit in a larger area in the Batticaloa municipality is one of the main reasons for the low-density development, which is similarly identified in the United States of America [24].

Figure 6 shows the built-up density in the Batticaloa municipal council area by Grama Niladhari division. Based on this, higher density patterns were identified in the city centre from 2000 to 2020. However, nine divisions, which are Saththurukondan, Thiraimadu, Paalameenmadu, Panichalady, Kokkuvil, Thiruperunthurai, Thimilathivu, Veechikalmunai, and Navalady, mainly come under the low density in the divisions throughout these periods. The lack of space in the city centre for housing development has been limited to low-income residents where land value is most in demand. Further, nearly 150 low-income families living in the city had the rural characteristics identified in Sinna Oorani, Punnaicholai, Thiraimadu, Mamangam, Kokkuvil, and Saththurukondan areas. The developed built-in density map can be useful for identifying areas with low and very low density in the city. Based on this, an appropriate development plan can be adapted to this area to make the city more compact.

Table 6 shows the range of built-up density in the Batticaloa Municipality. The ranges were categorised from very low density to very high density. Based on this, around 0–4 buildings per hectare were identified in the very low-density areas and more than 22 buildings per hectare in the very high-density areas. The built-up density increased near the city centres, except in the northern and western part of the city, from 2000 to 2020. The main reason for the low density in these areas is the inadequate facilities such as accessibility to highways, commercial, and other services.

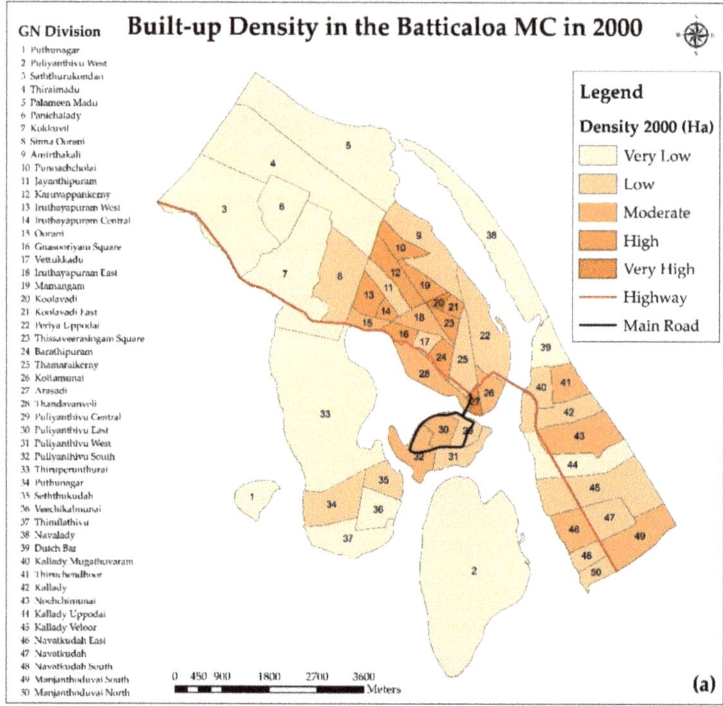

Figure 6. *Cont.*

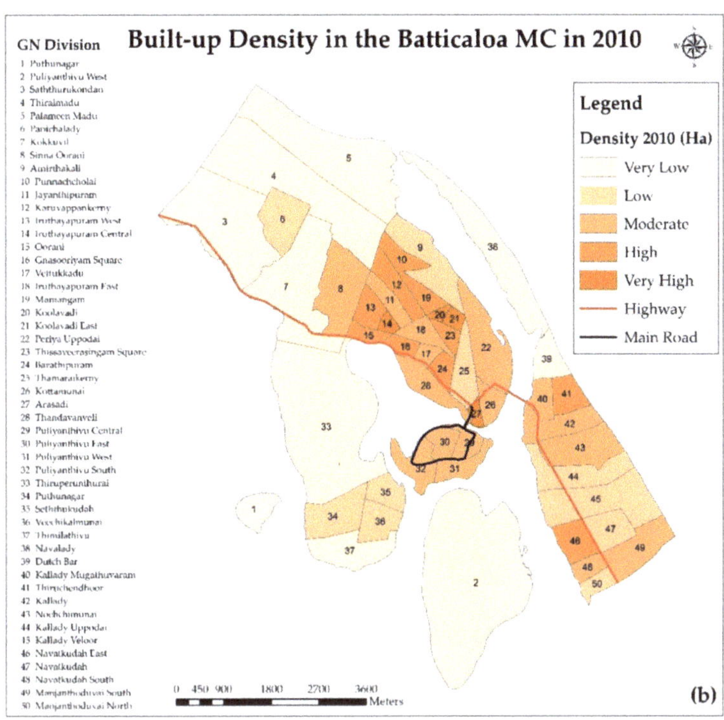

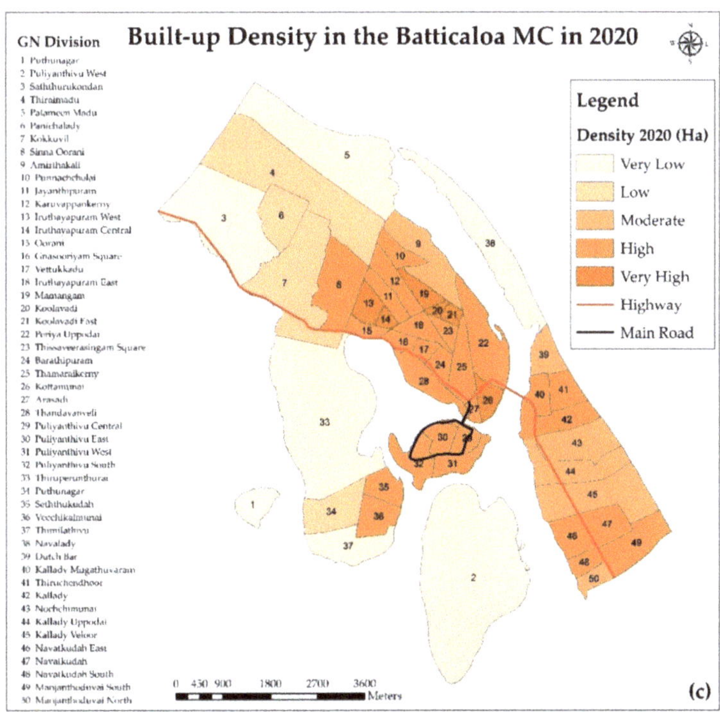

Figure 6. Built-up Density in the Batticaloa Municipal Council in (**a**) 2000, (**b**) 2010, and (**c**) 2020.

Table 6. The range of built-up density in Batticaloa Municipality.

Scale	Density (Buildings/Ha)		
	2000	**2010**	**2020**
Very Low	0.0–4.0	0.0–4.0	0.0–4.0
Low	4.1–8.0	4.1–8.0	4.1–8.0
Moderate	8.1–13.0	8.1–13.0	8.1–13.0
High	13.1–22.0	13.1–22.0	13.1–22.0
Very High	22.1–37.0	22.1–31.0	22.1–34.0

Most of the lands are used for single use, like individual housing, which created the low-density development in the Batticaloa municipality. For example, 88% of homes are single-storey separated houses, 9% of homes are two-storey separated houses, and 1% of homes are more than two-storey separated houses. These housing patterns show a rural characteristic in this city. In addition, these single housing developments are one of the main reasons for the low-density development in this area. Therefore, housing policy must be designed in accordance with Sustainable Development Goal 11 and the existing situation of Batticaloa municipality. Housing policy should be developed in consultation with stakeholders in Batticaloa Municipality who provide a clear view of all income earners and the different communities living in the city. The municipality then displays the decision for public responses that provide different ideas for improving the policy before it is implemented. Furthermore, reporting the municipality's policy in public can also make the right decisions in all activities by people, including building houses and maintaining the land.

In addition, building codes must be provided to track building types and the location of buildings in the city. This method can help to quickly identify a specific building in all situations and demolish illegal constructions. These activities primarily help control the future sprawling growth of the city. Further, the developed built-up pattern and density maps are helpful to identify the additional unregistered buildings in each Grama Niladhari division. For example, a Grama Niladhari division already has 25 buildings registered in the municipality, but the map shows 28 buildings in the same division. By this, the constructions can be understood as illegal development in the area in question. The municipality can take the necessary measures against them and also minimise the sprawling growth in the city. In addition, a monitoring unit should be set up to review housing policy and the necessary strategies for building construction in the city. An online platform should be developed to guide public discussions and consultations. This continuous monitoring activity can control illegal housing development in the city.

The population density is measured by the ratio of people inhabiting a specific region in persons per square kilometre or hectare. A city that occupies a smaller land area is considered more compact and less sprawled, and that with more extensive land occupied by less population implies low density and a more sprawled characteristic [41]. Table 7 shows the range of population density in the Batticaloa Municipality. Population density ranges from very low to very high density. Areas with 0 to 10 persons per hectare are known as very low-density areas. Areas with over 100 people per hectare were identified as very high-density areas in 2001, and with more than 84 people per hectare were very high-density areas in 2012 and 2019. The population density is almost high in the city centre and close to commercial areas.

Table 7. The range of population density in Batticaloa Municipality.

Scale	Density (Persons/Ha)		
	2001	**2012**	**2019**
Very Low	0.00–10.00	0.00–10.00	0.00–10.00
Low	10.01–34.00	10.01–29.00	10.01–29.00
Moderate	34.01–60.00	29.01–53.00	29.01–53.00
High	60.01–100.00	53.01–84.00	53.01–84.00
Very High	100.01–151.00	84.01–124.00	84.01–131.00

The consumption of the land is faster than the population growth, which revealed low-density development. As shown in Figure 7, the population density in the Batticaloa Municipality is identified by the Grama Niladhari divisions. Based on this, developed areas such as Puliyanthivu, Arasady, Thandavanveli, Oorani, Kallay, Iruthayapuram, Mamangam, Thiruchendhoor, and Navakkudah are almost high built-up density areas. However, those areas have not populated much when compared with built-up. However, the highest population density was identified in the Arasady, Koolavady and Iruthayapuram areas. Low density was registered in the edge areas, but the core city and the highway area only showed higher population density and built-up. Paalameenmadu, Navalady, Saththurukondan, Puliyanthivu west, and Thiruperunthurai areas were identified with the lowest population density patterns in the Batticaloa municipality. The main reason for the low density is poor accessibility to highway and other services. In addition, Seththukudah, Vechukalmunai, and Thimilaithivu areas were under military control during the civil war period. Thus, people did not desire to make their settlement in those areas.

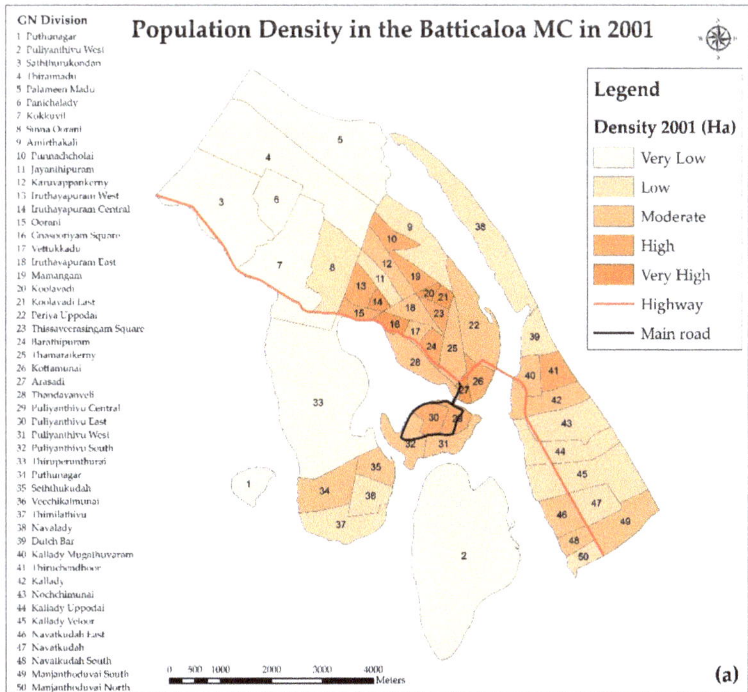

Figure 7. *Cont.*

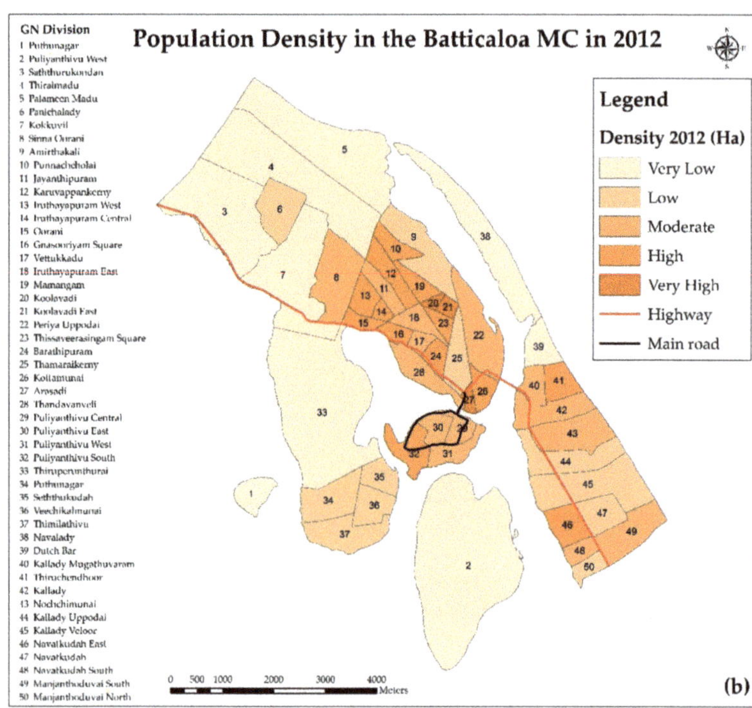

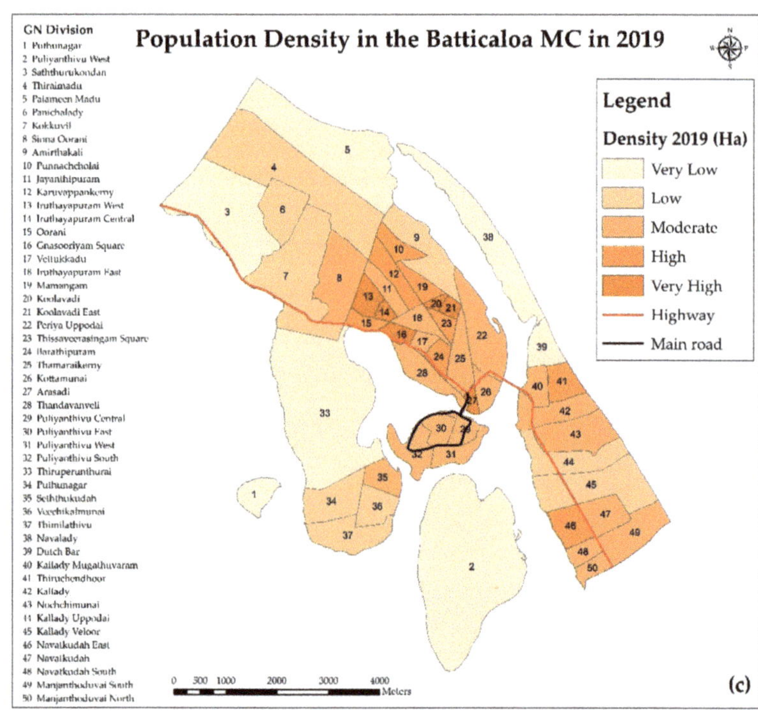

Figure 7. Population Density by Grama Niladhari Divisions in (**a**) 2001, (**b**) 2012, and (**c**) 2019.

Additionally, Batticaloa Domestic Airport is located in the Thirupperunthurai division, which is closest to the Puthunagar and Sethukkudah areas. Residential buildings are banned around the airport areas such as Puthunagar, Thirupperunthurai, and Sethukudah due to the airport expansion project. In addition, a water supply system was launched in the high-density populated areas such as the city centre and the nearby areas. In other areas, people use well water for their needs. Further, 38 schools are located in the Batticaloa municipality area. Of these, seven schools are national schools with a high level of education, located in the city centre and the nearest areas, which is one of the reasons for the high population density.

Further, the population of this city has a high growth rate of 3.92% in the period 1990–2001 and around 2.07% in the period 2001–2010. This rate changes to around 2.27% in the period 2010–2019, which indicates the fastest growth of cities in Sri Lanka. In addition, the population growth rate in the future is expected to be 2.5% to 3.0% in 2030. The minimum expected population is 127,291 persons and a maximum of 170,714 people in 2030 [37]. This rapid population growth can lead to more sprawling development when political influences disrupt municipality development plans and regulations. Therefore, the rules and regulations must be strictly followed to become a sustainable city in the future.

4.2.2. Leapfrog Development

A discontinuous irregular pattern on developed land is widely recognised within the city limits. This type of development makes it costly to provide essential services like water and drainage. This development consumed a wide range of land and created an arbitrary development pattern that destroys urban beauty. Figure 8 shows the leapfrog development, which creates more changes in the land use pattern, leading to the urban sprawl established in the visual map. This development can identify a very inefficient land use pattern, which is one of the most extreme examples of urban sprawl. Such growth affects the development of the city directly, including infrastructure and services.

The leapfrog development creates less housing and population density due to the undeveloped land, such as the urban fringe. This density is higher than the individual homes, which affects sustainable development [21]. The vacant land in specific areas such as Thiraimadu, Paalameenmadu, Kokkuvil, Panichalady, and Navalady are good examples of leapfrog growth with low density in the Batticaloa municipality area. Fundamental accessibilities such as public transportation, telecommunication, and water supply are comparatively poor, leading to less population growth. These people live in small housing units, which means one-room or two-room houses built using bricks or clay, but the land extent is larger than the houses. The people living in these areas are from low-income classes. The municipality should introduce more housing schemes, incorporating with the National Housing Development Authority, Sri Lanka. This development can reduce the leapfrog development within the city limits. In addition, the municipality should raise people's awareness about the leapfrog development and its impacts on the city through community programmes. People's understanding of this issue can regulate built-up development.

The leapfrog pattern develops due to the spatial heterogeneity of agricultural facilities [42], which is the case in the Batticaloa municipality area, such as Thiruperunthurai and Puthunagar. Thus, this discontinuous pattern forms the vacant land between built-up areas, making it difficult to afford facilities by the municipality. The transformation of non-urban land influences leapfrog and edge development into built-up land that increases faster than the growth of population in the Chinese cities [9]. However, this transformation was not identified in the Batticaloa municipality area during the selected periods. Rather, built-up discontinuous patterns were mainly identified in this city, which cause the rising cost of infrastructure.

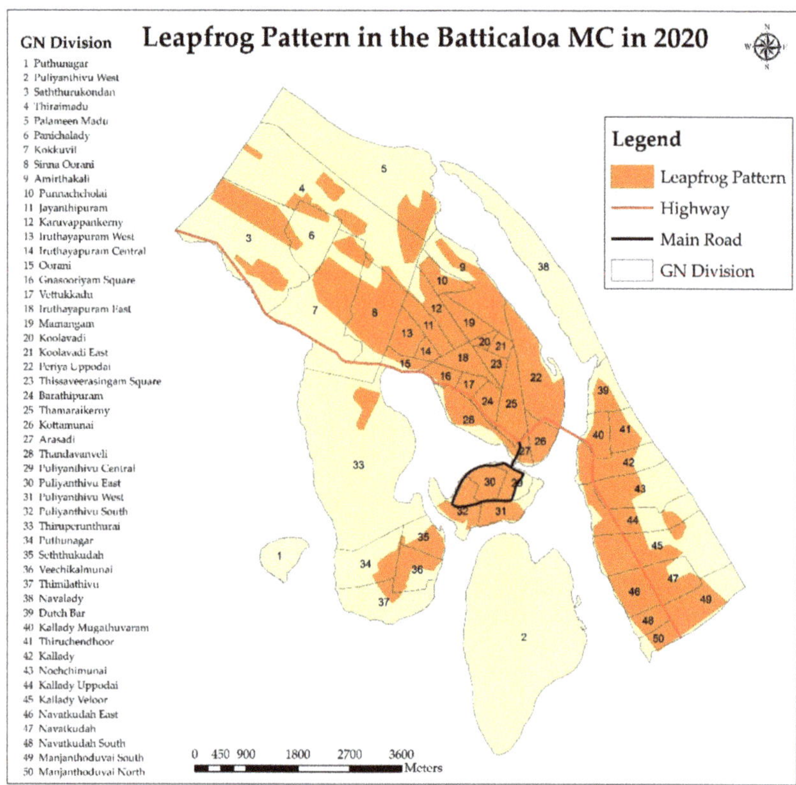

Figure 8. A leapfrog development pattern in the Batticaloa MC—2020.

4.2.3. Scattered Development

Scattered development also provides an inaccessible pattern on the urban edge, like undeveloped areas, creating sprawling. Figure 9 illustrates the scattered patterns in the study area that grows in the urban limit. The built-up growth develops in a dispersed way, which creates a considerable change in the city. This pattern was mainly identified in Thiruperunthurai and Navalady areas in the Batticaloa municipality. One reason for the scattered growth in the Thiruperunthurai area is that the land is mainly used for agriculture purposes. Municipal open spaces assigning a convenience value is one of the ways to account for scattered development. Therefore, people can be willing to spend more money on owning a home in these spacious areas, even if they are further away from the central business district, such as in the European cities [42]. This activity is sometimes experienced in Batticaloa city as well, in recent decades.

Further, the settlement areas of the core city developed by the individual homes in a vast land are mostly composed of single-storey buildings and some of double-storey houses. The individual housing preference of these people induces them to occupy the spacious land for building their dream house. Most families here are nuclear families rather than the extended family needed to build many individual houses for each family. For example, a nuclear family has four members, such as a father, mother, and two daughters. The parents must build two houses for these two daughters to marry them. Parents must build an individual house for each daughter, even if they have five daughters, because of the tradition in Batticaloa. A better solution for this, parents can build a low-rise building like three or four floors with all facilities and assign each floor to each daughter, which can minimise the number of individual single homes in the city. This practice should be included in municipality policy, which makes the city more compact.

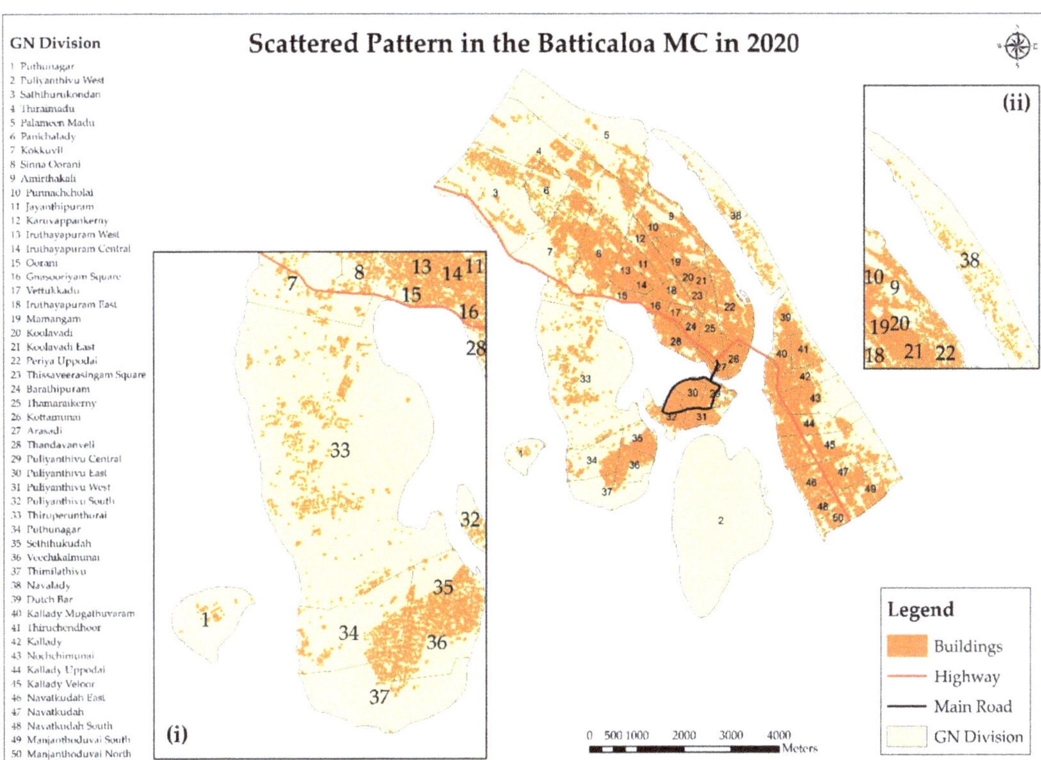

Figure 9. A scattered development pattern in the Batticaloa MC—2020, (i) and (ii) shows the scattered pattern in some GN divisions in the city.

4.2.4. Commercial Strip or Ribbon Development

Commercial development, along with highways, is another characteristic of urban sprawl called ribbon development that threatens sustainable urban growth in the city. Commercial buildings are mostly built along the main transport corridors in the core city and outside of the downtown area (see Figure 10). This type of development caused an increase in the value of the land near the highways. These urbanised areas use a mixed mode, such as commercial and residential, affecting urban land use. Many buildings utilise only the ground floor for commercial purposes, though in some buildings the second and third floors are also used as commercial spaces. Nevertheless, in the rest, all second or third levels are mainly used for residential purposes. People want to buy things such as clothes, grocery items, and electronic products for different purposes simultaneously, and they have to walk long distances to buy these things. This distance affects people's continuous shopping and time consumption.

Figure 10. Commercial Ribbon Development in the Batticaloa MC—2020.

Further, this commercial ribbon pattern causes several problems in the city's functions. For example, approximately 75,000 to 100,000 people commute to Batticaloa municipality from 6 a.m. to 8 p.m. every day. The reasons for this commuting are to access services such as the railway station, the teaching hospital, the Faculty of Healthcare Sciences at the Eastern University, the Swami Vipulananda Institute of Aesthetic Studies at the Eastern University, the Open University, the district court, the airport, the technical college, the financial institutions, and other government institutions. Most of these institutions are located in commercial areas. Thus, this continuous commuting activity causes traffic congestion and frequent accidents in the city. Mainly Koddamaunai Bridge and the new bridge areas face traffic congestion daily. The main reason is that the traffic lights are not fixed in many areas of this city. Further, about 500 private buses and 320 public buses operate in the Batticaloa municipality area. This bus service starts from the Batticaloa's central bus station, which is located in the city centre, and has caused overcrowding. This ribbon development pattern is one of the main reasons for congestion in the city.

However, all these characteristics have created similar and different effects on the Batticaloa municipality. Leapfrog and scattered developments are the discontinuous and dispersed built-up growth that have less connectivity between the buildings. Meanwhile, commercial ribbon development with shopping complexes, restaurants, and banks built along the street of the core city generally depends on the highways for the developments. Various development projects were implemented in this area at the end of the civil war, which caused more changes in the city. Thus, land use classes were analysed to extract the area through characteristic changes in urban sprawl during the years 2000, 2010, and 2020.

The building patterns are not growing in a planned manner, such as housing, administration, shopping complexes, schools, police station, health office, and playgrounds. The police station and its quarters are built between the commercial area and also the core city. At the same time, administration buildings and playgrounds are also built in the core city. The suitability for the development of each sector was not considered until now. The main reason is less development planning in this area because it was affected by the civil war for three decades (1983–2009).

Further, the transportation network in the city was also not well designed. The primary and minor roads were not planned based on the network tracking method, which is more familiar for making road maps in ArcGIS. The road network of Batticaloa is not well mapped to identify the closest route between two locations in order to avoid traffic-related problems. For example, it is difficult to reach the general hospital of Batticaloa because of the lack of the closest facility in an emergency during traffic congestion. Closest facilities must consider tracking the places as quickly as possible to reach the location, and this must be a consideration when developing the transport pattern.

As a developing city in Sri Lanka, in Batticaloa the remaining categories that cause sprawl, such as auto-dependent or car-dependent development, uncontrolled growth, and uncoordinated growth, are not highly identified in this study area. The lower population of the city is the reason for these characteristics not having grown during the selected periods. Therefore, based on the identified characteristics of the study area, urban sprawl refers to the urban expansion [9,43] with low density beyond the built-up area [1,5,9,12,22,23], leapfrog development [5,12,13,26], a commercial ribbon development along the highways [12,13], and scattered growth [11,22,25].

4.3. Future Plans and Regulations in the Batticaloa Municipality

Plans and regulations regarding the construction of buildings, especially the construction of homes in current and future development, are the most important to minimise the urban sprawl development in the future. The development plan for Batticaloa municipality in 2030 is proposed in nine major zones, which are: residential zone, commercial zone, information technology zone, environmental conservation zone, airport-related activity zone, mixed development zone, administrative zone, fort conservation zone, and agricultural zone. These zones are separated by boundary lines such as roads or railways, or canals. These zones will only be used for the specific development for which the areas were designated. However, Batticaloa as a tourist city encourages activities related to tourism in any area of the municipality, depending on the suitability of the tourist development.

Further, the minimum land extent for the residential building is 6 perches, mentioned in the regulations of Batticaloa municipality. However, the maximum land area is not defined. Therefore, there are no obstacles to buying a large land and building a single house on this large land. Some people prefer the large spacious land to build their home today. The land for their preference is mostly available in the peripheral areas of the city. This development is the reason for the low density and scattered development patterns in the future as well.

Further, all buildings within 300 m of the coastal zone must be constructed with the prior approval of the Department of Coastal Conservation, Sri Lanka. A green belt was developed in the coastal strips to control the loss of biodiversity and the barriers from disasters such as tsunami and cyclone. Any building constructions in proximity to these disaster-prone areas should obtain clearance from the Urban Development Authority, Batticaloa. These regulations are the reasons for the low population and built-up density in the coastal areas of the city.

In addition, environmentally sensitive areas such as scrubland and mangrove forest are considered conservation areas, which do not allow the built-up development in the municipality areas, showing low density and leapfrog characteristics. Even though people can get approval to build houses in some environmental conservation areas, the buildings should be designed to retain natural beauty rather than obstruct the open green spaces.

Therefore, people should follow the regulations mentioned by the municipality to control the urban sprawl development.

However, Land Value Capture and Taxation are favourable for affordable housing development in the city with lower land value. Proper activation of this system can increase revenues and fix up downtown buildings. The significant increase in land value is due to the locational preferences as commercial areas and a transport hub. This tax system can control the increase of land value by the community preferences and use these tax revenues to develop public spaces such as bus stands, roads, and public spaces. Thus, the Municipality should strictly follow this policy in this city and encourage people to adopt this system to control land value, since people who pay tax can force the authorities to take care of their land and control the illegal occupancy of land in the city.

In addition, the urban development plan must also be open to feedback from stakeholders and experts who can make the development plan stronger. Most of the time, the municipality does not follow this practice when trying to implement a policy. The public is largely ignored in this process, which affects the sustainable planning of the city. Thus, the municipality should consider development activities in all aspects, including public participation. Therefore, proper regulations should be implemented based on the Sustainable Development Goal 11 and New Urban Agenda.

5. Conclusions

Urban sprawl has been increasing due to rapid construction development, especially housing in the Batticaloa municipal council. Urban fringe in the study area showed different sprawling characteristics, which are low density, leapfrog development, and scattered growth. However, a characteristic such as commercial ribbon development along the highways has been identified in the core city areas. Rapid population growth causes more sprawling development in the city, which was increasingly identified after 1990 in this area. In addition, people's preferences, land value, education, employment, income level, illegal construction, and permanent and temporary migrations are the main reasons for the sprawling development of the city. Based on these factors, it is clear that urban sprawl is a socioeconomic phenomenon that should be focused more on socioeconomic aspects in future studies.

Further, political influence interrupts the municipality's development plans and regulations. The rules and regulations need to be followed carefully in this area to develop itself as a sustainable city in the future. Maximum land extent for housing development should be stated in the regulations, which are the most important to minimise the pattern of low density and scattered development in the city. The municipality must identify existing illegal buildings, and when the buildings have not been built following the regulations, they must be demolished by the municipality. In addition, the illegal landfill in wetland areas such as scrubland areas must be punished and charged a penalty by the municipality. These regulations can control illegal activities in future. People have a lack of awareness of the urban sprawl development in this city; thus, it is important to educate them about the urban sprawl development and its effects through research, mapping, and community programmes.

Finally, this study identified the particular characteristics of urban sprawl such as low density, leapfrog development, scattered growth, and commercial ribbon development in the Batticaloa municipality. This finding empirically contributes to understanding the patterns of uncontrolled urban sprawl in Batticaloa city and other cities of Sri Lanka, and other developing countries in the future. This study can help to formulate strategic policies to minimise sprawling growth in the Batticaloa Municipal Council. Despite the above, this study, nevertheless, has limitations. Although the current low-resolution satellite images are deemed adequate in terms of accuracy, the findings could be enriched via a more high-resolution images analysis to identify micro-level changes and built-up patterns/forms. That may, in turn, provide more interesting explanations about urban sprawl characteristics. In addition, these sprawl characteristics creating impacts to land use/land cover patterns are associated with physical and socioeconomic influences, but they are not considered in this study. Therefore, future studies should consider these influencing factors to obtain a holistic understanding and then form a predictive model curbing urban sprawl more effectively.

Author Contributions: Conceptualisation, methodology, M.S. and N.R.; software, M.S.; validation, M.S., N.R., G.H.T.L. and I.S.; formal analysis, M.S.; writing—original draft preparation, M.S.; writing—review and editing, M.S., N.R., G.H.T.L. and I.S.; supervision, N.R., G.H.T.L. and I.S.; funding acquisition, N.R. and G.H.T.L. All authors have read and agreed to the published version of the manuscript.

Funding: This research received no external funding.

Institutional Review Board Statement: Not applicable.

Informed Consent Statement: Not applicable.

Acknowledgments: We are very grateful to the Center for Innovative Planning and Development (CIPD) and the Faculty of Built Environment and Surveying, Universiti Teknologi Malaysia for the encouragement and financial support for this study, and our gratitude also goes to the reviewers who provided constructive comments and insights on this manuscript.

Conflicts of Interest: The authors declare no conflict of interest.

References

1. Hamad, R. A remote sensing and GIS-based analysis of urban sprawl in Soran District, Iraqi Kurdistan. *SN Appl. Sci.* **2020**, *2*, 24. [CrossRef]
2. Jat, M.K.; Garg, P.K.; Khare, D. Monitoring and modelling of urban sprawl using remote sensing and GIS techniques. *Int. J. Appl. Earth Obs. Geoinf.* **2008**, *10*, 26–43. [CrossRef]
3. Jain, M.; Dimri, A.; Niyogi, D. Urban sprawl patterns and processes in Delhi from 1977 to 2014 based on remote sensing and spatial metrics approaches. *Earth Interact.* **2016**, *20*, 1–29. [CrossRef]
4. Amarawickrama, S.; Singhapathirana, P.; Rajapaksha, N. Defining urban sprawl in the Sri Lankan context: With special reference to the Colombo Metropolitan Region. *J. Asian Afr. Stud.* **2015**, *50*, 590–614. [CrossRef]
5. Grigorescu, I.; Kucsicsa, G.; Mitrică, B. Assessing spatio-temporal dynamics of urban sprawl in the Bucharest metropolitan area over the last century. *Land Use/Cover Chang. Sel. Reg. World* **2015**, *10*, 19–27.
6. Wolny, A.; Dawidowicz, A.; Źróbek, R. Identification of the spatial causes of urban sprawl with the use of land information systems and GIS tools. *Bull. Geography. Socio-Econ. Ser.* **2017**, *35*, 111–122. [CrossRef]
7. Lv, Z.-q.; Dai, F.-q.; Sun, C. Evaluation of urban sprawl and urban landscape pattern in a rapidly developing region. *Environ. Monit. Assess.* **2012**, *184*, 6437–6448. [CrossRef] [PubMed]
8. Masakorala, P.; Dayawansa, N. Spatio-temporal Analysis of Urbanization, Urban Growth and Urban Sprawl Since 1976–2011 in Kandy City and Surrounding Area using GIS and Remote Sensing. *Bhumi Plan. Res. J.* **2015**, *4*, 26–44. [CrossRef]
9. Yue, W.; Liu, Y.; Fan, P. Measuring urban sprawl and its drivers in large Chinese cities: The case of Hangzhou. *Land Use Policy* **2013**, *31*, 358–370. [CrossRef]
10. Siedentop, S.; Fina, S. Monitoring urban sprawl in Germany: Towards a GIS-based measurement and assessment approach. *J. Land Use Sci.* **2010**, *5*, 73–104. [CrossRef]
11. Tamilenthi, S.; Punithavathi, J.; Baskaran, R.; ChandraMohan, K. Dynamics of urban sprawl, changing direction and mapping: A case study of Salem city, Tamilnadu, India. *Arch. Appl. Sci. Res.* **2011**, *3*, 277–286.
12. Paul, S.; Dasgupta, A. Spatio-temporal analysis to quantify urban sprawl using geoinformatics. *Int. J. Adv. Remote Sens. GIS* **2012**, *1*, 234–248.
13. Farooq, S.; Ahmad, S. Urban sprawl development around Aligarh city: A study aided by satellite remote sensing and GIS. *J. Indian Soc. Remote Sens.* **2008**, *36*, 77–88. [CrossRef]
14. Howley, P.; Scott, M.; Redmond, D. An examination of residential preferences for less sustainable housing: Exploring future mobility among Dublin central city residents. *Cities* **2009**, *26*, 1–8. [CrossRef]
15. Antalyn, B.; Weerasinghe, V. Assessment of Urban Sprawl and Its Impacts on Rural Landmasses of Colombo District: A Study Based on Remote Sensing and GIS Techniques. *Asia-Pac. J. Rural Dev.* **2020**, *30*, 139–154. [CrossRef]
16. Mathanraj, S.; Ratnayake, R.; Rajendram, K. A GIS-Based Analysis of Temporal Changes of Land Use Pattern in Batticaloa MC, Sri Lanka from 1980 to 2018. *World Sci. News* **2019**, *137*, 210–228.
17. Partheepan, K.; Manobavan, M.; Dayawansa, N. Assessment of landuse changes in the Batticaloa district (2000–2003/2005) for the preparation of a (spatial) zonation plan to aid in decision making for development. *JSc-East. Univ. Sri Lanka* **2008**, *5*, 19–31.
18. Prakasa, D.T.; Soemardiono, B.; Defiana, I. Theoretical Review Effect and Solution of Urban Sprawl. *J. Eng. Sci.* **2018**, *7*, 47–58.

19. Galster, G.; Hanson, R.; Ratcliffe, M.R.; Wolman, H.; Coleman, S.; Freihage, J. Wrestling sprawl to the ground: Defining and measuring an elusive concept. *Hous. Policy Debate* **2001**, *12*, 681–717. [CrossRef]
20. Sudhira, H.; Ramachandra, T. Characterising urban sprawl from remote sensing data and using landscape metrics. In Proceedings of the 10th International Conference on Computers in Urban Planning and Urban Management, Iguassu Falls, Brazil, 11–13 July 2007.
21. Ottensmann, J.R. An alternative approach to the measurement of urban sprawl. *SSRN Electron. J.* **2018**, 1–39. [CrossRef]
22. Shirkhanloo, N. Analysing the Effects of Urban Sprawl on the Physical Environment in the Case of Kyrenia. Master's Thesis, Eastern Mediterranean University (EMU)-Doğu Akdeniz Üniversitesi (DAÜ), Gazimağusa, Cyprus, 2013.
23. Nikolov, P. A survey of Bulgarian (national) planning and regulation acts and documents concerning urban sprawl. In Proceedings of the 2nd International Scientific Conference, Riga, Latvia, 11–12 April 2013. RESPAG 2013.
24. Sinha, S.K. Causes of Urban Sprawl: A comparative study of Developed and Developing World Cities. *Res. Rev. Int. J. Multidiscip.* **2018**, *3*, 1–5.
25. Bhatta, B. *Analysis of Urban Growth and Sprawl from Remote Sensing Data*; Springer Science & Business Media: Berlin/Heidelberg, Germany, 2010.
26. Osman, S.; Nawawi, A.H.; Abdullah, J. Urban sprawl and its financial cost:-a conceptual framework. *Asian Soc. Sci.* **2008**, *4*, 39–50. [CrossRef]
27. Johnson, M.P. Environmental impacts of urban sprawl: A survey of the literature and proposed research agenda. *Environ. Plan. A* **2001**, *33*, 717–735. [CrossRef]
28. Pichler-Milanović, N. European urban sprawl: Sustainability, cultures of (anti) urbanism and "hybrid cityscapes". *Dela* **2007**, *27*, 101–133. [CrossRef]
29. OECD. Rethinking Urban Sprawl: Moving Towards Sustainable Cities (9264189823). 2018. Available online: http://dx.doi.org/10.1787/9789264189881-en (accessed on 14 September 2020).
30. Green, C.T. The Origins of Urban Sprawl. Master's Thesis, The University of Georgia, Athens, GA, USA, 2006.
31. Frenkel, A.; Ashkenazi, M. Measuring urban sprawl: How can we deal with it? *Environ. Plan. B Plan. Des.* **2008**, *35*, 56–79. [CrossRef]
32. Noor, N.M.; Rosni, N.A. Determination of spatial factors in measuring urban sprawl in Kuantan using remote sensing and GIS. *Procedia-Soc. Behav. Sci.* **2013**, *85*, 502–512. [CrossRef]
33. Gavrilidis, A.A.; Niţă, M.R.; Onose, D.A.; Badiu, D.L.; Năstase, I.I. Methodological framework for urban sprawl control through sustainable planning of urban green infrastructure. *Ecol. Indic.* **2019**, *96*, 67–78. [CrossRef]
34. Aurambout, J.-P.; Barranco, R.; Lavalle, C. Towards a simpler characterisation of urban sprawl across urban areas in Europe. *Land* **2018**, *7*, 33. [CrossRef]
35. Abidin, S.Z.; Zamani, N.A.M.; Aliman, S. A computerised tool based on cellular automata and modified game of life for urban growth region analysis. In Proceedings of the 4th International Conference on Soft Computing in Data Science, Bangkok, Thailand, 15–16 August 2018.
36. Chettry, V.; Surawar, M. Urban sprawl assessment in Raipur and Bhubaneswar urban agglomerations from 1991 to 2018 using geoinformatics. *Arab. J. Geosci.* **2020**, *13*, 1–17. [CrossRef]
37. Statistic Office, District Secretariat. *Statistical Hand Book*; Statistic Office, District Secretariat: Batticaloa, Sri Lanka, 2019.
38. Department of Census and Statistic. *Population and Housing*; Department of Census and Statistic: Batticaloa, Sri Lanka, 2019.
39. Earth Explorer. United States Geological Survey (USGS). Available online: https://earthexplorer.usgs.gov/ (accessed on 26 April 2021).
40. Verma, P.; Raghubanshi, A.; Srivastava, P.K.; Raghubanshi, A. Appraisal of kappa-based metrics and disagreement indices of accuracy assessment for parametric and nonparametric techniques used in LULC classification and change detection. *Modeling Earth Syst. Environ.* **2020**, *6*, 1–15. [CrossRef]
41. Angel, S.; Parent, J.; Civco, D. Urban sprawl metrics: An analysis of global urban expansion using GIS. In Proceedings of the ASPRS 2007 Annual Conference, Tampa, FL, USA, 7–11 May 2007.
42. Oueslati, W.; Alvanides, S.; Garrod, G. Determinants of urban sprawl in European cities. *Urban Stud.* **2015**, *52*, 1594–1614. [CrossRef] [PubMed]
43. Inostroza, L.; Baur, R.; Csaplovics, E. Urban sprawl and fragmentation in Latin America: A dynamic quantification and characterisation of spatial patterns. *J. Environ. Manag.* **2013**, *115*, 87–97. [CrossRef] [PubMed]

Article

Monitoring and Modeling the Patterns and Trends of Urban Growth Using Urban Sprawl Matrix and CA-Markov Model: A Case Study of Karachi, Pakistan

Muhammad Fahad Baqa [1], Fang Chen [1], Linlin Lu [1,*], Salman Qureshi [2], Aqil Tariq [3], Siyuan Wang [4], Linhai Jing [1], Salma Hamza [5] and Qingting Li [6]

1. Key Laboratory of Digital Earth Science, Aerospace Information Research Institute, Chinese Academy of Sciences, Beijing 100094, China; 2252293808@mails.ucas.edu.cn (M.F.B.); chenfang@radi.ac.cn (F.C.); jinglh@aircas.ac.cn (L.J.)
2. Institute of Geography, Humboldt University of Berlin, 12489 Berlin, Germany; salman.qureshi@geo.hu-berlin.de
3. State Key Laboratory of Information Engineering in Surveying, Mapping and Remote Sensing (LIESMARS), Wuhan University, Wuhan 430079, China; aqiltariq@whu.edu.cn
4. Research Center for Eco-Environmental Sciences, State Key Laboratory of Urban and Regional Ecology, Chinese Academy of Sciences, Beijing 100085, China; wangsy@rcees.ac.cn
5. Department of Earth and Environmental Sciences, Bahria University Karachi Campus, Karachi 75300, Pakistan; salma.bukc@bahria.edu.pk
6. Airborne Remote Sensing Center, Aerospace Information Research Institute, Chinese Academy of Sciences, Beijing 100094, China; liqt@radi.ac.cn
*   Correspondence: lull@radi.ac.cn

Citation: Baqa, M.F.; Chen, F.; Lu, L.; Qureshi, S.; Tariq, A.; Wang, S.; Jing, L.; Hamza, S.; Li, Q. Monitoring and Modeling the Patterns and Trends of Urban Growth Using Urban Sprawl Matrix and CA-Markov Model: A Case Study of Karachi, Pakistan. *Land* **2021**, *10*, 700. https://doi.org/10.3390/land10070700

Academic Editors: Luca Salvati, Iwona Cieślak and Andrzej Biłozor

Received: 17 June 2021
Accepted: 1 July 2021
Published: 2 July 2021

Publisher's Note: MDPI stays neutral with regard to jurisdictional claims in published maps and institutional affiliations.

Abstract: Understanding the spatial growth of cities is crucial for proactive planning and sustainable urbanization. The largest and most densely inhabited megapolis of Pakistan, Karachi, has experienced massive spatial growth not only in the core areas of the city, but also in the city's suburbs and outskirts over the past decades. In this study, the land use/land cover (LULC) in Karachi was classified using Landsat data and the random forest algorithm from the Google Earth Engine cloud platform for the years 1990, 2000, 2010, and 2020. Land use/land cover classification maps as well as an urban sprawl matrix technique were used to analyze the geographical patterns and trends of urban sprawl. Six urban classes, namely, the primary urban core, secondary urban core, sub-urban fringe, scatter settlement, urban open space, and non-urban area, were determined for the exploration of urban landscape changes. Future scenarios of LULC for 2030 were predicted using a CA–Markov model. The study found that the built-up area had expanded in a considerably unpredictable manner, primarily at the expense of agricultural land. The increase in mangroves and grassland and shrub land proved the effectiveness of afforestation programs in improving vegetation coverage in the study area. The investigation of urban landscape alteration revealed that the primary urban core expanded from the core districts, namely, the Central, South, and East districts, and a new urban secondary core emerged in Malir in 2020. The CA–Markov model showed that the total urban built-up area could potentially increase from 584.78 km^2 in 2020 to 652.59 km^2 in 2030. The integrated method combining remote sensing, GIS, and an urban sprawl matrix has proven invaluable for the investigation of urban sprawl in a rapidly growing city.

Keywords: urban sprawl; Landsat; CA–Markov model; SDG 11; urban sustainable development

1. Introduction

Urbanization is a complex socioeconomic process that shifts the distribution of a population from dispersed rural settlements to dense urban settlements [1]. In spatial terms, the urbanization process is manifested in the physical development of urban settlements and the transition of landscapes into urban forms [2,3]. In the Global South, rapid and unplanned urban sprawl leads to problems such as fragmented landscape, reduction in arable land, increase in urban poverty, and environmental degradation, which pose a huge threat to sustainable development in these regions [4–6]. By 2030, Sustainable Development Goal 11 of the United Nations intends to make cities and human settlements more inclusive, safe, resilient, and sustainable [7]. Building policies to promote the sustainable development of cities, especially in developing countries, need accurate and timely monitoring and understanding of the spatial growth of urban settlements [8].

Geospatial techniques have enabled the analysis and forecasting of urban growth at regional and global scales. These methods are useful for observing and understanding the dynamics of urban landscapes [6,9,10]. Previously, efforts have been made to model and analyze urban spatial growth and patterns using methods such as cellular automata [11–13], the artificial neural network [14,15], the Markov chain [16,17], geographical weighted regression [18], the non-ordinal and Sleuth model [19–21], the analytic hierarchy process [22], machine learning models [23,24], and an urban sprawl matrix [25,26]. Batty demonstrated how cellular and agent-based models have the ability to clearly incorporate spatial interaction and mobility [27]. Considering the limitation of basic logistic regression models, Arsanjani et al. used a hybrid model to uncover the interaction of various environmental and socioeconomic variables that cause urban expansion [28]. By combining the CA model's benefit of modeling spatial variation in complex systems with the Markov model's advantage of long-term prediction, the CA–Markov model was developed, which is an effective method for simulating LULC transformation. It has been widely applied to examining and measuring urbanization and landscape dynamics [29]. The Markov model predicts the future status of a land use based on its current rate [30]. Cellular automata (CA) detects the geographic location of changes, whereas the Markov chain predicts future change based on the past [30].

Karachi, Pakistan's largest city, has seen massive urban growth in recent decades not only in the city's center, but also in the surrounding suburbs [6]. If the urban land expansion rate is higher than the population increase rate, the population density in the urban area will significantly decline, and the phenomena of urban sprawl will occur. Due to institutional inefficiency and governance failure, rural lands have been converted into residential and industrial areas without considering the urban planning schemes in Karachi [31]. The massive conversion of rural lands for urban areas has caused the sprawl phenomenon since 2000, which has led to loss of agricultural lands, an increase in commuting costs, and flooding [31]. The unplanned urban sprawl has also resulted in a range of social problems such as a lack of health care, shortage of education facilities and infrastructures, an increase in criminal incidents, and sociocultural imbalance [32–34]. The introduction of new urban forms and structures that adapt to climate change issues can mitigate the environmental problems caused by dispersed urban area growth and create more efficient urban economies [35]. Therefore, the spatiotemporal modeling of urban sprawl is crucial to better understand the changing urban patterns of Karachi divisions, thus helping local governments in prioritizing the demands of the local population and formulating strategies and practical solutions to achieve the goal of urban sustainable development.

Previous studies have attempted to use remote sensing data to analyze the general pattern of urban land cover changes and urban suitability in Karachi [36,37]. Although land use land cover changes were significant based on the analysis using satellite imagery, the landscape changes during the urbanization process were not fully investigated. Moreover, the simulation and prediction of future LULC scenarios in the growing city have barely been reported. To fill such gaps, this study aimed to thoroughly analyze the LULC changes and the spatiotemporal dynamics of urban expansion in Karachi using satellite data from 1990 to 2020. The future LULC scenarios and urban expansion were also simulated using a CA–Markov model in the city for the year 2030.

2. Study Area and Datasets

2.1. Study Area

Karachi, the provincial capital of Sindh, is Pakistan's largest and most densely populated megacity. It is the principal industrial center, seaport, and financial and commercial hub. Karachi Urban Agglomeration (Karachi UA), extending over 3527 km^2, is located on the coastline of the Arabian Sea, between 24°45′ N to 25°15′ N and 66°37′ E to 67°37′ E (Figure 1). Karachi is mainly made up of flat rolling plains with hills on the western and northern boundaries. The southern and southeastern banks of the Malir River have a contagious linear concentration of urban settlements [38].

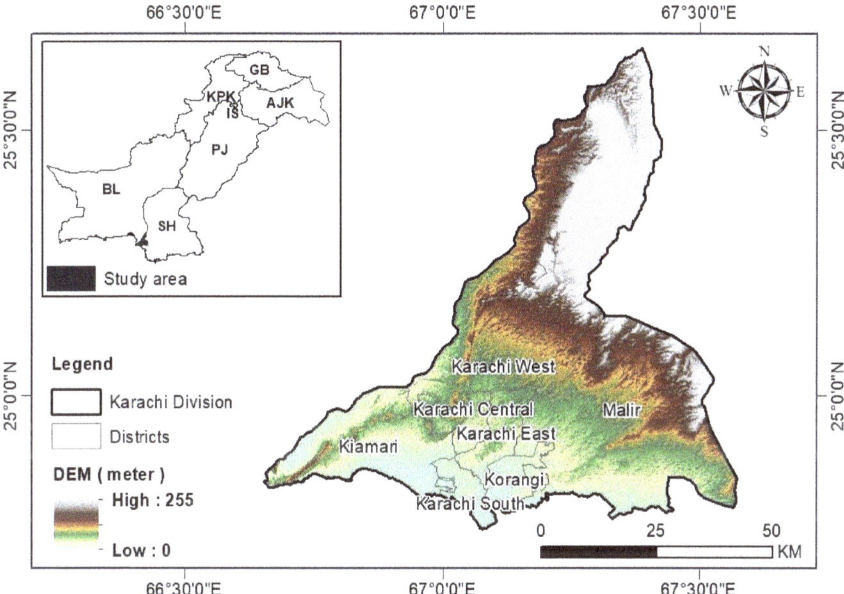

Figure 1. Study area. PJ = Punjab; SH = Sindh; BL = Balochistan; IS = Islamabad; AJK = Azad Jammu Kashmir; GB = Gilgit-Baltistan; KPK = Khyber Pakhtunkhwa.

According to the 2017 Census Report [39], more than 16 million people live in Karachi, and the population will increase to more than 20 million by 2025 with a density of 4115 persons per square kilometer [40,41]. The city consists of seven districts, which can be further divided into 31 sub-divisions [39]. As an increasing metropolitan city in a developing country, Karachi faces unplanned urban expansion, inappropriate essential infrastructure and facilities, crises in drinking water and solid-waste management services, inconvenient public transport, environmental pollution, and poor governance [42]. Of the total population, nearly 40% live in slum areas [43,44].

2.2. Datasets

The primary data source for measuring urban spatial patterns and analyzing the trend of urban growth in Karachi was Landsat Thematic Mapper (TM) pictures from 1990, 2000, and 2010 as well as Landsat 8 OLI images from 2020 from Google Earth Engine (Table 1). The atmospheric correction technique LaSRC was used to correct the available Landsat Surface Reflectance Tier 1 data in Google Earth Engine. The CFMASK algorithm was used to mask cloud, shadow, and water regions in these images. The entire study area covered three Landsat tiles (152_042, 152_043, and 153_043). The atmospherically corrected and cloud removed images with a ten-year interval were used to perform the initial LULC classification. As supplementary features for land cover classification, the normalized difference vegetation index (NDVI) and the normalized built up index (NDBI) were computed for each decadal image [45].

Table 1. Details of datasets used in this study.

Data	Details	Period	Source
Landsat images	30 m resolution, Path/Row 152,042, 152/043, 153/043	1990–2020	USGS
Base map	scale 1:25,000	2000	SOP
Population	divisional level	1990–2020	COP
DEM	SRTM DEM and Slope	2015	USGS
Roads	Road Network	2018	OSM

Several datasets were used as supplementary data in our study (Table 1). To distinguish LULC classes between plain and hilly areas, SRTM digital elevation model (DEM) data were employed. To evaluate the accuracy of LULC, high spatial resolution images with multiple acquisition dates collected from Google Earth and topographical maps published by the Survey of Pakistan, Government of Pakistan were used as reference data. District-level population data were gathered for the years 1990, 2000, 2010, and 2020 from the official census and Pakistan Bureau of Statistics [39]. The road network data were used to train the CA–Markov model for the LULC scenario simulation.

3. Methods

The workflow was primarily comprised of three steps: classification of land use/land cover, analysis of urban expansion, and modeling of future LULC scenarios. Figure 2 depicts the entire data processing workflow adopted in this study.

3.1. Land Use/Land Cover Classification

We used the Google Earth Engine's random forest classification technique to produce land use/land cover maps for the years 1990, 2000, 2010, and 2020 in the study area [46]. The overall accuracy (OA), producers' accuracy (PA), and users' accuracy of the classification results were measured using the confusion matrix [46].

3.2. Urban Landscape Change Analysis

The post-classification change matrix methodology was used to create a land use/land cover change map from 1990 to 2020. To analyze land use/land cover changes, a transition model was developed using cross-tabulation in the GIS module. The transition matrix indicates that the study area had experienced major alterations.

An urban sprawl matrix was utilized to examine urban expansion dynamics and measure urban spatial patterns in Karachi [47]. For the categorization of urban spatial patterns, matrix functions based on urban pixels were used. Using the urban sprawl matrix, the study area was divided into six classes, namely, the urban primary core, urban secondary core, suburban fringe, scatter settlement, urban open space, and non-urban area (Table 2 and Figure 3).

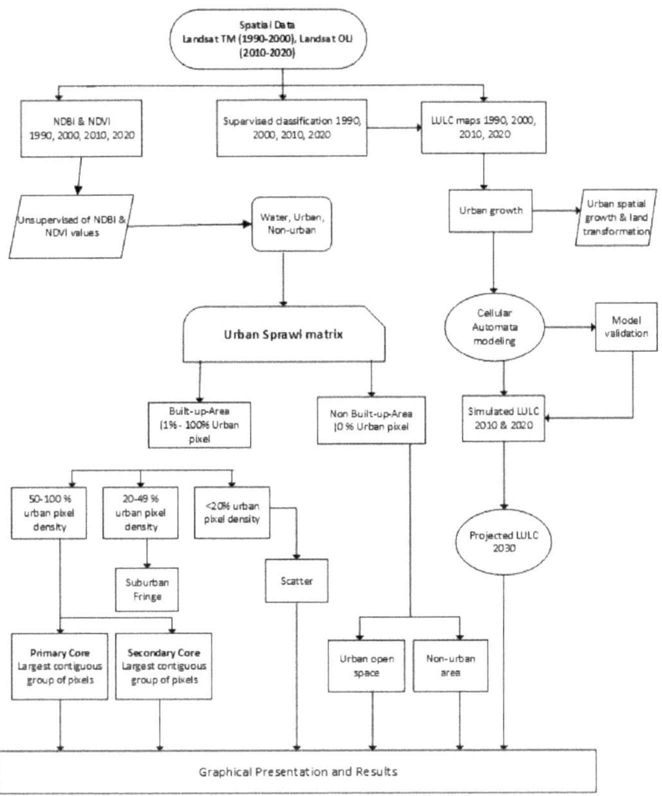

Figure 2. The data processing workflow in this study.

Table 2. Classification of the study area into urban spatial patterns.

Urban Spatial Patterns	Criteria
Urban primary core	The most densely packed set of pixels in which at least 50% of the surrounding neighborhood is densely populated.
Urban secondary core	It is at least 50% built-up in the same way as urban primary core although it is not part of it.
Suburban fringe	The built-up pixels with a 30–50% urbanness surrounded by primary and secondary core.
Scatter settlement	The built-up pixel is less than 20% built up and is located apart from the primary and secondary cores.
Urban open space	The non-urban regions encircled the primary and secondary urban cores.
Non-urban area	Apart from the primary and secondary urban cores, non-urban area.

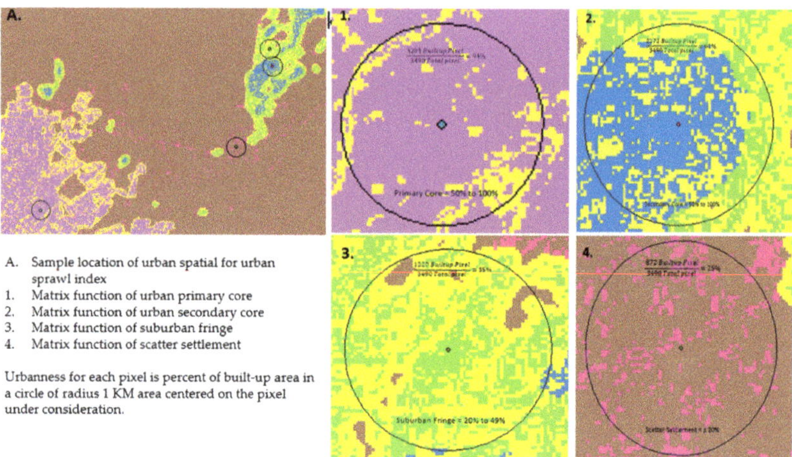

A. Sample location of urban spatial for urban sprawl index
1. Matrix function of urban primary core
2. Matrix function of urban secondary core
3. Matrix function of suburban fringe
4. Matrix function of scatter settlement

Urbanness for each pixel is percent of built-up area in a circle of radius 1 KM area centered on the pixel under consideration.

Figure 3. Depiction of the function of the urban sprawl matrix and urban spatial pattern. The percentage of the built-up area in a 1000 m radius circle centered on the pixel under examination is the pixel's urbanness.

3.3. LULC Simulation

CA–Markov simulation was implemented using several steps: (a) the generation of LULC maps with the same time interval (1990, 2000, 2010, and 2020); (b) the calculation of transition probability matrices based on LULC maps; (c) the generation of transition suitability maps using driving factors such as distance to water body, distance to main roads, distance to built-up areas, and slope [4,31]; (d) the evaluation of the model's ability to simulate future changes using a kappa index of agreement (KIA) approach; and (e) the simulation of LULC maps for the predicted year (here, 2030). The projections of LULC change in the study area were performed using the land change modeler (LCM) within the TerrSet software (Clarke Labs 2019, https://clarklabs.org (accessed on 10 December 2020) [48].

As an input to the CA–Markov model, the Markov chain model was employed to produce a transition probability matrix between an initial state and a final state. The transitional probability matrices were generated using LULC information from 2010 to 2020 in order to investigate how each land cover class was expected to change. The Markov model can be described using the following equation:

$$S\,(t+1) = P_{ij} \times S(t) \tag{1}$$

where S represents the land use condition at time t; S (t + 1) represents the land use status at time t + 1; and P_{ij} is the transition probability matrix in a certain state, which is calculated using the following equations [49]:

$$\|P_{ij}\| = \left\|\begin{array}{ccc} P_{1,1} & P_{1,2} & P_{1,N} \\ P_{2,1} & P_{2,2} & P_{2,N} \\ P_{N,1} & P_{N,2} & P_{N,N} \end{array}\right\| \tag{2}$$

$$(0 \leq P_{ij} \leq 1) \tag{3}$$

where P refers to the transition probability; P_{ij} refers to the probability of changing from state i to state j in the next time; and PN refers to the state probability of any time. The low transition probability is close to 0, and the high transition probabilities is close to 1 [49].

Using the multi-criteria evaluation (MCE) module, suitability maps, which show the suitability of cell transformation for a particular land cover type, were created for the application of the CA model. The characteristics of LULC types were taken into consideration. For example, the built-up area cannot be converted into a water body [50,51]. As an inherent part for geospatial modeling, the kappa index of agreement (KIA) representing the model's simulation accuracy was used here to evaluate the model's ability to simulate the spatial pattern of land use [52,53]. The KIA was calculated with the following equation:

$$KIA = Pr(a) - Pr(e)/1 - Pr(e) \qquad (4)$$

where Pr (a) refers to the observed agreement, and Pr(e) refers to chance agreement. The kappa coefficients (K-no, K location, and K-standard as well as the overall kappa coefficient) were used to compare the simulated and the LULC map based on remote sensing data of 2020. The kappa coefficient values were calculated using TerrSet IDRISI software.

4. Results and Discussion

4.1. LULC Change

In the study area, six LULC classes were identified: bare land, built-up area, cultivated land, grassland and shrub land, water body, and mangroves (Figure 4). According to the accuracy assessment results, the overall classification accuracies were 89, 91, 91, and 89% for 1990, 2000, 2010, and 2020, respectively. The kappa coefficient values were 0.86, 0.90, 0.89, and 0.87 for 1990, 2000, 2010, and 2020, respectively.

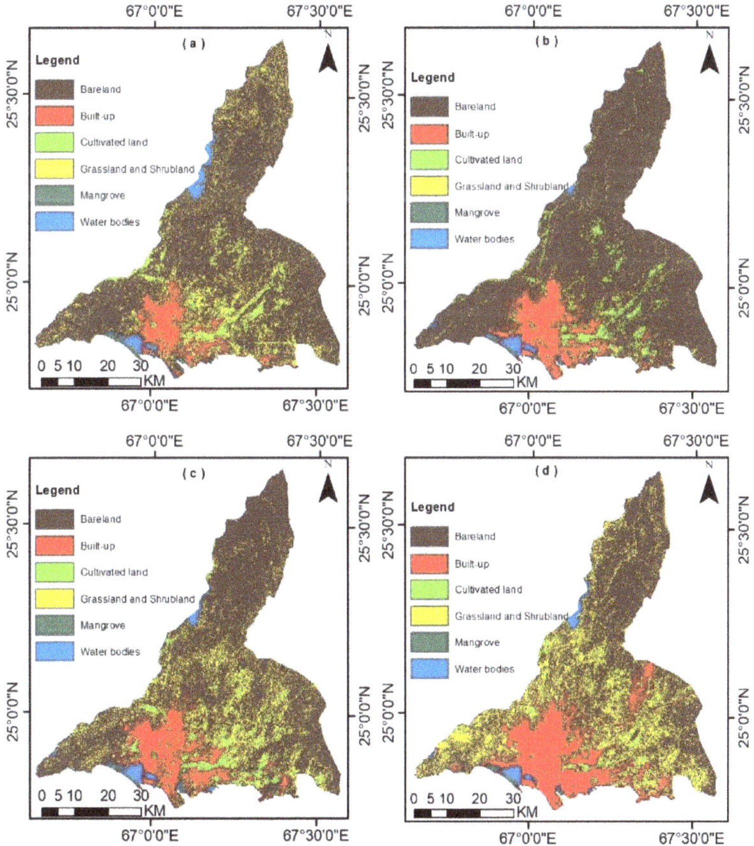

Figure 4. LULC map of Karachi in (**a**) 1990, (**b**) 2000, (**c**) 2010, and (**d**) 2020.

Over the last three decades, the land use/land cover in the study region has changed dramatically (Table 3). Between 1990 and 2020, the area covered by built-up area and grassland and shrub land expanded, while the area occupied by agricultural land, mangroves, and open bare ground declined. Divergent changing trends were revealed in the time periods before 2000 and after 2000 for cultivated land, grassland, and shrub land, and mangroves (Table 3). The increase in the area of mangroves and grassland and shrub land since 2000 indicates that afforestation programs have played a positive role in improving vegetation coverage in the study area. The Sindh Forest Department made great efforts to restore and plant endangered mangrove species. With the help of local communities, they planted more than 800,000 saplings of Rhizophora mucronata mangroves in the coastal zone of Pakistan in 2013 [54]. The decrease in cultivated land was observed near the built-up area, which indicates urban expansion at the cost of cultivated land (Figure 4).

Table 3. Areal changes in each land use land cover type in Karachi.

LULC Classes	Area (sq.km)				Change Rate (%)		
	1990	2000	2010	2020	1990–2000	2000–2010	2010–2020
Bare land	2663.7	2811	2491.7	2156.6	5.53	−11.35	−13.44
Built-up area	221.1	358.7	424.3	573.9	62.23	18.28	35.25
Cultivated land	112.2	159.3	148.4	81.5	41.97	−6.84	−45.08
Grassland and shrub land	534.3	370.5	563	867.7	−30.65	51.95	54.12
Mangroves	65.9	13.8	14.2	17	−79.05	2.89	19.71
Water bodies	23.8	46.5	54	56.8	95.37	16.12	5.18

The increase in urban areas in different districts of the study area is illustrated in Figure 5. It was observed that districts near coast and far from the core area (Karachi Central, South, and East districts) had a record high urban growth from 1990 to 2020, particularly in the Malir (417.92%), West (279.38%), and Kiamari (257.05%) districts (Table 4). Among the core areas, the East district of Karachi experienced a higher increase in the built-up area than that in the Central and South districts of Karachi. The central city's congestion caused outgrowth at the periphery of the megacity during the study period. As a main driver of built-up area growth, the density of the population in Karachi has constantly been increasing over the last three decades. The population of central city has remained highly concentrated, and its population increased from 1.8 million in 1990 to 3.09 million in 2020. Simultaneously, the population of the suburban Malir and West districts increased from 0.8 million and 0.7 million in 1990 to 2.8 million and 2.23 million in 2020, respectively [39].

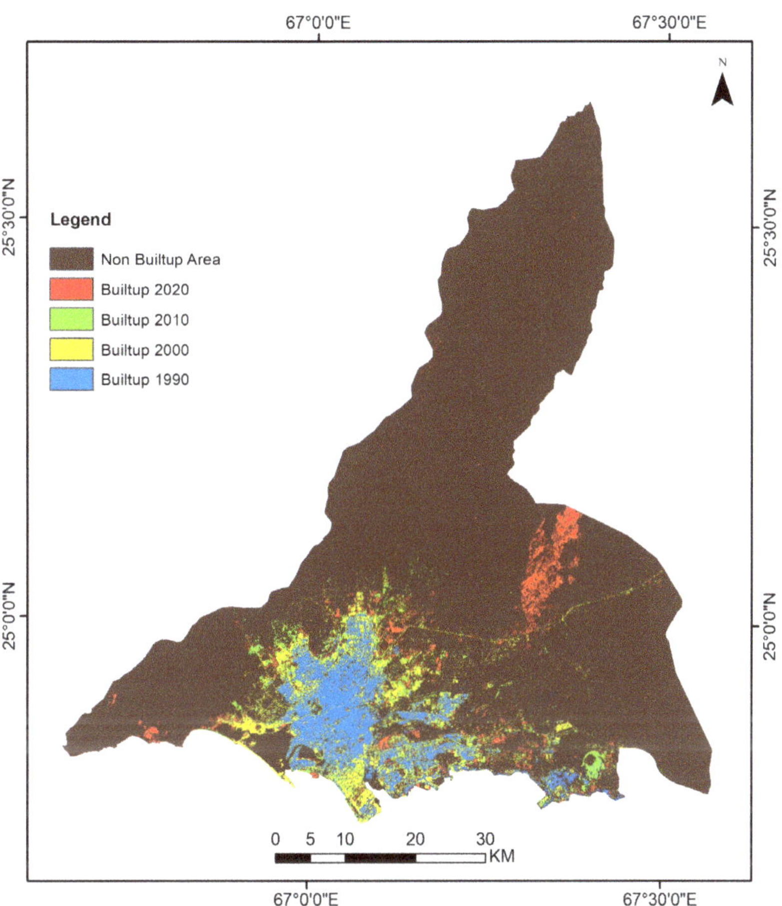

Figure 5. Built-up area expansion in Karachi in 1990, 2000, 2010, and 2020.

Table 4. Built-up area increases in each district of Karachi (in percentage).

District	1990–2000	2000–2010	2010–2020	1990–2020
Kiamari	118.49	3.08	14.13	257.05
Central	29.67	−3.95	3.11	128.43
South	29.07	−0.77	11.64	142.97
West	96.63	26.78	12.07	279.38
East	47.61	3.95	22.57	188.07
Korangi	29.91	10.75	28.98	185.59
Malir	66.16	19.13	111.12	417.92

Figure 6 shows the land transformation in various districts and time periods induced by the process of urbanization. The majority of areas converted to urban land at the expense of open bare land, grassland and shrub land, and agricultural land (Table 5).

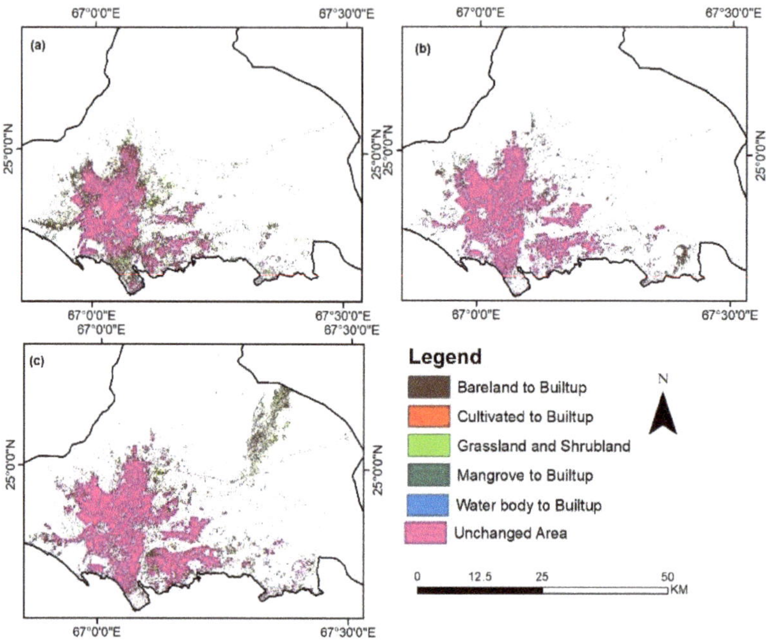

Figure 6. Spatial pattern of land transformation in Karachi during the periods of (**a**) 1990–2000, (**b**) 2000–2010, and (**c**) 2010–2020.

Table 5. Land transformation from other LULC classes to built-up area during the periods of 1990–2000, 2000–2010, and 2010–2020.

Urban Land Transformation	Area (km^2)			Change Rate (%)			
	1990–2000	2000–2010	2010–2020	1990–2000	2000–2010	2010–2020	1990–2020
Bare land to Built-up	114	68.7	134.2	4.28	2.44	5.38	117.71
Cultivated land to Built-up	8.4	9.7	12.5	3.79	2.70	2.94	148.81
Grassland and Shrub land to Built-up	42.6	13.1	30.2	37.96	8.22	20.35	70.89
Water to Built-up	3	0.1	0.9	0.56	0.02	0.16	30.00
Mangroves to Built-up	2.1	2	3.2	3.18	14.49	22.53	152.38
Expansion of Built-up Area	170.1	93.6	181	714.70	201.29	335.18	106.40

4.2. Urban Landscape Change

The urban sprawl matrix was used to create urban landscape maps in the study area. The area of urban primary core increased from 145.9 square kilometers in 1990 to 363.5 square kilometers in 2020 (Table 6). In 1990, changes in the area of the primary core were registered in the areas that comprise the CBD area, namely, the South, East, and Central districts, and later in 2020, the urban primary core expended further into the suburban districts of Karachi such as the Malir, West, and Kiamari districts (Figure 7). The area of the urban secondary core also changed from 25.9 sq.km 1990 to 22.3 sq.km in 2020 (Table 6). In 1990, the urban secondary core was observed only in the districts of Malir and Korangi, while later in 2020, the urban secondary core could be observed in other suburban areas of Karachi such as the districts of West and Kiamari. The observed urban secondary core areas in 1990 merged with the urban primary core in 2020 due to rapid expansion, and a new urban secondary core area emerged in the suburban areas of Karachi (Figure 7).

Table 6. Urban landscape changes in Karachi during the periods of 1990–2000, 2000–2010, and 2010–2020.

Urban Spatial Patterns	Area (sq.km)				Change Rate (%)		
	1990	2000	2010	2020	1990–2000	2000–2010	2010–2020
Urban Primary Core	145.9	248	255	363.5	69.979	2.823	42.549
Urban Secondary Core	25.9	29.3	40	22.3	13.127	36.519	−44.250
Suburban Fringe	25.2	22	22.09	35.8	−12.698	0.409	62.064
Scatter Settlement	16.7	32.8	33.9	42.2	96.407	3.354	24.484
Urban Open Space	121.8	152.5	168.2	234.3	25.205	10.295	39.298
Non-Urban Open Space	3272	3131	3089	2911	−4.309	−1.341	−5.762

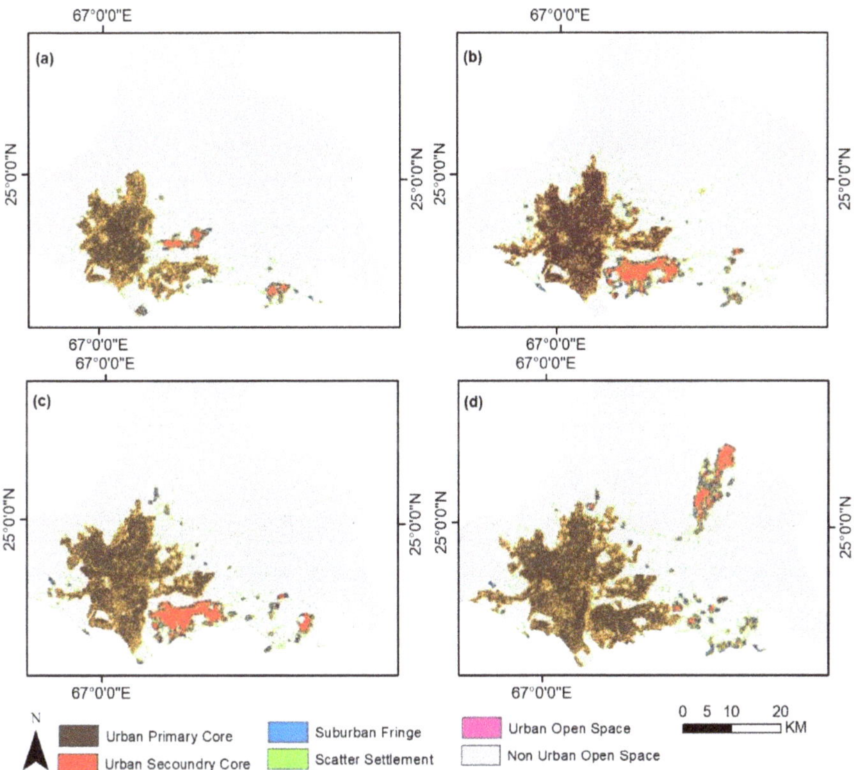

Figure 7. Urban landscape in Karachi in (**a**) 1990, (**b**) 2000, (**c**) 2010, and (**d**) 2020.

The area of the suburban fringe and scatter settlements showed a marginal increase (Table 6). The urban open space increased from 121.8 sq.km in 1990 to 243.3 ssq.km in 2020, which indicates an increase in green space under the urbanized area. Most of this increase was observed in the core of districts of East, Korangi, and Kiamari. A drastic decrease from 3272 sq.km in 1990 to 2911 sq.km in 2020 in the area of non-urban open space can also be observed.

The changes in each urban spatial pattern class within districts were analyzed (Table 7). Between 1990 and 2020, the Kiamari, East, West, and Korangi districts had rapid growth in the primary urban core. From 1990 to 2020, no urban secondary core was found in the district of Kiamari, Central, South, West, or East, while this was observed in the districts of Malir and Korangi. The urban secondary core in the Korangi district merged with the urban primary core in 2020. The newly developed Malir district experienced high urban secondary core growth due to the large number of commercial and residential developmental activities over the last two decades. The suburban fringe increased in the districts of Kiamari, West, East, and Malir, while the urban open space decreased within the Central and South districts. The decrease in open space in the CBD area might be attributed to the conversion of open space to residential and commercial lands.

Table 7. Urban landscape changes in different districts of Karachi (in percentage).

District	Primary Core			Secondary Core			Suburban Fringe			Urban Open Space		
	Phase 1	Phase 2	Phase 3	Phase 1	Phase 2	Phase 3	Phase 1	Phase 2	Phase 3	Phase 1	Phase 2	Phase 3
Kiamari	103.66	1.80	12.44	0.00	0.00	0.00	0.00	100.00	29.45	73.26	−0.89	13.18
Central	30.77	−3.86	3.25	0.00	0.00	0.00	0.00	0.00	0.00	−52.39	24.26	−15.33
South	44.86	−5.02	15.44	0.00	0.00	0.00	0.00	0.00	−100.00	−12.23	60.64	−7.21
West	98.47	16.87	11.42	0.00	0.00	2100.00	0.00	0.00	−9.18	33.73	37.25	33.09
East	40.37	8.52	26.99	0.00	−100.00	0.00	427.78	76.62	50.00	103.99	10.64	22.85
Korangi	5728.57	7.78	251.51	100.10	25.38	−100.00	46.63	−114.47	−100.00	1.23	12.82	3.69
Malir	0	35.96	402.63	−350.61	69.31	30.39	122.99	18.69	144.37	59.99	4.67	148.41

4.3. Transition Probability Matrix Analysis

The transition probability matrix was generated for the time periods of 1990–2000, 2000–2010, and 2010–2020 to demonstrate the probability that each land cover type was projected to change (Table 8). The values on the diagonal of the matrix represent the possibility of a land cover type maintaining its original state, and the values on the non-diagonal represent the possibility of a land cover type converting to other types. From 1990 to 2000, bare land was the most stable class with 0.77 probabilities, while the most dynamic class was cultivated land with transition probabilities of 0.32. From 2000 to 2010, water bodies were the most stable class with 0.67 probabilities, and grassland and shrub land were most dynamic with 0.20 probabilities. Similarly, from 2010 to 2020, mangroves were the most stable class with 0.76 probabilities, and cultivated land was the most dynamic class with 0.20 probabilities. The transition probability matrix from 2010 to 2020 was used to simulate the LULC map for Karachi city in 2030.

Table 8. Transition probability matrix of LULC classes in Karachi from 1990 to 2000, 2000 to 2010, and 2010 to 2020.

Class	Time Period	Bare Land	Built-Up	Cultivated Land	Grassland and Shrub Land	Mangroves	Water Bodies
Bare land	Phase 1	0.77	0.10	0.40	0.06	0.00	0.00
	Phase 2	0.68	0.05	0.02	0.23	0.00	0.00
	Phase 3	0.61	0.05	0.01	0.32	0.00	0.00
Built-up	Phase 1	0.14	0.70	0.05	0.08	0.00	0.00
	Phase 2	0.14	0.76	0.03	0.05	0.00	0.00
	Phase 3	0.03	0.85	0.03	0.03	0.03	0.03
Cultivated land	Phase 1	0.39	0.88	0.32	0.18	0.00	0.00
	Phase 2	0.25	0.12	0.29	0.30	0.00	0.01
	Phase 3	0.27	0.05	0.20	0.46	0.00	0.00
Grassland and shrub land	Phase 1	0.30	0.30	0.30	0.85	0.03	0.03
	Phase 2	0.54	0.14	0.10	0.20	0.00	0.00
	Phase 3	0.55	0.04	0.04	0.34	0.00	0.00
Mangroves	Phase 1	0.09	0.14	0.02	0.01	0.41	0.30
	Phase 2	0.04	0.05	0.00	0.00	0.63	0.25
	Phase 3	0.07	0.04	0.00	0.02	0.76	0.08
Water bodies	Phase 1	0.49	0.38	0.04	0.02	0.00	0.39
	Phase 2	0.12	0.12	0.00	0.00	0.05	0.67
	Phase 3	0.13	0.05	0.01	0.02	0.09	0.66

4.4. LULC Simulation Results

The validation results showed strong agreement with the simulation map (Table 9). The kappa values indicate that the CA–Markov model used is suitable for simulating future LULC maps in the study area.

Table 9. Validation results of the CA–Markov model.

Year	Kappa Parameters				
2020	K-location	K-no	K-location Strata	K-standard	Overall Kappa Value
	0.91	0.91	0.91	0.87	0.87

The simulated LULC maps for Karachi city in 2030 are shown in Figure 8, and the changes for each LULC type are tabulated in Table 10. The simulation results show that the bare land area will significantly decrease in 2030 due to its conversion to a built-up area. The districts of Malir, South, and Kiamari are seeing the most growth in terms of urban built-up area. The spatial pattern of the predicted LULC indicated that the city's new residents would settle in sub-urban fringes surrounding the urban cores. Living in these areas allows them to be closer to work and facilitates a more convenient commute. Grassland and shrub land covered about 838.42 km^2 in 2020 and are expected to gradually increase to 999.06 km^2 in 2030.

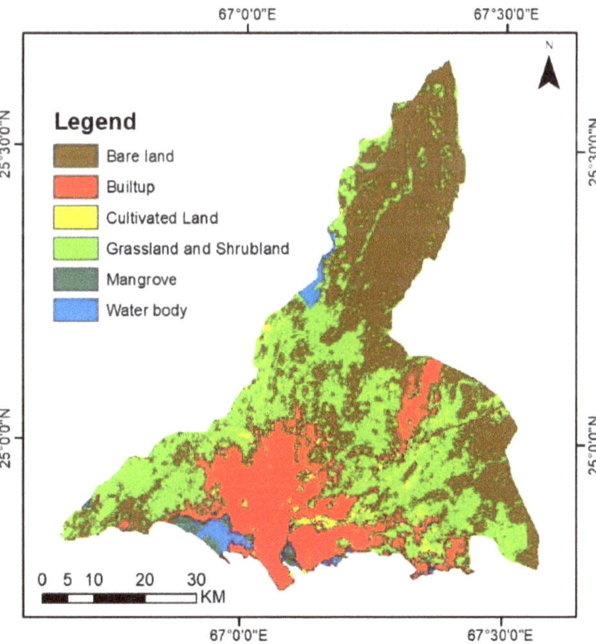

Figure 8. LULC prediction results for Karachi city in 2030.

Table 10. Predicted LULC changes for Karachi city in 2030.

LULC Class	Area (sq.km)		Change	
	2020	2030	Sq.km	%
Bare land	2031.74	1750.84	−280.9	−86.17%
Built-up	584.78	652.59	67.81	111.60%
Cultivated land	78.39	99.46	21.07	126.88%
Grassland and shrub land	838.42	999.06	160.64	119.16%
Mangroves	15.61	33.36	17.75	213.71%
Water bodies	53.35	66.97	13.62	125.53%

Although the validation results showed that the CA-Markov model was a reliable method for simulating land use change, there are several limitations in our study. Socioeconomic factors are among the most important variables influencing land use changes. Our study was unable to investigate several potential socioeconomic causes of urban expansion due to a lack of spatial data. Moreover, more sophisticated models can be developed to simulate urban growth in different areas of the study area [55]. Landsat images with a resolution of 30 m were used to construct the land-use/cover maps for LULC modeling. High-resolution satellite data may be employed in the future to generate more detailed observations of specific agricultural and urban covers.

5. Conclusions

An urban sprawl matrix methodology was used in this study to analyze changes in urban spatial patterns in Karachi over three decadal epochs (1990–2000, 2000–2010, and 2010–2020). The utilization of the urban sprawl matrix provided an accurate and effective assessment of Karachi's urban expansion tendencies. Future land cover changes in the study area were predicted using a CA–Markov model for 2030. The results indicate that the built-up area had expanded in a considerably unpredictable manner, which was mainly at the expense of agricultural land. The increase in mangroves and grassland and shrub land demonstrated the effectiveness of afforestation programs in improving vegetation coverage in the study area. Fast urban development was recorded in districts including Malir, West, and Kiamari from 1990 to 2020. The primary urban core expanded from the core districts, namely, the Central, South, and Eastern districts, and a new urban secondary core was observed in Malir in 2020. The LULC simulation results for 2030 revealed a significant increase in urban built-up area of 111.6% compared with that in 2020, mainly distributed in sub-urban fringes.

This study proved remote sensing and GIS techniques to be valuable tools in tracking and assessing changes in urban spatial patterns. The findings of the analysis can provide policy implications for future urban land transformation management and planning in order to achieve the Sustainable Development Goals. Future research could explore the forces that drive urban sprawl and examine how they interact with social, economic, and environmental repercussions in fast growing cities.

Author Contributions: Conceptualization, L.L. and F.C.; Methodology, M.F.B.; Software, M.F.B.; Validation, M.F.B.; Formal analysis, M.F.B.; Investigation, S.Q.; Resources, L.L.; Data curation, A.T.; Writing—original draft preparation, M.F.B.; Writing—review and editing, L.L., S.Q. and S.W.; Visualization, S.H.; Supervision, L.J.; Project administration, Q.L.; Funding acquisition, Q.L. All authors have read and agreed to the published version of the manuscript.

Funding: This research was funded by the National Key Research and Development Program of China (No. 2019YFD1100803).

Institutional Review Board Statement: Not applicable.

Informed Consent Statement: Not applicable.

Data Availability Statement: The data presented in this study are available on request from the corresponding author.

Acknowledgments: The authors are grateful for the comments from anonymous reviewers and the editors.

References

1. United Nations. *World Population Prospects 2019: Highlights*; Department of Economic and Social Affairs, Population Division: New York, NY, USA, 2019; Available online: https://population.un.org/wpp/Publications/Files/WPP2019_10KeyFindings.pdf (accessed on 10 March 2021).
2. Foley, J.A.; DeFries, R.; Asner, G.P.; Barford, C.; Bonan, G.; Carpenter, S.R.; Chapin, F.S.; Coe, M.T.; Daily, G.C.; Gibbs, H.K. Global consequences of land use. *Science* **2005**, *309*, 570–574. [CrossRef]
3. Grimm, N.B.; Faeth, S.H.; Golubiewski, N.E.; Redman, C.L.; Wu, J.; Bai, X.; Briggs, J.M. Global change and the ecology of cities. *Science* **2008**, *319*, 756–760. [CrossRef]
4. Thapa, R.B.; Murayama, Y. Urban growth modeling of Kathmandu metropolitan region, Nepal. *Comput. Environ. Urban Syst.* **2011**, *35*, 25–34. [CrossRef]
5. Pauleit, S.; Breuste, J.; Qureshi, S.; Sauerwein, M. Transformation of rural-urban cultural landscapes in Europe: Integrating approaches from ecological, socio-economic and planning perspectives. *Landsc. Online* **2010**, *20*, 1–10. [CrossRef]
6. Lu, L.; Weng, Q.; Guo, H.; Feng, S.; Li, Q. Assessment of urban environmental change using multi-source remote sensing time series (2000–2016): A comparative analysis in selected megacities in Eurasia. *Sci. Total Environ.* **2019**, *684*, 567–577. [CrossRef] [PubMed]
7. United Nations. *Transforming Our World: The 2030 Agenda for Sustainable Development*; United Nations, Department of Economic and Social Affairs: New York, NY, USA, 2015.
8. Inostroza, L.; Hamstead, Z.; Spyra, M.; Qureshi, S. Beyond urban–rural dichotomies: Measuring urbanisation degrees in central European landscapes using the technomass as an explicit indicator. *Ecol. Indic.* **2019**, *96*, 466–476. [CrossRef]
9. Hashem, N.; Balakrishnan, P. Change analysis of land use/land cover and modelling urban growth in Greater Doha, Qatar. *Ann. GIS* **2015**, *21*, 233–247. [CrossRef]
10. Lu, L.; Guo, H.; Corbane, C.; Li, Q. Urban sprawl in provincial capital cities in China: Evidence from multi-temporal urban land products using Landsat data. *Sci. Bull.* **2019**, *64*, 955–957. [CrossRef]
11. Li, X.; Liu, X.; Yu, L. A systematic sensitivity analysis of constrained cellular automata model for urban growth simulation based on different transition rules. *Int. J. Geogr. Inf. Sci.* **2014**, *28*, 1317–1335. [CrossRef]
12. Ke, X.; Qi, L.; Zeng, C. A partitioned and asynchronous cellular automata model for urban growth simulation. *Int. J. Geogr. Inf. Sci.* **2016**, *30*, 637–659. [CrossRef]
13. Mondal, B.; Das, D.N.; Bhatta, B. Integrating cellular automata and Markov techniques to generate urban development potential surface: A study on Kolkata agglomeration. *Geocarto Int.* **2017**, *32*, 401–419. [CrossRef]
14. Pijanowski, B.C.; Pithadia, S.; Shellito, B.A.; Alexandridis, K. Calibrating a neural network-based urban change model for two metropolitan areas of the Upper Midwest of the United States. *Int. J. Geogr. Inf. Sci.* **2005**, *19*, 197–215. [CrossRef]
15. Maithani, S. A neural network based urban growth model of an Indian city. *J. Indian Soc. Remote Sens.* **2009**, *37*, 363–376. [CrossRef]
16. Tang, J.; Wang, L.; Yao, Z. Spatio-temporal urban landscape change analysis using the Markov chain model and a modified genetic algorithm. *Int. J. Remote Sens.* **2007**, *28*, 3255–3271. [CrossRef]
17. Firozjaei, M.K.; Kiavarz, M.; Alavipanah, S.K.; Lakes, T.; Qureshi, S. Monitoring and forecasting heat island intensity through multi-temporal image analysis and cellular automata-Markov chain modelling: A case of Babol city, Iran. *Ecol. Indic.* **2018**, *91*, 155–170. [CrossRef]
18. Mondal, B.; Das, D.N.; Dolui, G. Modeling spatial variation of explanatory factors of urban expansion of Kolkata: A geographically weighted regression approach. *Model. Earth Syst. Environ.* **2015**, *1*, 1–13. [CrossRef]
19. Clarke, K.C.; Gaydos, L.J. Loose-coupling a cellular automaton model and GIS: Long-term urban growth prediction for San Francisco and Washington/Baltimore. *Int. J. Geogr. Inf. Sci.* **1998**, *12*, 699–714. [CrossRef]
20. Dadashpoor, H.; Nateghi, M. Simulating spatial pattern of urban growth using GIS-based SLEUTH model: A case study of eastern corridor of Tehran metropolitan region, Iran. *Environ. Dev. Sustain.* **2017**, *19*, 527–547. [CrossRef]
21. Bihamta, N.; Soffianian, A.; Fakheran, S.; Gholamalifard, M. Using the SLEUTH Urban Growth Model to Simulate Future Urban Expansion of the Isfahan Metropolitan Area, Iran. *J. Indian Soc. Remote Sens.* **2015**, *43*, 407–414. [CrossRef]
22. Park, S.; Jeon, S.; Choi, C. Mapping urban growth probability in South Korea: Comparison of frequency ratio, analytic hierarchy process, and logistic regression models and use of the environmental conservation value assessment. *Landsc. Ecol. Eng.* **2012**, *8*, 17–31. [CrossRef]
23. Xian, G.; Crane, M.; McMahon, C. Quantifying multi-temporal urban development characteristics in Las Vegas from Landsat and ASTER data. *Photogramm. Eng. Remote Sens.* **2008**, *74*, 473–481. [CrossRef]

24. Bununu, Y.A. Integration of Markov chain analysis and similarity-weighted instance-based machine learning algorithm (SimWeight) to simulate urban expansion. *Int. J. Urban Sci.* **2017**, *21*, 217–237. [CrossRef]
25. Angel, S.; Parent, J.; Civco, D. Urban sprawl metrics: An analysis of global urban expansion using GIS. In Proceedings of the ASPRS 2007 Annual Conference, Tampa, FL, USA, 7–11 May 2007.
26. Sharma, R.; Joshi, P.K. Monitoring Urban Landscape Dynamics Over Delhi (India) Using Remote Sensing (1998–2011) Inputs. *J. Indian Soc. Remote Sens.* **2013**, *41*, 641–650. [CrossRef]
27. Batty, M. Agents, Cells, and Cities: New Representational Models for Simulating Multiscale Urban Dynamics. *Environ. Plan. A* **2005**, *37*, 1373–1394. [CrossRef]
28. Jokar Arsanjani, J.; Helbich, M.; Kainz, W.; Darvishi Boloorani, A. Integration of logistic regression, Markov chain and cellular automata models to simulate urban expansion. *Int. J. Appl. Earth Obs. Geoinf.* **2013**, *21*, 265–275. [CrossRef]
29. Keshtkar, H.; Voigt, W. Potential impacts of climate and landscape fragmentation changes on plant distributions: Coupling multi-temporal satellite imagery with GIS-based cellular automata model. *Ecol. Inform.* **2016**, *32*, 145–155. [CrossRef]
30. Rimal, B.; Zhang, L.; Keshtkar, H.; Haack, B.; Rijal, S.; Zhang, P. Land Use/Land Cover Dynamics and Modeling of Urban Land Expansion by the Integration of Cellular Automata and Markov Chain. *ISPRS Int. J. Geo-Inf.* **2018**, *7*, 154. [CrossRef]
31. Akhtar, S.; Dhanani, M.R. Urban Sprawl in Karachi. *Glob. Adv. Res. J. Geogr. Reg. Plan.* **2013**, *2*, 160–171.
32. Schetke, S.; Qureshi, S.; Lautenbach, S.; Kabisch, N. What determines the use of urban green spaces in highly urbanized areas?—Examples from two fast growing Asian cities. *Urban For. Urban Green.* **2016**, *16*, 150–159. [CrossRef]
33. Qureshi, S. The fast growing megacity Karachi as a frontier of environmental challenges: Urbanization and contemporary urbanism issues. *J. Geogr. Reg. Plan.* **2010**, *3*, 306–321.
34. Hamza, S.; Khan, I.; Lu, L.; Liu, H.; Burke, F.; Nawaz-ul-Huda, S.; Baqa, M.F.; Tariq, A. The Relationship between Neighborhood Characteristics and Homicide in Karachi, Pakistan. *Sustainability* **2021**, *13*, 5520. [CrossRef]
35. D'Acci, L. A new type of cities for liveable futures. Isobenefit Urbanism morphogenesis. *J. Environ. Manag.* **2019**, *246*, 128–140. [CrossRef] [PubMed]
36. Shaikh, A.; Gotoh, K. A Satellite Remote Sensing Evaluation of Urban Land Cover Changes and Its Associated Impacts on Water Resources in Karachi, Pakistan. *J. Jpn. Soc. Photogramm. Remote Sens.* **2006**, *45*, 41–55. [CrossRef]
37. Mahboob, M.A.; Atif, I.; Iqbal, J. Remote Sensing and GIS Applications for Assessment of Urban Sprawl in Karachi, Pakistan. *Sci. Technol. Dev.* **2015**, *34*, 179–188. [CrossRef]
38. City District Government Karachi. *Karachi Strategic Development Plan 2020*; CDGK (City District Government Karachi): Karachi, Pakistan, 2007; Available online: https://urckarachi.org/2020/07/19/karachi-strategic-development-plan-2020/ (accessed on 15 January 2021).
39. Government of Pakistan. *Pakistan Bureau of Statistics*; 2017. Available online: https://www.pbs.gov.pk/ (accessed on 15 December 2020).
40. Ahmed, Q.I.; Lu, H.; Ye, S. Urban transportation and equity: A case study of Beijing and Karachi. *Transp. Res. Part A Policy Pract.* **2008**, *42*, 125–139. [CrossRef]
41. Government of Pakistan. Pakistan Economic Survey; 2017–2018. Available online: http://www.finance.gov.pk/survey_1718.html (accessed on 15 December 2020).
42. World Bank. *Transforming Karachi into a Livable and Competitive Megacity: A City Diagnostic and Transformation Strategy*; The World Bank: Washington, DC, USA, 2018; Available online: https://openknowledge.worldbank.org/bitstream/handle/10986/29376/211211ov.pdf?sequence=8&isAllowed=y (accessed on 15 January 2021).
43. Haq, U. The Rise of Karachi as a Mega-City: Issues and Challenges. Human Development Centre. Available online: http://www.mhhdc.org (accessed on 9 January 2021).
44. Hasan, A.; Ahmed, N.; Raza, M.; Sadiq, A.; Ahmed, S.; Sarwar, M.B. *Land Ownership, Control and Contestation in Karachi and Implications for Low-Income Housing*; Human Settlements Group, International Institute for Environment and Development: London, UK, 2013.
45. Lu, L.; Kuenzer, C.; Wang, C.; Guo, H.; Li, Q. Evaluation of Three MODIS-Derived Vegetation Index Time Series for Dryland Vegetation Dynamics Monitoring. *Remote Sens.* **2015**, *7*, 7597–7614. [CrossRef]
46. Lu, L.; Weng, Q.; Xiao, D.; Guo, H.; Li, Q.; Hui, W. Spatiotemporal Variation of Surface Urban Heat Islands in Relation to Land Cover Composition and Configuration: A Multi-Scale Case Study of Xi'an, China. *Remote Sens.* **2020**, *12*, 2713. [CrossRef]
47. Parent, J.; Civco, D.; Angel, S. *Urban Growth Analysis: Calculating Metrics to Quantify Urban Sprawl*; University of Connecticut: Storrs, CT, USA, 2008; Available online: https://proceedings.esri.com/library/userconf/proc08/papers/papers/pap_1692.pdf (accessed on 9 March 2021).
48. Mas, J.-F.; Kolb, M.; Paegelow, M.; Olmedo, M.T.C.; Houet, T. Inductive pattern-based land use/cover change models: A comparison of four software packages. *Environ. Model. Softw.* **2014**, *51*, 94–111. [CrossRef]
49. Kumar, S.; Radhakrishnan, N.; Mathew, S. Land use change modelling using a Markov model and remote sensing. *Geomat. Nat. Hazards Risk* **2014**, *5*, 145–156. [CrossRef]
50. Guan, D.; Li, H.; Inohae, T.; Su, W.; Nagaie, T.; Hokao, K. Modeling urban land use change by the integration of cellular automaton and Markov model. *Ecol. Model.* **2011**, *222*, 3761–3772. [CrossRef]
51. Tariq, A.; Shu, H. CA-Markov Chain Analysis of Seasonal Land Surface Temperature and Land Use Land Cover Change Using Optical Multi-Temporal Satellite Data of Faisalabad, Pakistan. *Remote Sens.* **2020**, *12*, 3402. [CrossRef]

52. Lillesand, T.; Kiefer, R.W.; Chipman, J. *Remote Sensing and Image Interpretation*; John Wiley & Sons: Hoboken, NJ, USA, 2015.
53. Shawul, A.A.; Chakma, S. Spatiotemporal detection of land use/land cover change in the large basin using integrated approaches of remote sensing and GIS in the Upper Awash basin, Ethiopia. *Environ. Earth Sci.* **2019**, *78*, 1–13. [CrossRef]
54. Mahar, G.A.; Hussain, B.; Wagan, R.A.; Hussain, A.T.; Khan, S.H. Remote sensing (RS) monitoring of mangroves plantation against the guinness world record (GWR) of maximum plantation in indus delta. *Pak. J. Bot.* **2020**, *52*, 1497–1503. [CrossRef]
55. Liu, Y.; He, Q.; Tan, R.; Liu, Y.; Yin, C. Modeling different urban growth patterns based on the evolution of urban form: A case study from Huangpi, Central China. *Appl. Geogr.* **2016**, *66*, 109–118. [CrossRef]

Article

Characterizing Urban Expansion Combining Concentric-Ring and Grid-Based Analysis for Latin American Cities

Su Wu [1], Neema Simon Sumari [2], Ting Dong [3], Gang Xu [4,5,*] and Yanfang Liu [1]

[1] School of Resource and Environmental Sciences, Wuhan University, 129 Luoyu Road, Wuhan 430079, China; wusu027@whu.edu.cn (S.W.); yfliu610@163.com (Y.L.)

[2] Department of Mathematics, Informatics and Computational Sciences, Solomon Mahlangu College of Science and Education, Sokoine University of Agriculture, P.O. Box 3038 Morogoro, Tanzania; neydsumari@sua.ac.tz

[3] School of Geography and Tourism, Anhui Normal University, 189 Jiuhua South Road, Wuhu 241000, China; dongtingdt@whu.edu.cn

[4] School of Remote Sensing and Information Engineering, Wuhan University, 129 Luoyu Road, Wuhan 430079, China

[5] Key Laboratory of Urban Land Resources Monitoring and Simulation, Ministry of Natural Resources, 8007 West Hongli Road, Shenzhen 518034, China

* Correspondence: xugang@whu.edu.cn

Abstract: Spatio-temporal characterization of urban expansion is the first step towards understanding how cities grow in space. We summarize two approaches used in urban expansion measurement, namely, concentric-ring analysis and grid-based analysis. Concentric-ring analysis divides urban areas into a series of rings, which is used to quantify the distance decay of urban elements from city centers. Grid-based analysis partitions a city into regular grids that are used to interpret local dynamics of urban growth. We combined these two approaches to characterize the urban expansion between 2000–2014 for five large Latin American cities (São Paulo, Brazil; Mexico City, Mexico; Buenos Aires, Argentina; Bogotá, Columbia; Santiago, Chile). Results show that the urban land (built-up area) density in concentric rings decreases from city centers to urban fringe, which can be well fitted by an inverse S curve. Parameters of fitting curves reflect disparities of urban extents and urban form among these five cities over time. Grid-based analysis presents the transformation of population from central to suburban areas, where new urban land mostly expands. In the global context, urban expansion in Latin America is far less rapid than countries or regions that are experiencing fast urbanization, such as Asia and Africa. Urban form of Latin American cities is particularly compact because of their rugged topographies with natural limitations.

Keywords: urban expansion; concentric-ring analysis; grid-based analysis; invers S curve; Latin America

Citation: Wu, S.; Sumari, N.S.; Dong, T.; Xu, G.; Liu, Y. Characterizing Urban Expansion Combining Concentric-Ring and Grid-Based Analysis for Latin American Cities. *Land* **2021**, *10*, 444. https://doi.org/10.3390/land10050444

Academic Editor: Iwona Cieślak

Received: 17 March 2021
Accepted: 19 April 2021
Published: 22 April 2021

Publisher's Note: MDPI stays neutral with regard to jurisdictional claims in published maps and institutional affiliations.

1. Introduction

More and more people are now living in cities, driving the persistent expansion of urban land across the world [1]. Rapid expansion of urban land swallows cultivated land and natural land, threatening biodiversity and exacerbating environmental degradation [2–4]. Numerous studies have investigated spatio-temporal characteristics of urban land expansion and its environmental consequences [5,6]. Remote sensing and geographical information science (GIS) have greatly contributed to quantifying land use changes [7]. Remotely sensed imagery provides the first-hand data used to monitor urban growth in space [8]. GIS-based technologies provide a wealth of tools for urban expansion measurement [9].

The first step to measure urban expansion is its speed and intensity, for instance, the annual growth rate of built-up areas [10,11]. The intensity of urban expansion refers to proportions of urban land use changes to the total land area in a defined region [10,12]. Beyond the statistical description of urban land use changes, the spatial perspective is employed to answer the spatial pattern of urban form and where the urban expansion happens [13]. Thirdly, the temporal process cannot be separated when we characterize urban expansion [14]. Landscape metrics are broadly used to measure dynamics and heterogeneities in the temporal process of urban expansion [15]. Finally, urban theories behind urban land use changes are proposed based on these quantitative measurements, such as the diffusion-coalescence theory [16–18], pattern-process interrelationships, distance-decay of densities [19,20], types of urban expansion (infilling, extension, and leapfrog) [21–24], and driving forces of urban expansion [25], etc.

As far as the spatial unit is concerned, there are two approaches to quantify urban expansion, namely, the concentric-ring analysis and grid-based analysis [13,26]. The concentric-ring analysis, also known as gradient analysis, divides urban areas into a series of concentric rings from the city center, which usually is the point of origin or central business district (CBD) of the city [27,28]. The concentric-ring analysis is often used in conjunction with landscape metrics (such as the percent of landscape, patch density, and many other indicators) [12,22]. They are used to analyze spatial patterns (distance-decay) of urban landscape. The distance-decay of urban land density (built-up density) in concentric rings reflects the gradient of urban development intensity [19]. Some studies divide concentric rings into multiple sectors to measure the heterogeneity of urban growth in different directions [15]. Considering irregular urban forms or spatial constrains, such as UK's 'green belts' and large water bodies in Wuhan, China, an improved partitioning method for concentric-ring analysis was proposed [12].

The grid-based analysis divides urban areas into regular grids; for example, 1 km grids [26,29–31]. Each grid is a sample and the urban dynamics in each grid are different, which determines the overall spatial pattern of land-use transformation. Grid-based analysis allows the correlation analysis between urban land and other occupied land, and it also builds the bridge between urban land expansion and population growth [26]. The growth in population fundamentally drives the expansion of urbanized land [32]. Numerous studies reported the faster growth rate of urban land than that of population over time, resulting in a decline in the urban population density [33–40].

In this study, we attempt to combine the concentric-ring analysis and grid-based analysis to quantify spatio-temporal characteristics of urban expansion. Our case study area is Latin America, defined as the Americas south of the United States. Latin America consists of 20 countries and 13 dependencies, with more than 640 million population in 2016 and an area of approximately 19,000,000 km^2. In many countries of Latin America, the economy developed rapidly after the middle of the 20th century, and the level of urbanization has continued to increase. Latin America is one of the most urbanized continents in the world with almost 84% of the total population living in cities [41]. For example, Brazil's urbanization level is more than 85%, and Argentina's urbanization level is even more than 90% [42].

Although the urbanization level in Latin America is relatively high, compared with the extensive research of urban expansion in the United States and Europe, there are few studies on the temporal and spatial characteristics of urban expansion in Latin America [41]. In particular, there is a lack of comparative analysis of urban expansion and spatial dynamics among cities in Latin America [13,41,43]. Investigating the dynamics of urban land expansion in Latin America not only has regional significance, but also has international comparative value. It also provides precedents for developing countries that are undergoing rapid urbanization, such as Africa and Asia [44,45].

2. Materials and Methods

2.1. Study Area

New York University, United Nations-Habitat, and Lincoln Institute of Land Policy have published *The Atlas of Urban Expansion* (2016 Edition) of 200 cities around the world, which includes 26 cities in Latin America and the Caribbean [46,47]. Considering the representativeness of large cities and the evenness of their spatial distributions, we selected five large cities with a population of more than five million in 2014 from 26 cities, namely,

São Paulo (Brazil), Mexico City (Mexico), Buenos Aires (Argentina), Bogotá (Colombia), and Santiago (Chile) (Figure 1, Table 1).

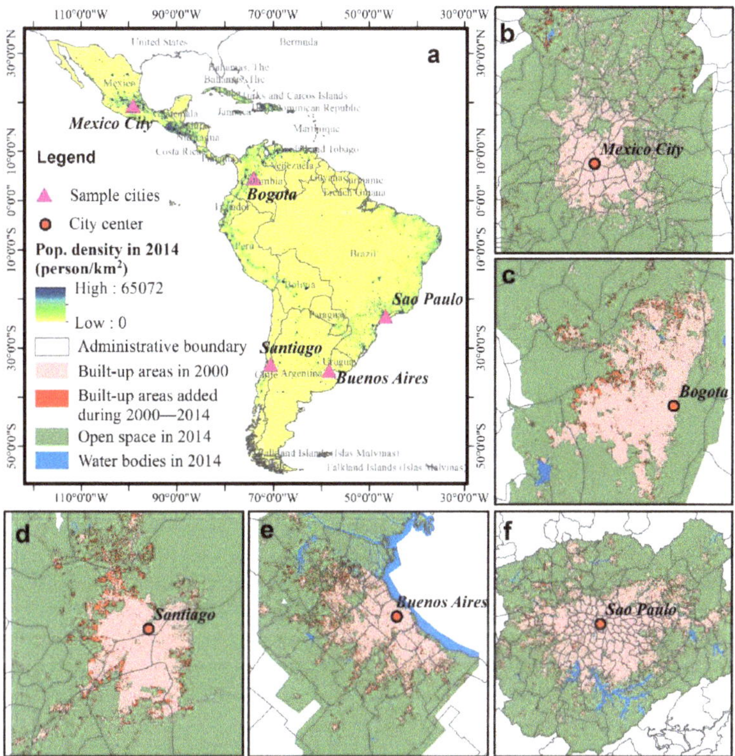

Figure 1. Spatial distributions of five major cities in Latin America with their urban land use in 2000 and 2014.

Table 1. Urban population, built-up area, and population density in 2014 and their annual growth rates from 2000 to 2014 in five large Latin American cities.

City	City Center	Urban Population in 2014 (Million) [#]	AGR * of Urban Population (2000–2014, %)	Built-up Area in 2014 (km²) [#]	AGR * of Built-up Area (2000–2014, %)	Population Density in 2014 (Person/km²) [#]	AGR * of Population Density (2000–2014, %)
São Paulo, Brazil	Catedral Metropolitana de São Paulo	19.61	1.08	1724	0.70	11,372	0.38
Mexico City, Mexico	Palacio Nacional	17.77	2.53	1618	3.56	10,978	−0.99
Buenos Aires, Argentina	Parque centenario	13.88	1.46	1473	1.88	9421	−0.41
Bogotá, Colombia	Museo Nacional	7.80	2.29	319	1.26	24,460	1.02
Santiago, Chile	Plaza Baquedano	6.49	1.32	604	1.76	10,742	−0.43

Data source: The atlas of urban expansion (2016 Edition) http://www.atlasofurbanexpansion.org/ (accessed on 20 April 2021). * AGR (annual growth rate) is calculated as $AGR = \left(\sqrt[n]{P_{2014}/P_{2000}} - 1 \right) \cdot 100\%$, where P is the population, built-up area, or population density of a city, and n is the interval in years.

São Paulo is the largest city in Brazil and the economic, cultural, and technological center. Nearly 20 million people lived in São Paulo in 2014 with the total built-up area being around 1700 km^2. Mexico City is the capital of Mexico and is located in the valley of the plateau of central Mexico, at an altitude over 2000 m. It is the political, economic, cultural and transportation center of Mexico. Buenos Aires is the capital and largest city of Argentina and it is a coastal city on the southern bank of the La Plata River. Bogotá is the capital and largest city in Colombia and it is located in the center of Colombia, on a high plateau known as the Bogotá savanna. Due to the eastern mountains, Bogotá can only expand to the west (Figure 1c). Santiago is the capital and largest city in Chile, which is entirely located in the country's central valley. We identified city centers for the five cities using road maps and high resolution images from Google Map (Figure 1, Table 1).

2.2. Land Use and Population Data

We collected urban land use maps of five Latin American cities in 2000 and 2014 from *The Atlas of Urban Expansion* (*2016 Edition*), which were interpreted using Landsat imagery (30 m resolution) [46]. They classified land use into three categories, namely, built-up area, open space, and water bodies, using the unsupervised classification method (clustering analysis) with the ISODATA (Iterative Self-Organizing Data Analysis Technique) algorithm. The classification result has a high accuracy with verification using Google Earth high resolution imagery. The user's accuracy for the built-up area is 91% and the producer's accuracy is 89.3% [46]. Detailed information on the land use classification procedures can be found in Hurd [48]. The data from *The Atlas of Urban Expansion* (*2016 Edition*) has been successfully used in related studies on urban expansion and urban form [49–51]. The spatial distributions of urban land in 2000 and 2014 and open space and water bodies in 2014 of five Latin American cities are shown in Figure 1.

The population data in Table 1 also comes from *The Atlas of Urban Expansion* (*2016 Edition*), but it is statistical data without information on its spatial distribution. We further collected *LandScan* population data for the five sample cities in 2000 and 2014. *LandScan* is a community standard for global population distribution data at approximately 1 km (30″ × 30″) spatial resolution, and it represents an ambient population (average over 24 h) distribution [52–54]. The spatial distribution of population data supports the correlation analysis between urban expansion and population growth using grid-based analysis.

2.3. Concentric Ring and Grid-Based Analysis

We generate a series of 1-km equidistant buffer rings to cover almost the entire built-up areas in 2014 (Figure 2a). Previous studies calculated landscape metrics in concentric rings and analyze their spatial variations [15]. In this study, we calculate the urban land density (built-up density) in each concentric ring, and then fit its distance decay using the inverse *S* curve (See Section 2.4). The urban land density is defined as the proportion of urban land to the area of buildable land in each ring (Equation (1)).

$$Dens = \frac{S_{urban\ land}}{S_{buildable\ land}} \tag{1}$$

where *Dens* is the urban land density in a concentric ring (doughnut). $S_{urban\ land}$ is the area of urban land (built-up area) in a ring, and $S_{buidable\ land}$ is the area of buildable land in a ring. The buildable land is the total land area excluding water bodies and mountains in a ring. The concentric rings are developed and urban land density in each ring are calculated in ArcGIS 10.6 (ESRI, Inc., Redwoodds, CA, USA).

The unit of grid-based analysis is 1 km, which is consistent with the resolution of *LandScan* population data. We first clip the *LandScan* population data in 2000 and 2014 for each city using the spatial extent of land use map of the same city (Figure 2b). We convert the population raster file to a vector file in 1 km × 1 km grid segments. We calculate the area of built-up area in each grid for two years, 2000 and 2014, using zonal statistics, and then calculate the change of built-up area in each grid from 2000 to 2014. Finally, we use map algebra to calculate the change in population from 2000 to 2014 in each grid of each city and convert the result to a vector file. We combine the changes in built-up areas and population for the 2000–2014 period based on the ID of each grid. All of these spatial analytics, such as format conversion, zonal statistics, and map algebra, are conducted in ArcGIS 10.6 (ESRI, Inc., Redwoodds, CA, USA).

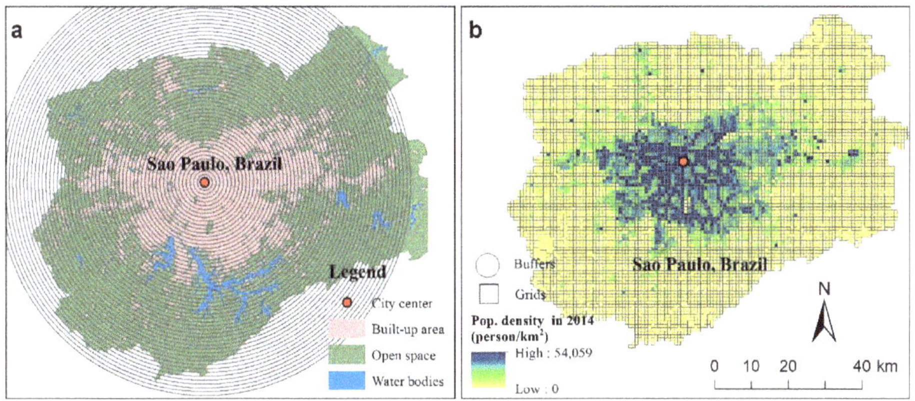

Figure 2. An example of concentric ring analysis (**a**) and grid-based analysis (**b**) in São Paulo, Brazil.

2.4. Inverse S Curve

Jiao (2015) proposed a function with an inverse S curve (Figure 3) to characterize the spatial distribution of the urban land density within a city, as shown in Equation (2) [19]:

$$f(r) = \frac{1-c}{1 + e^{\alpha((2r/D)-1)}} + c \tag{2}$$

where f is the urban land density (built-up density) in each concentric ring that is calculated using Equation (1), r is the distance to the city center, e is the Euler's number, and α, c and D are parameters, which vary across cities and over time.

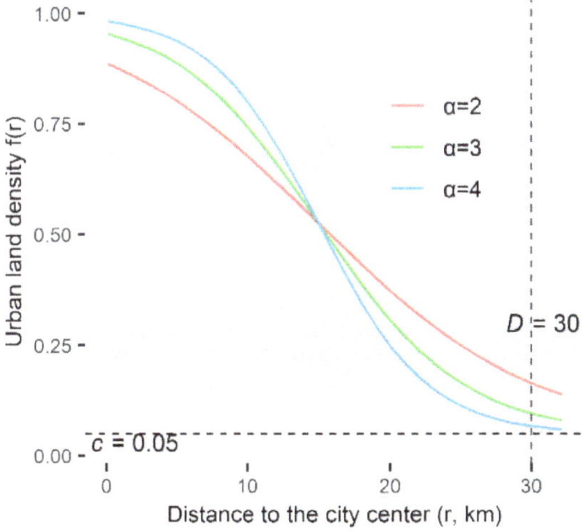

Figure 3. Curves of the urban land density function with different α parameters.

Parameters in the inverse S curve have explicit physical meanings. The α parameter controls the form of the fitting curve and a higher α indicates a more compact urban form (Figure 3). The D parameter denotes the approximate boundary of an urban extent. The c

parameter represents the background value of the urban land density. The D parameter generally increases over time because the urban extent expands. Detailed explanations of these three parameters can be found in Jiao (2015) [19].

The inverse S curve has been an effective and quantitative tool in urban studies. It can be used to not only fit the distance decay of urban land density but also can be applied to other urban indicators, such as the population density, road density, and even land surface temperature (LST) [19,55,56]. The inverse S curve was first proposed using Chinese cities and then it has been applied to cities in other countries across the world [57–60].

3. Results

3.1. Distance Decay of Urban Land Density

We calculate the urban land density in each concentric ring using Formula (1). Overall, the urban land density decreases slowly around the city center, and then decreases quickly to a relatively low level, and finally decreases slowly again to the background level of urban land density, showing an inverse S-shape (Figure 4). The inverse S curve can fit the distance decay of urban land density very well in all cities in both 2000 and 2014. Parameters for fitted curves are shown in Table 2. The urban physical size varies among the five cities. The radius of the urban extent (represented by the D parameter) is the largest for Buenos Aires, which is over 50 km. The radii of urban extents of São Paulo and Mexico City are around 45 km and 42 km in 2014, respectively. The space size is relatively small for Bogotá and Santiago with urban extent radii of 27 km and 26 km, respectively, in 2014.

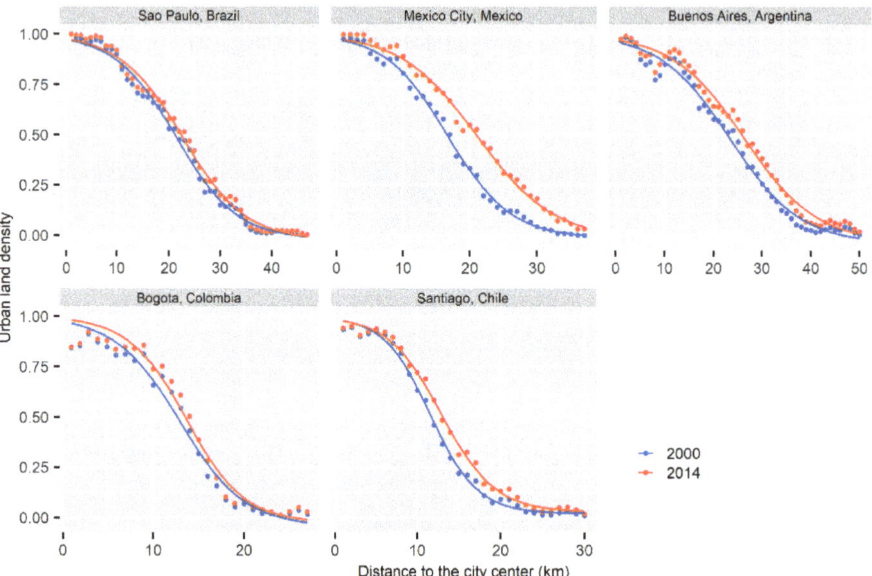

Figure 4. Distance decay of urban land density and fitted results using inverse S curve in concentric rings in five Latin American cities in 2000 and 2014.

The α parameter reflects the compactness of urban form and a higher α means a more compact urban form. Bogotá has the highest value of α (=4.49) in 2014, indicating the most compact urban form among the five cities. The urban population density in Bogotá in 2014 is also the highest among the five cities (24,460 person/km^2, Table 1), further supporting its compact form. Generally, the α parameter increases from 2000 to 2014 in four cities except for Santiago, which has a small decline. Cross-sectionally, Buenos Aires has the lowest value of α, indicating that it has the most relatively dispersed urban form.

The urban population density in Buenos Aires is also the lowest among the five cities (9421 person/km^2, Table 1).

Table 2. Parameters of the inverse S curve for five Latin American cities in 2000 and 2014.

City	Year	Num. of Rings	α	c	D	Adjusted R^2
São Paulo	2000	47	3.70	0.01	42.11	0.993
	2014	47	3.83	0.03	44.51	0.992
Mexico city	2000	37	3.65	0.02	33.12	0.996
	2014	37	3.78	0.03	41.73	0.994
Buenos Aires	2000	50	3.39	0.01	45.38	0.987
	2014	50	3.70	0.04	51.43	0.988
Bogotá	2000	27	3.87	0.03	25.13	0.974
	2014	27	4.49	0.05	26.95	0.976
Santiago	2000	30	4.11	0.02	23.17	0.996
	2014	30	4.03	0.03	25.82	0.996

The gap between the two curves of the same city for 2000 and 2014 reflects the dynamics of the urban expansion, which varies among the five cities. Mexico City has the largest gap between the two curves, indicating the most obvious urban expansion in this city. Comparatively, the two curves of São Paulo nearly overlap, implying a limited urban expansion in this city. The gap between the two curves also reflects the spatial disparities in urban expansion. The new urban land is mainly added around 10 km from the city center in Bogotá from 2000 to 2014, while in Santiago the urban expansion took place in its outskirts. This is why the urban form of Santiago has become more dispersed during 2000–2014 (Table 2). The new urban land is evenly distributed in space in Buenos Aires (Figure 4).

3.2. Directional Heterogeneity in Urban Form

Urban expansion is generally not homogeneous, but there is usually a directional heterogeneity due to the influence of natural conditions or urban planning [15,61]. Taking São Paulo as an example, we divided the concentric ring into four directions, namely East, South, West, and North, as shown in Figure 5a. Due to the limited urban expansion from 2000 to 2014 in São Paulo (Figure 4), we only take the land use information in 2014 to present the heterogeneity of urban form in different directions. Overall, São Paulo has an urban form with a longer-extent in the east-west direction but a more limited extent in a north-south direction, which is restricted by the macro physical geography (mountains in the north and water bodies in the south). The built-up area in the easterly direction is 717 km^2, which is the largest among the four directions, taking 42% of the whole built-up area (1724 km^2, Table 1). The built-up area in the northern direction is only 248 km^2.

We calculate the urban land density in four directions in São Paulo using concentric rings (Figure 5a). There are obvious disparities in the distance decay of urban land density and fitted curves in the four directions (Figure 5b). The parameters for the inverse S curve in the four directions are shown in Table 3. The fitted urban extents (parameter D) are 55 km, 47 km, 47 km, and 27 km in the East, South, West, and North, respectively. Unsurprisingly, the urban extent in the east is the largest, which is more than twice of that in the north. Remember that the α parameter indicates the compactness of urban form. Urban expansion in the south is restricted by water bodies, resulting in the most compact form in this direction ($\alpha = 8.81$), while the urban form is relatively dispersed in the west ($\alpha = 4.37$). There is an obvious bump of the urban land density in the north at around 26 km from the city center, which is caused by leapfrogging urban patches in the north of the city center (Figure 5b).

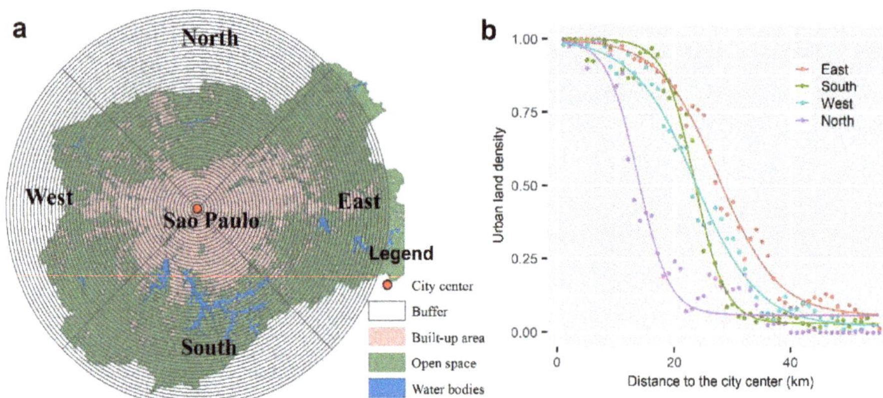

Figure 5. The concentric ring in four directions (**a**) and the spatial attenuation of urban land density in four directions of São Paulo in 2014 (**b**).

Table 3. Parameters of inversed-S shape function in four directions of São Paulo.

Directions	α	c	D	Adjusted R^2
East	5.43	0.06	55.02	0.990
South	8.81	0.03	46.83	0.989
West	4.37	0.02	47.05	0.989
North	5.16	0.06	27.36	0.963

3.3. Correlations between Land Expansion and Population Growth

We use grid-based analysis to investigate urban land expansion, population growth and their correlations. Changes in population and built-up areas in 1 km grid squares for 2000 and 2014 are presented in Figure 6. There are apparent disparities in space for both land expansion and population growth. Population increases in some places while it decreases in other locations. Taking São Paulo as an example, the urban population decreased around the city center and urban core area, while it increased in the urban fringe and suburban areas (Figure 6a), reflecting the process of suburbanization [62]. Mexico City, Buenos Aires, and Santiago also experienced the transformation of population from the city center to suburban areas, but with different characteristics (Figure 6b,c,h). In Mexico City, urban population transferred from the east to the west during 2000–2014 (Figure 6b). Although the central part of Buenos Aires had an increased population, the urban population obviously decreased near the city center, forming a doughnut with a reduced population around the center (Figure 6c). The urban population change in Santiago shows the strongest heterogeneity, which declined in the main corridors passing through the city center. Different from the other four cities, there is no obvious decline in population around the city center in Bogotá (Figure 6g).

Generally, the expansion of urban land shares a similar pattern as seen in the population growth. The new built-up areas mainly occurred at the urban fringe and suburban areas, while a limited newly built-up area was found around the city center (Figure 6d–f,i,j). In São Paulo, the urban expansion hotspots are located in the east and west from 2000 to 2014 (Figure 6b). The urban land spread evenly around the city center in Mexico City, particularly in the north and southeast (Figure 6e). As a coastal city, the new urban land can only spread to the inland side of Buenos Aires and most of these lie along main roads (Figure 6f). Bogotá shows a massive urban expansion northwest of the city center (Figure 6i). The urban expansion surrounded the city center of Santiago with more new urban land located in the northwest of the city (Figure 6j).

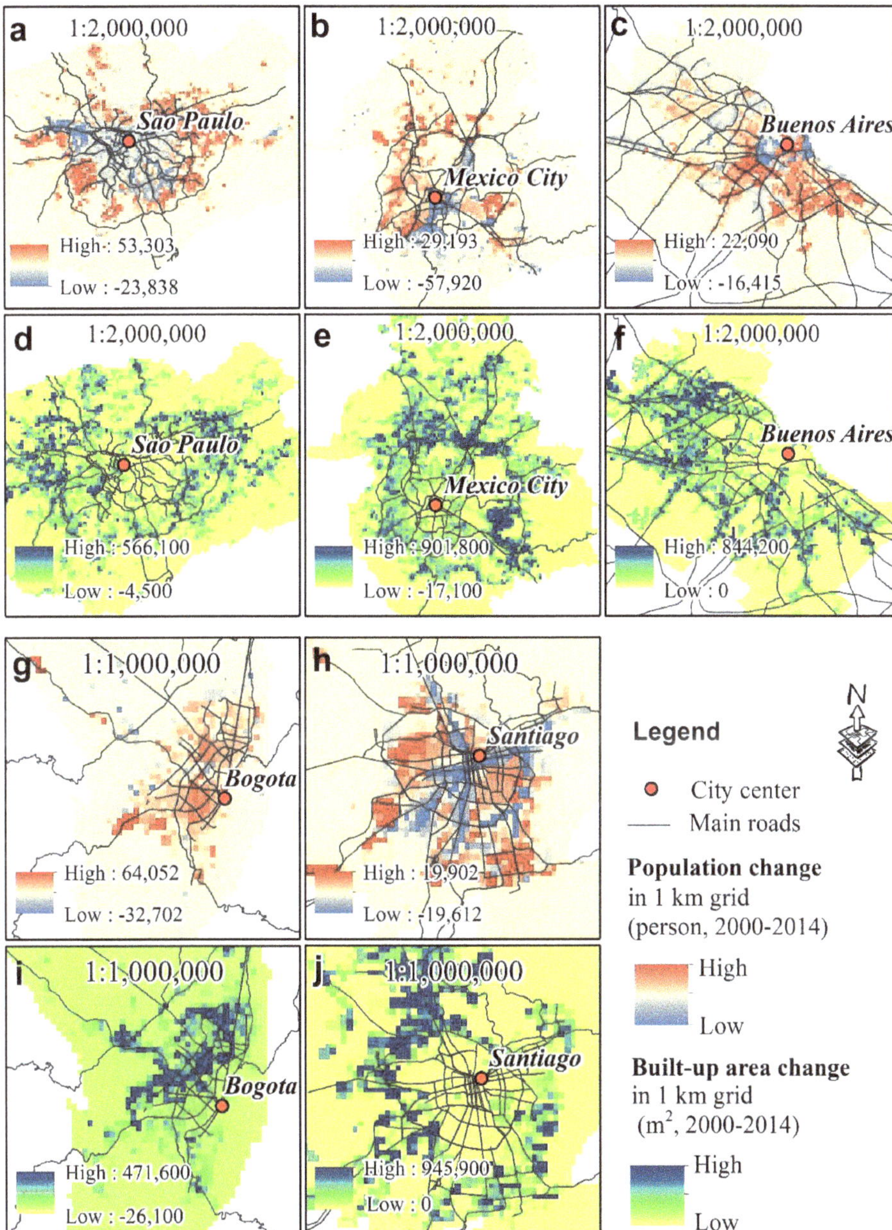

Figure 6. Spatial distributions of population changes (**a–c,g,h**) and built-up area changes (**d–f,i,j**) from 2000 to 2014 in five Latin American cities. Only three types of roads (trunk, primary, motorway) were preserved from the OSM road net (https://www.openstreetmap.org) (accessed on 20 April 2021).

The frequency distributions of changes in population and built-up area from 2000 to 2014 in 1 km grid squares are shown in Figure 7a,b, and their correlations are presented in Figure 7c. Population and built-up areas remain unchanged in most grids during 2000–2014. The mean values of population change in the 1 km grid are 563 persons

(São Paulo), 536 persons (Mexico City), 518 persons (Buenos Aires), 944 persons (Bogotá), and 200 person (Santiago). The mean values of changes in the built-up areas using the 1 km grid are 2.49 hm^2 (São Paulo), 9.88 hm^2 (Mexico City), 4.64 hm^2 (Buenos Aires), 1.96 hm^2 (Bogotá), and 3.15 hm^2 (Santiago). From the mean values for each grid, Bogotá experiences the highest population increase but with the least expansion in use of urban land, verifying its high population density (Tables 1 and 2) and compact urban form. Scatter plots show that the population change and the built-up area change are positively correlated (Figure 7c).

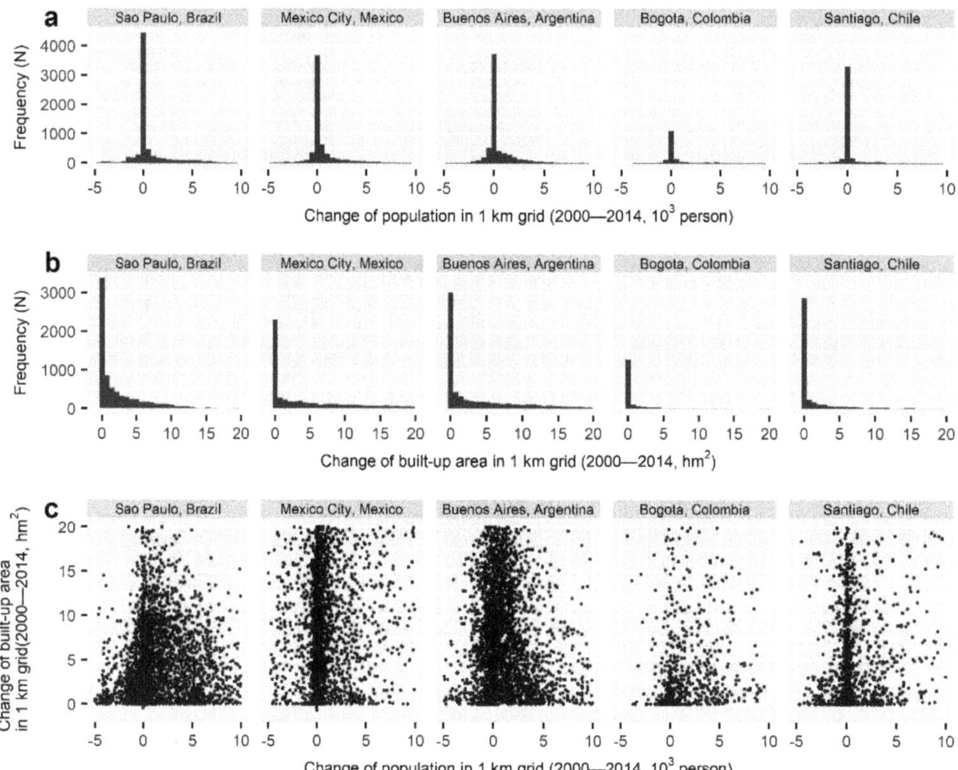

Figure 7. Frequency distribution and correlation analysis between the population growth and the urban land expansion from 2000 to 2014 in five Latin American cities. (**a**) Frequency of population change; (**b**) Frequency of built-up area change; (**c**) Scatter plots of population change and built-up area change in each 1 km grid.

4. Discussion

The city is composed of different spatial units. In addition to administrative divisions, the two most commonly used units in urban studies are concentric-rings and grids. Concentric-ring analysis, also called gradient analysis, is developed based on the monocentric model of cities, which does well in measuring macro pattern (urban-rural gradient) of urban expansion and urban form. Grid-based analysis has advantages in portraying the local microscopic dynamics of urban expansion. Grid-based analysis is usually used to build correlations between urban land and population changes. Urban expansion is a process from micro dynamics to macro patterns. Every land use change occurred in a local area, but the accumulation of local micro-processes shaped the macro pattern of urban expansion. The combined use of concentric-ring and grid-based analysis can measure urban expansion from both macro and micro perspectives. Although the concentric-ring analysis and grid-based analysis have been used in previous studies, they are all used in a single way. We propose for the first time that these are two different strategies for measuring urban expansion.

We combined these two approaches to quantify urban expansion for five major Latin America cities from 2000 to 2014. The urban land density (built-up density) in concentric rings decreases from the city center to suburban areas, presenting an inverse S-shape (Figure 4). Partitioning concentric rings into different sectors can effectively quantify the directional heterogeneity of urban expansion (Figure 5). The grid-based analysis particularly presents the process of suburbanization where population transferred from central areas to suburban areas from 2000 to 2014 (Figure 6). Meanwhile, the newly added urban land mainly expands in the urban fringe and suburban areas to accommodate the increasing population there (Figure 6).

Parameters of the inverse S curve reveal the urban extents and compactness of urban form (Figure 4, Table 2). Although São Paulo has the largest built-up area, Buenos Aires has a larger urban extent than São Paulo. Buenos Aires experienced the most dispersed urban form, while the urban form is the most compact in Bogotá. The compactness of the urban form quantified by the inverse S curve is consistent with urban population densities that is the highest in Bogotá but the lowest in Buenos Aires (Table 1). The calculation of population change and built-up change using the grid-based analysis also confirms that the compact growth in Bogotá, where the urban population averagely increase by 944 person at a cost of 1.96 hm^2 for each 1 km grid during 2000–2014 (Figure 7).

In the global context, the speed of urban land expansion in Latin American cities is relatively low. The intensity of the urban expansion is reflected by the gaps between the two fitting curves using the inverse S curve. The gaps between the two curves in 2000 and 2014 are very close and even overlapped in the five major Latin American cities (Figure 4). Not to mention the fast urban expansion in China and other countries in Asia and Africa, even the expansion of urban land in the United States and Europe is greater than in Latin America [19,44,45,63]. On the other hand, the urban form of Latin American cities is more compact than for cities in other regions, which is quantified by the α parameter of the inverse S curve. The α parameter for the five major Latin American cities varies in 3.39–4.49, while that for Chinese and European cities is around 2.5, and is less than 2 for cities in the United States for the same two years (2000 and 2014) [6,20,44]. Many Latin American cities grow in rugged topographies which constrains expansion, resulting in the overall compact urban form [43].

The distance-decay of urban population densities from the city center has long been revealed in urban geography and urban economics [64]. Classical models include the Clark's negative exponential model [64], the Batty's inverse power function [65] and many other models [66]. The urban land density, defined as the proportion of built-up areas to the buildable land in concentric rings, also declines with distance to the city center, but there are limited quantitative models to measure this process [28]. Urban land density declines slowly around the city center followed by a quick decline outward, and then declines slowly again in the urban periphery, showing an inverse S shape as a whole [19]. Inspired by the widely used sigmoid function with an S curve, Jiao (2015) proposed an inverse S curve to quantify the distance-decay of urban land density from the city center [19]. The urban land density function (inverse S curve) has been not only used to quantify past urban expansion but also to guide the spatial allocation of future urban land, which has potential applications in urban planning [67]. Recently, a geographic micro-process model was proposed to explain why the distance-decay of urban land density presents an inverse S shape [20].

This study also has limitations. This study only quantifies the growth of cities from a spatial perspective, neglecting the underlying driving forces of urban transformation [25]. Secondly, we chose the city center based on the road network and we only identified one city center for each city. With the development of cities, those large cities tend to be polycentric. There are slight bumps of urban land densities around 12 km from the city center in Buenos Aires (Figure 4). These bumps are closely related to the polycentric structure of cities, which can be controlled using polycentric scenarios in concentric ring partitioning [19,45]. In addition, the intervals (1 km, 2 km, and so on) in concentric-ring analysis and the size of grids (1 × 1 km, 2 × 2 km, and so on) have an effect of scale [68], which is not further discussed in this study.

5. Conclusions

This study contributes to the characterization of urban expansion by summarizing two approaches: Concentric-ring analysis and grid-based analysis. Concentric-ring analysis is developed based on the monocentric model of cities and it is easy to present the distance decay of urban elements from the city center to suburban areas. Grid-based analysis partitions a city into regular grids, which reflects local dynamics of urban growth.

Taking five major cities in Latin America as examples, this study combines concentric-ring analysis and grid-based analysis to quantify their urban expansion and population growth from 2000 to 2014. Urban land density in concentric rings declines with the distance from the city center. The inverse S curve not only can well fit the distance decay of densities but also reflects disparities in the urban extent and compactness of the urban form. Grid-based analysis characterizes the transformation of population from central areas to suburban areas. It is suggested to combine concentric-ring and grid-based analysis to fully quantify urban expansion from both perspectives of macro patterns and micro dynamics in other global cities.

Author Contributions: Conceptualization, G.X. and S.W.; methodology, S.W., T.D., and G.X.; software, S.W.; data curation, S.W.; writing—original draft preparation, S.W. and G.X.; writing—review and editing, N.S.S., T.D., G.X. and Y.L. All authors have read and agreed to the published version of the manuscript.

Funding: This research was funded by the Open Fund of Key Laboratory of Urban Land Resources Monitoring and Simulation, Ministry of Natural Resources (KF-2019-04-036).

Institutional Review Board Statement: Not applicable.

Informed Consent Statement: Not applicable.

Data Availability Statement: Data used in this study is publicly available. Land use maps retrieved from Landsat images are downloaded from http://www.atlasofurbanexpansion.org/ (accessed on 20 April 2021). The spatial distributions of population are downloaded from https://landscan.ornl.gov/ (accessed on 20 April 2021).

Conflicts of Interest: The authors declare no conflict of interest.

References

1. Van Vliet, J. Direct and indirect loss of natural area from urban expansion. *Nat. Sustain.* **2019**, *2*, 755–763. [CrossRef]
2. Bren d'Amour, C.; Reitsma, F.; Baiocchi, G.; Barthel, S.; Guneralp, B.; Erb, K.H.; Haberl, H.; Creutzig, F.; Seto, K.C. Future urban land expansion and implications for global croplands. *Proc. Natl. Acad. Sci. USA* **2017**, *114*, 8939–8944. [CrossRef] [PubMed]
3. Yang, C.; Liu, H.; Li, Q.; Cui, A.; Xia, R.; Shi, T.; Zhang, J.; Gao, W.; Zhou, X.; Wu, G. Rapid Urbanization Induced Extensive Forest Loss to Urban Land in the Guangdong-Hong Kong-Macao Greater Bay Area, China. *Chin. Geogr. Sci.* **2021**, *31*, 93–108. [CrossRef]
4. Yang, Q.; Huang, X.; Yang, J.; Liu, Y. The relationship between land surface temperature and artificial impervious surface fraction in 682 global cities: Spatiotemporal variations and drivers. *Environ. Res. Lett.* **2021**, *16*, 024032. [CrossRef]
5. Xia, C.; Zhang, A.; Wang, H.; Zhang, B.; Zhang, Y. Bidirectional urban flows in rapidly urbanizing metropolitan areas and their macro and micro impacts on urban growth: A case study of the Yangtze River middle reaches megalopolis, China. *Land Use Policy* **2019**, *82*, 158–168. [CrossRef]
6. Dong, T.; Jiao, L.; Xu, G.; Yang, L.; Liu, J. Towards sustainability? Analyzing changing urban form patterns in the United States, Europe, and China. *Sci. Total Environ.* **2019**, *671*, 632–643. [CrossRef]
7. Hailemariam, S.N.; Soromessa, T.; Teketay, D. Land Use and Land Cover Change in the Bale Mountain Eco-Region of Ethiopia during 1985 to 2015. *Land* **2016**, *5*, 41. [CrossRef]
8. Liu, X.; Hu, G.; Chen, Y.; Li, X.; Xu, X.; Li, S.; Pei, F.; Wang, S. High-resolution multi-temporal mapping of global urban land using Landsat images based on the Google Earth Engine Platform. *Remote Sens. Environ.* **2018**, *209*, 227–239. [CrossRef]

9. Xiao, J.; Shen, Y.; Ge, J.; Tateishi, R.; Tang, C.; Liang, Y.; Huang, Z. Evaluating urban expansion and land use change in Shijiazhuang, China, by using GIS and remote sensing. *Landsc. Urban Plan.* **2006**, *75*, 69–80. [CrossRef]

10. Xu, X.; Min, X. Quantifying spatiotemporal patterns of urban expansion in China using remote sensing data. *Cities* **2013**, *35*, 104–113. [CrossRef]

11. Zhao, S.; Zhou, D.; Zhu, C.; Sun, Y.; Wu, W.; Liu, S. Spatial and temporal dimensions of urban expansion in China. *Environ. Sci. Technol.* **2015**, *49*, 9600–9609. [CrossRef] [PubMed]

12. Jiao, L.; Xu, G.; Xiao, F.; Liu, Y.; Zhang, B. Analyzing the Impacts of Urban Expansion on Green Fragmentation Using Constraint Gradient Analysis. *Prof. Geogr.* **2017**, *69*, 553–566. [CrossRef]

13. Schneider, A.; Woodcock, C.E. Compact, Dispersed, Fragmented, Extensive? A Comparison of Urban Growth in Twenty-five Global Cities using Remotely Sensed Data, Pattern Metrics and Census Information. *Urban Stud.* **2008**, *45*, 659–692. [CrossRef]

14. Herold, M.; Goldstein, N.C.; Clarke, K.C. The spatiotemporal form of urban growth: Measurement, analysis and modeling. *Remote Sens. Environ.* **2003**, *86*, 286–302. [CrossRef]

15. Ramachandra, T.V.; Aithal, B.H.; Sanna, D.D. Insights to urban dynamics through landscape spatial pattern analysis. *Int. J. Appl. Earth Obs. Geoinf.* **2012**, *18*, 329–343.

16. He, Q.; Song, Y.; Liu, Y.; Yin, C. Diffusion or coalescence? Urban growth pattern and change in 363 Chinese cities from 1995 to 2015. *Sustain. Cities Soc.* **2017**, *35*, 729–739. [CrossRef]

17. Dietzel, C.; Oguz, H.; Hemphill, J.J.; Clarke, K.C.; Gazulis, N. Diffusion and coalescence of the Houston Metropolitan Area: Evidence supporting a new urban theory. *Environ. Plan. B Plan. Des.* **2005**, *32*, 231–246. [CrossRef]

18. Herold, M.; Hemphill, J.; Dietzel, C.; Clarke, K. Remote sensing derived mapping to support urban growth theory. In Proceedings of the 3rd International Symposium Remote Sensing and Data Fusion over Urban Areas (URBAN 2005) and 5th International Symposium Remote Sensing of Urban Areas (URS 2005), Tempe, AZ, USA, 1 March 2005.

19. Jiao, L. Urban land density function: A new method to characterize urban expansion. *Landsc. Urban Plan.* **2015**, *139*, 26–39. [CrossRef]

20. Jiao, L.; Dong, T.; Xu, G.; Zhou, Z.; Liu, J.; Liu, Y. Geographic micro-process model: Understanding global urban expansion from a process-oriented view. *Comput. Environ. Urban Systems* **2021**, *87*, 101603. [CrossRef]

21. Liu, X.; Li, X.; Chen, Y.; Tan, Z.; Li, S.; Ai, B. A new landscape index for quantifying urban expansion using multi-temporal remotely sensed data. *Landsc. Ecol.* **2010**, *25*, 671–682. [CrossRef]

22. Liu, J.; Jiao, L.; Zhang, B.; Xu, G.; Yang, L.; Dong, T.; Xu, Z.; Zhong, J.; Zhou, Z. New indices to capture the evolution characteristics of urban expansion structure and form. *Ecol. Indic.* **2021**, *122*, 107302. [CrossRef]

23. Jiao, L.; Liu, J.; Xu, G.; Dong, T.; Gu, Y.; Zhang, B.; Liu, Y.; Liu, X. Proximity Expansion Index: An improved approach to characterize evolution process of urban expansion. *Comput. Environ. Urban Syst.* **2018**, *70*, 102–112. [CrossRef]

24. Jiao, L.; Mao, L.; Liu, Y. Multi-order landscape expansion index: Characterizing urban expansion dynamics. *Landsc. Urban Plan.* **2015**, *137*, 30–39. [CrossRef]

25. Brueckner, J.K. Urban sprawl: Diagnosis and remedies. *Int. Reg. Sci. Rev.* **2000**, *23*, 160–171. [CrossRef]

26. Bagan, H.; Yamagata, Y. Landsat analysis of urban growth: How Tokyo became the world's largest megacity during the last 40 years. *Remote Sens. Environ.* **2012**, *127*, 210–222. [CrossRef]

27. Seto, K.C.; Fragkias, M. Quantifying spatiotemporal patterns of urban land-use change in four cities of China with time series landscape metrics. *Landsc. Ecol.* **2005**, *20*, 871–888. [CrossRef]

28. Guérois, M.; Pumain, D. Built-Up Encroachment and the Urban Field: A Comparison of Forty European Cities. *Environ. Plan. A Econ. Space* **2008**, *40*, 2186–2203. [CrossRef]

29. Bagan, H.; Yamagata, Y. Land-cover change analysis in 50 global cities by using a combination of Landsat data and analysis of grid cells. *Environ. Res. Lett.* **2014**, *9*, 0640155. [CrossRef]

30. Maimaitijiang, M.; Ghulam, A.; Sandoval, J.O.; Maimaitiyiming, M. Drivers of land cover and land use changes in St. Louis metropolitan area over the past 40 years characterized by remote sensing and census population data. *Int. J. Appl. Earth Obs. Geoinf.* **2015**, *35*, 161–174. [CrossRef]

31. Hou, H.; Estoque, R.C.; Murayama, Y. Spatiotemporal analysis of urban growth in three African capital cities: A grid-cell-based analysis using remote sensing data. *J. Afr. Earth Sci.* **2016**, *123*, 381–391. [CrossRef]

32. Colsaet, A.; Laurans, Y.; Levrel, H. What drives land take and urban land expansion? A systematic review. *Land Use Policy* **2018**, *79*, 339–349. [CrossRef]

33. Angel, S.; Parent, J.; Civco, D.L.; Blei, A.M. *The Persistent Decline in Urban Densities: Global and Historical Evidence of Sprawl*; Working Paper; Lincoln Institute of Land Policy: Cambridge, MA, USA, 2010.

34. Seto, K.C.; Fragkias, M.; Guneralp, B.; Reilly, M.K. A meta-analysis of global urban land expansion. *PLoS ONE* **2011**, *6*, e23777. [CrossRef]

35. Haase, D.; Kabisch, N.; Haase, A. Endless urban growth? On the mismatch of population, household and urban land area growth and its effects on the urban debate. *PLoS ONE* **2013**, *8*, e66531. [CrossRef]

36. Kasanko, M.; Barredo, J.I.; Lavalle, C.; McCormick, N.; Demicheli, L.; Sagris, V.; Brezger, A. Are European cities becoming dispersed? A comparative analysis of 15 European urban areas. *Landsc. Urban Plan.* **2006**, *77*, 111–130. [CrossRef]

37. Güneralp, B.; Reba, M.; Hales, B.U.; Wentz, E.A.; Seto, K.C. Trends in urban land expansion, density, and land transitions from 1970 to 2010: A global synthesis. *Environ. Res. Lett.* **2020**, *15*, 044015. [CrossRef]

38. Xu, G.; Jiao, L.; Yuan, M.; Dong, T.; Zhang, B.; Du, C. How does urban population density decline over time? An exponential model for Chinese cities with international comparisons. *Landsc. Urban Plan.* **2019**, *183*, 59–67. [CrossRef]

39. Xu, G.; Zhou, Z.; Jiao, L.; Zhao, R. Compact Urban Form and Expansion Pattern Slow Down the Decline in Urban Densities: A Global Perspective. *Land Use Policy* **2020**, *94*, 104563. [CrossRef]

40. Song, X.; Feng, Q.; Xia, F.; Li, X.; Scheffran, J. Impacts of changing urban land-use structure on sustainable city growth in China: A population-density dynamics perspective. *Habitat Int.* **2021**, *107*, 102296. [CrossRef]

41. Inostroza, L.; Baur, R.; Csaplovics, E. Urban sprawl and fragmentation in Latin America: A dynamic quantification and characterization of spatial patterns. *J. Environ. Manag.* **2013**, *115*, 87–97. [CrossRef] [PubMed]

42. WorldBank. Data bank. Available online: https://data.worldbank.org/indicator (accessed on 16 March 2021).

43. Duque, J.C.; Lozano-Gracia, N.; Patino, J.E.; Restrepo, P.; Velasquez, W.A. Spatio-Temporal Dynamics of Urban Growth in Latin American Cities: An Analysis Using Nighttime Lights Imagery. *Landsc. Urban Plan.* **2019**, *191*. [CrossRef]

44. Xu, G.; Dong, T.; Cobbinah, P.B.; Jiao, L.; Sumari, N.S.; Chai, B.; Liu, Y. Urban expansion and form changes across African cities with a global outlook: Spatiotemporal analysis of urban land densities. *J. Clean. Prod.* **2019**, *224*, 802–810. [CrossRef]

45. Xu, G.; Jiao, L.; Liu, J.; Shi, Z.; Zeng, C.; Liu, Y. Understanding urban expansion combining macro patterns and micro dynamics in three Southeast Asian megacities. *Sci. Total Environ.* **2019**, *660*, 375–383. [CrossRef]

46. Angel, S.; Blei, A.M.; Parent, J.; Lamson-Hall, P.; Sanchez, N.G. *Atlas of Urban. Expansion—2016 Edition*; Areas and Densities; New York University: New York, NY, USA; UN-Habitat: Nairobi, Kenya; Lincoln Institute of Land Policy: Cambridge, MA, USA, 2016; Volume 1.

47. Atlas of Urban Expansion. Available online: http://www.atlasofurbanexpansion.org/ (accessed on 16 March 2021).

48. Hurd, J. *Atlas of Global Expansion 2015 Edition Cities Classification Procedures Manual*; University of Connecticut: Storrs, CT, USA, 2015.

49. Angel, S.; Lamson-Hall, P.; Blei, A.; Shingade, S.; Kumar, S. Densify and Expand: A Global Analysis of Recent Urban Growth. *Sustainability* **2021**, *13*, 3835. [CrossRef]

50. Angel, S.; Arango Franco, S.; Liu, Y.; Blei, A.M. The shape compactness of urban footprints. *Prog. Plan.* **2020**, *139*, 100429. [CrossRef]

51. Lemoine-Rodríguez, R.; Inostroza, L.; Zepp, H. The global homogenization of urban form. An assessment of 194 cities across time. *Landsc. Urban Plan.* **2020**, *204*, 103949. [CrossRef]

52. Landscan. Available online: https://landscan.ornl.gov/ (accessed on 16 March 2021).

53. Bright, E.A.; Coleman, P.R. *LandScan 2000*; Oak Ridge National Laboratory: Oak Ridge, TN, USA, 2001.

54. Bright, E.A.; Rose, A.N.; Urban, M.L. *LandScan 2014*; Oak Ridge National Laboratory: Oak Ridge, TN, USA, 2015.

55. Xiao, R.; Tian, Y.; Xu, G. Spatial gradient of urban green field influenced by soil sealing. *Sci.Total Environ.* **2020**, *735*, 139490. [CrossRef] [PubMed]

56. Bonafoni, S.; Keeratikasikorn, C. Land surface temperature and urban density: Multiyear modeling and relationship analysis using MODIS and Landsat data. *Remote Sens.* **2018**, *10*, 1471. [CrossRef]

57. Sumari, N.S.; Cobbinah, P.B.; Ujoh, F.; Xu, G. On the absurdity of rapid urbanization: Spatio-temporal analysis of land-use changes in Morogoro, Tanzania. *Cities* **2020**, *107*, 102876. [CrossRef]

58. Sumari, N.S.; Xu, G.; Ujoh, F.; Korah, P.I.; Ebohon, O.J.; Lyimo, N.N. A Geospatial Approach to Sustainable Urban Planning: Lessons for Morogoro Municipal Council, Tanzania. *Sustainability* **2019**, *11*, 6508. [CrossRef]

59. Govind, N.R.; Ramesh, H. Exploring the relationship between LST and land cover of Bengaluru by concentric ring approach. *Environ. Monit. Assess.* **2020**, *192*, 1–25. [CrossRef] [PubMed]

60. Keeratikasikorn, C. A comparative study on four major cities in Northeastern Thailand using urban land density function. *Geo-Spat. Inf. Sci.* **2018**, *21*, 93–101. [CrossRef]

61. Li, X.; Zhang, L.; Liang, C. A GIS-based buffer gradient analysis on spatiotemporal dynamics of urban expansion in Shanghai and its major satellite cities. *Procedia Environ. Sci.* **2010**, *2*, 1139–1156. [CrossRef]

62. Mieszkowski, P.; Mills, E.S. The causes of metropolitan suburbanization. *J. Econ. Perspect.* **1993**, *7*, 135–147. [CrossRef]

63. Chai, B.; Seto, K.C. Conceptualizing and characterizing micro-urbanization: A new perspective applied to Africa. *Landsc. Urban Plan.* **2019**, *190*, 103595. [CrossRef]

64. Clark, C. Urban population densities. *J. R. Stat. Soc.* **1951**, *114*, 490–496. [CrossRef]

65. Batty, M.; Sik Kim, K. Form follows function: Reformulating urban population density functions. *Urban Stud.* **1992**, *29*, 1043–1069. [CrossRef]

66. Zielinski, K. Experimental analysis of eleven models of urban population density. *Environ. Plan. A* **1979**, *11*, 629–641. [CrossRef] [PubMed]

67. Wang, W.; Jiao, L.; Dong, T.; Xu, Z.; Xu, G. Simulating urban dynamics by coupling top-down and bottom-up strategies. *Int. J. Geogr. Inf. Sci.* **2019**, *33*, 2259–2283. [CrossRef]

68. Zhang, B.; Xu, G.; Jiao, L.; Liu, J.; Dong, T.; Li, Z.; Liu, X.; Liu, Y. The scale effects of the spatial autocorrelation measurement: Aggregation level and spatial resolution. *Int. J. Geogr. Inf. Sci.* **2019**, *33*, 945–966. [CrossRef]

land

Article

Exploring the Patterns and Drivers of Urban Expansion in the Texas Triangle Megaregion

Jiani Guo and Ming Zhang *

School of Architecture, University of Texas at Austin, Austin, TX 78712, USA; jiani_guo@utexas.edu
* Correspondence: zhangm@austin.utexas.edu

Abstract: As the world becomes increasingly urbanized, it is vital for planners and policy-makers to understand the patterns of urban expansion and the underlying driving forces. This study examines the spatiotemporal patterns of urban expansion in the Texas Triangle megaregion and explores the drivers behind the expansion. The study used data from multiple sources, including land cover and imperviousness data from the National Land Cover Database (NLCD) 2001–2016, transportation data from the Texas Department of Transportation (TxDOT), and ancillary socio-demographic data from the U.S. Census Bureau. We conducted spatial cluster analysis and mixed-effect regression analysis. The results show that: (1) urban expansion in the Texas Triangle between 2001 and 2016 showed a decreasing trend, and 95% of the newly urbanized land was in metropolitan areas, especially at the periphery of the central cities; (2) urban expansion in non-metropolitan areas displayed a scattered pattern, comparing to the clustered form in metro areas; (3) the expansion process in the Texas Triangle exhibited a pattern of increased development compactness and intensity; and (4) population and economic growth played a definitive role in driving the urban expansion in the Texas Triangle while highway density also mattered. These results suggest a megaregion-wide emerging trend deviating from the sprawling development course known in Texas' urban growth history. The changing trend can be attributed to the pro-sustainability initiatives taken by several anchor cities and metropolitan planning agencies in the Texas Triangle.

Keywords: megaregion; urban expansion; spatiotemporal patterns; driving forces; the Texas Triangle

check for
updates

Citation: Guo, J.; Zhang, M. Exploring the Patterns and Drivers of Urban Expansion in the Texas Triangle Megaregion. *Land* **2021**, *10*, 1244. https://doi.org/10.3390/land10111244

Academic Editors: Iwona Cieślak, Andrzej Biłozor and Luca Salvati

Received: 5 October 2021
Accepted: 10 November 2021
Published: 13 November 2021

Publisher's Note: MDPI stays neutral with regard to jurisdictional claims in published maps and institutional affiliations.

1. Introduction

Urbanization has long interested academia, policy-makers, and international agencies. One important aspect of urbanization pertains to urban land expansion. Urban land expansion (or urban expansion in short) is a process of creating urban land for the needs of urban population and activities [1]. According to the U.S. Census Bureau, about 80% of Americans lived in urban areas in 2018. Urban expansion is often accompanied by many ecological and environmental challenges [2], for example, ecosystem damage, traffic pollutions [3], climate change [4], and resource depletion. These challenges also adversely affect people and environment in both urban and rural areas [2]. Furthermore, massive and aggressive urban expansion has resulted in worsening social issues such as inequality, urban and rural poverty, and housing unaffordability [5].

There are many different perspectives to understand urban expansion. The neoclassical perspective of urban expansion emphasizes the role of free market in deciding the land to be developed for urban functions. This perspective holds that land price, transportation cost, income, and population distribution are predominant driving forces of urban expansion [6]. Researchers in this domain have developed sophisticated statistical models to explain and quantify the extent to which these forces drive urban expansion. On the other hand, the institutional perspective pays close attentions to the importance of institutional factors such as land use control, capital investments, and organizational capacities in the urban expansion process [7].

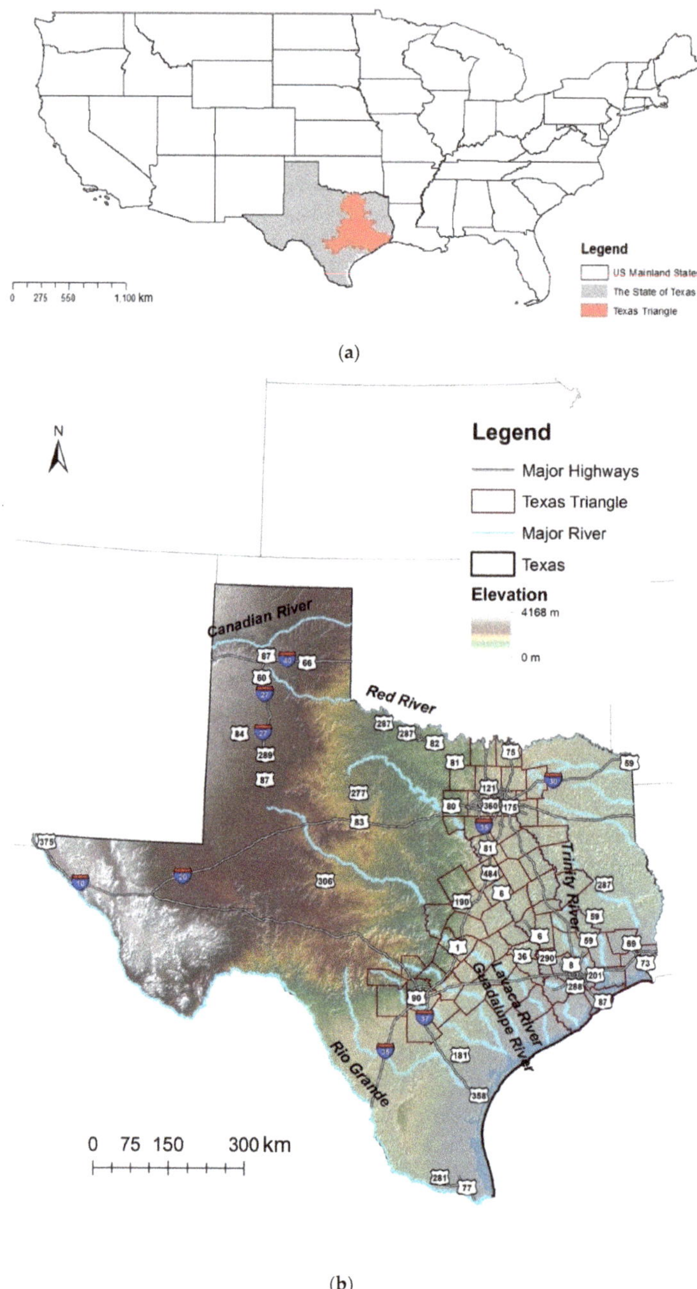

Figure 1. The Texas Triangle. (**a**): The location of the Texas Triangle in the U.S.; (**b**) the location of the Texas Triangle in the State of Texas. The elevation information is from the Texas Water Department Broad [26].

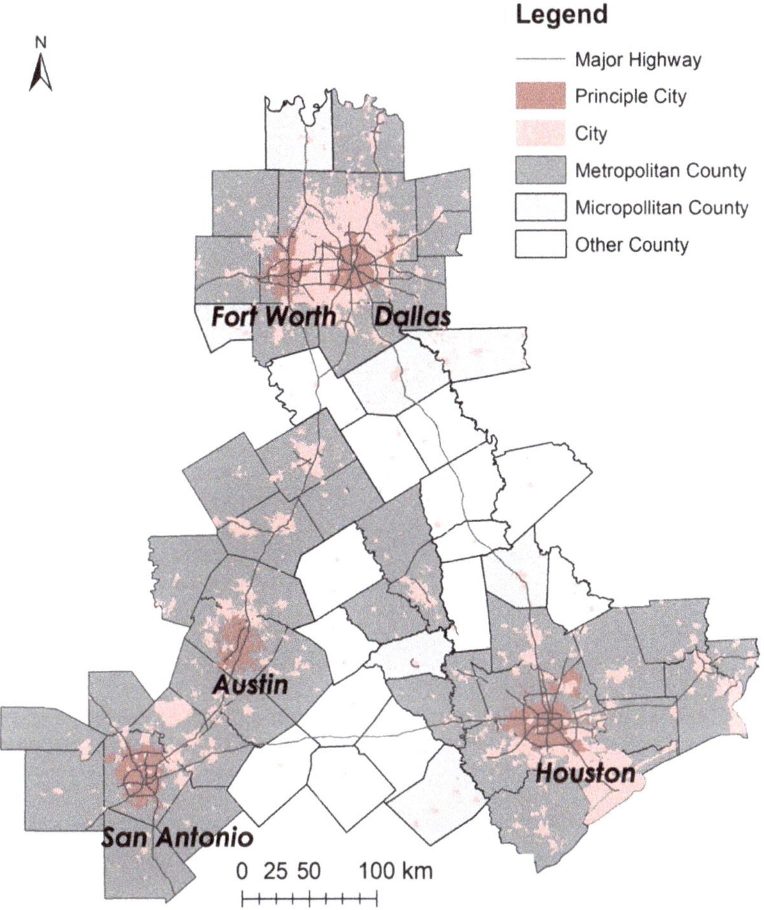

Figure 2. The Texas Triangle.

The Dallas-Fort Worth-Arlington MSA is the most populous metropolitan area in Texas and the fourth populous metropolitan area in the nation. It consists of 11 counties and a total area of 9286 square miles. This metropolis is home to 25 Fortune 500 companies, only behind New York City and Chicago. Houston-The Woodlands-Sugar Land is the second largest MSA in Texas and fifth most populous MSA in the U.S. This MSA includes nine counties (Harris, Fort Bend, Montgomery, Brazoria, Galveston, Liberty, Waller, Chambers, and Austin County) with a total area of over 10,000 square miles. Besides, this metropolitan area, is one of the fastest-growing MSA in the country. San Antonio-New Braunfels is an 8-county metropolitan area (Atascosa, Bandera, Bexar, Comal, Guadalupe, Kendall, Medina, and Wilson County), which covers a total area of 7387 square miles. As a famous historical city, the city of San Antonio is one of the top tourist cities in the U.S. Lastly, the Austin-Round Rock MSA includes six counties, Travis County, Bastrop County, Williamson County, Caldwell County, and Hays County. The Austin-Round Rock MSA is another rising metropolitan in which the population has increased from less than 300 thousand in 1970 to over 2 million in 2016 [29]. Austin was established in 1839 as the capital city of the Republic of Texas. The city is now a major education, technology, and economic center in the state, home to a flagship public university and world-renowned technology companies such as IBM (Endicott, NY, USA), Dell (Austin, TX, USA), and Apple (Los Altos, CA, USA).

2.2. Data

First, we retrieved the developed land from the National Land Cover Dataset (NLCD), U.S. Geographic Survey (USGS), to quantify the urbanized area. NLCD is a multi-year pre-prepared remote sensing data with a resolution of 30 m. The performance of the developed strategies and methods were tested in twenty World Reference System-2 path/row throughout the conterminous U.S. An overall agreement ranging from 71% to 97% between land cover classification and reference data was achieved for all tested areas and all years [30]. This remote sensing data has been used as a valuable data source in urban expansion research because of its broad and consistent area coverage and the virtue of being repeatedly updated regularly [31]. Rifat and Liu used NLCD and Coastal Change Analysis Program (C-CAP) datasets to study the urban expansion in the Miami Metropolitan area [32]. Terando et al. also used NLCD data to predict future urban sprawl in the Southern megapolis region [33]. In this research, the four categories of developed areas are treated as the urbanized area: developed open space, developed low density, developed medium density, and developed high density. The urbanized areas are calculated as the sum of the four types of developed land cover.

Moreover, we used the imperviousness data layer to retrieve the weighted urbanized area. Imperviousness data present impervious urban surfaces, representing the percentage of the developed surface. In the NLCD imperviousness dataset, each pixel is from 0% to 100%, where 80% to 100% pixels were classified as high intensity developed area. Several studies also used impervious information to measure the intensity of urban land [21,34]. We considered each pixel's imperviousness as an urbanized area's intensity weight. The weighted urbanized area is calculated as Equation (1):

$$\text{Weighted urbanized area} = \sum \text{pixel}_i * \text{impreviousness} \qquad (1)$$

Because of the data availability, we used the NLCD data layers for the years of 2001, 2006, 2011, and 2016. Other data in this research correspond to the four years.

Highway data were collected from the Texas Department of Transportation (TXDOT), Roadway Inventory 2019. This dataset provides all the roadways records in Texas up to 2019, including length, width, road type, start date, and traffic volume. We selected major highways (including interstate highway, state highway and U.S. highway) and calculated their density at the county level as the transportation indicator. The highway density was measured as the total length of the highway dividing the total area of each county.

To calculate the indicator for innovations and technological advances, we used the patent data which were collected from the U.S. Patent and Trademark Office (USPTO). USPTO posts the number of patents that were registered in the corresponding year and registration county. We retrieved county-level patent data in corresponding years as an indicator of technology level in those years. Finally, other ancillary social-demographic data were from the Bureau of Economic Analysis (BEA), including population, employment, and GDP. Table 1 presents the descriptive data of urban land change and the key drivers by major MSAs in the Texas Triangle from 2001 to 2016.

2.3. Methods

To answer the first research question, we conducted an Anselin Local Moran's I cluster and outlier analysis. For the second research question, we performed a mixed-effect regression analysis to determine the factors related to the Texas Triangle's urban expansion. Detailed descriptions of the methods are as follows.

Table 1. Descriptive data in the Texas Triangle, 2001–2016.

Metropolitan	Area (km²)			Population		Employment		GDP		Patent		Highway Length	
	Total	Urbanized Area in 2016	% Change Since 2001	2016 (Millions)	% Change Since 2001	2016 (Millions)	% Change Since 2001	2016 ($ Billion)	% Change Since 2001	2016	% Change Since 2001	2016 (km)	% Change Since 2001
Austin-Round Rock-Georgetown	11,085.37	1612.37	25.86	2.06	56.06	1.38	60.47	124.22	102.94	2701.00	55.86	1469.35	137.23
Beaumont-Port Arthur	6189.38	750.96	6.71	0.39	2.63	0.21	8.39	24.83	6.43	34.00	17.24	830.40	17.28
College Station-Bryan	5525.32	408.17	18.93	0.25	32.76	0.15	39.81	12.91	79.67	67.00	45.65	508.76	42.44
Dallas-Fort Worth-Arlington	23,328.57	5528.23	18.69	7.19	34.90	4.79	38.32	432.21	55.19	3028.00	42.09	6038.74	24.78
Houston-The Woodlands-Sugar Land	24,459.42	5531.00	21.98	6.81	41.29	4.04	40.40	446.78	50.18	3184.00	78.98	2959.86	56.31
Killeen-Temple	5554.46	498.58	17.00	4.16	30.82	0.22	26.61	15.81	47.07	21.00	5.00	452.14	67.33
San Antonio-New Braunfels	19,090.44	2162.71	17.66	2.42	38.44	1.41	41.86	108.63	62.50	413.00	73.53	3036.14	19.73
Sherman-Denison	2536.11	211.95	5.14	0.13	15.73	0.07	16.74	4.45	44.87	16.00	−33.33	485.67	10.31
Waco	4750.26	412.30	7.48	0.26	11.41	0.16	22.15	11.12	40.66	17.00	54.55	436.64	29.47
The Texas Triangle	**117,767.30**	**18,030.67**	**18.28**	**20.34**	**37.34**	**12.64**	**40.13**	**1195.98**	**56.13**	**9539.00**	**58.11**	**17,510.89**	**34.52**

2.3.1. Anselin Local Moran's I Cluster and Outlier Analysis

We identified hot-spot clusters and spatial outliners of urbanized land at the Census tract level through the Anselin Local Moran's I statistic [35]. This method is widely used in many fields, such as economics [36], demographics [37], and geography [38]. The Anselin local Moran's I cluster and outlier analysis (cluster analysis) is adopted because of its ability to capture the spatial patterns in not only their general but also abnormal trends. We used this method in our study to categorize four types of spatial clusters of urban expansion in the Texas Triangle. If the Local Moran's I test statistic turns out to be positive, this area belongs to a statistically significant cluster of either a high value (a high-high cluster) or a low value (a low-low cluster). On the contrary, if the test statistic is negative, this area is an outlier of either high value surrounded by low-value areas (a high-low outlier) or otherwise (a low-high outlier). We tested the absolute increase area of urbanized land at the census tract level in 2001 to 2006, 2006 to 2011, and 2011 to 2016 and to see if a place is the hotspot of urban expansion, or if this place is the outlier with the abnormal increasing urbanized land while its surrounding areas are not.

2.3.2. Regression Analysis of the Driving Forces of Urban Expansion

We estimated mixed-effect regression models to measure the relationship between urban expansion and its potential driving forces. The mixed-effect regression model can cancel out the unobserved the error from different geographic entities and other potential error. The time variables were fixed to test its relatively growth in different periods.

The dependent variable urban expansion is measured by the percentage of urbanized area and the percentage of the weighted urbanized area in the county. Besides selecting the urbanized area, we added the weighted urbanized area to model the intensity growth in this megaregion. The comparison of absolute urbanized area and weighted urbanized area can depict a more comprehensive urban expansion process beyond the horizontal land cover changes.

To model the urban expansion, we selected six widely discussed variables in the literature that can capture the most context at a higher level, such as the megaregion level used in this study. The details and rationales are elaborated and explained in the following paragraphs. Independent variables were categorized into five types, as is shown in Table 2.

Table 2. Selected major drivers of urban expansion.

Variable Category	Variable	Description	Sources
Social demographic factors	Population	The total population in the county	Bureau of Economic Analysis
Economic factors	Jobs	Employment in the county	Bureau of Economic Analysis
	GDP	GDP (millions) in the county	Bureau of Economic Analysis
Intellectual and technology innovation	Patents	Number of patents in the county	U.S. Patent and Trademark Office
Transportation infrastructure	Highway Density (kilometer per square kilometers)	The ratio of the length of highway to the total area in the county	TxDOT Roadway Inventory 2019
Institutional factor	1 if the county is part of a Metropolitan area; 0 otherwise	If this county is within/out of a Metropolitan area	The U.S. Census Bureau

We selected the population as one of the major predictors. It is indubitable that population growth and land expansion are two inseparable aspects of the urbanization process [39]. Research has shown that population migration from rural to urban areas is a major driving force of urban expansion [40]. Therefore, it was our expectation that the most important driving force of urban expansion was population growth, measured as the total population in the county in the corresponding years.

Second, we included the economic indicators in the model because urban economists think urban expansion results from the market and economy agglomeration and expansion [6]. Economic development and increasing economic activities have accelerated the urban expansion process in recent decades. On the other hand, urbanization also may, in return, promote economic development. The economic advantages in the urbanized area further attract more population and migration from rural to urban areas. Hence GDP and the number of employments widely serve as two indicators to measure the economic development in different counties.

The number of patents in the county was selected as an indicator of technology innovation. Technology development is wildly considered a significant factor of the prosperity of a region. The first technological revolution in the later 18th century in the United Kingdom is also the time the urbanization began. The second and the third technological revolution in the U.S. accelerated the urbanization process and urban expansion. From industrialization and informatization, technology innovation is always one of the central forces pushing the urban expansion process [41]. Friedman once argued that the technology is one of the reasons that the geographical location is less important nowadays [42]. In this research, the number of patents in the county is used as a proxy of technology innovation, as was used in Florida's research [41].

Highway density was chosen to represent the capital investments in transportation infrastructure. Early from the bid-rent theory, the distance to major transportation facilities is essential to location choice [43]. Later on, Dr. Adam's four stages model further emphasized how transportation infrastructure can shape the urban form and lead to urban expansion in different phases [44]. Moreover, transportation density is also an important indicator of built environments, influencing the urban expansion process. Therefore, the highway density is calculated to measure the supply of transportation infrastructure in each county.

The institutional perspective focuses on institutional or municipalities' role in the urban expansion [7]. The governmental policy is another factor for urban expansion. Due to the intricate and fragmented municipalities and governmental systems at the megaregion level, we considered being in a metropolitan area an institutional factor to investigate whether a county belongs to a larger administrative unit will make a difference in their urban land expansion. A metropolitan statistical area, defined by the U.S. Office of Management and Budget, is region with a principal city and its periphery containing more than 50,000 population. A micropolitan statistical area similarly is a place with population between 10,000 and 50,000.

The initial status (in the year 2001) of the urbanized area in each county varies; however, the growth rates of urbanized areas are relatively similar among different counties. Therefore, we did not use a growth model because of the low variation in slopes. In this case, we fixed the time effects, setting the initial year, 2001 as the baseline, while setting the geographic entities, that is each county, as random effects. The choice and form of variables are presented in Table 3.

The form of the equation is shown in Equations (2) and (3):

$$\begin{aligned}
\text{lglandpct}_i = \beta_{0i} \ &+ \beta_{1i}*\text{metro} + \beta_{2i}*\text{lgpop}_i + \beta_{3i}*\text{lgjob}_i + \beta_{4i}*\text{cpatent}_i \\
&+ \beta_{5i}*\text{lggdp}_i + \beta_{6i}*\text{lghwden}_i + \beta_{7i}*\text{year}_{2006i} + \beta_{8i} \\
&*\text{year}_{2011_i} + \beta_{9i}*\text{year}_{i_{2016}} + U_i + \varepsilon_i
\end{aligned} \tag{2}$$

$$\begin{aligned}
\text{lgimppct}_i = \beta_{0i} &+ \beta_{1i}*\text{metro} + \beta_{2i}*\text{lgpop}_i + \beta_{3i}*\text{lgjob}_i + \beta_{4i}*\text{cpatent}_i + \beta_{5i} \\
&*\text{lggdp}_i + \beta_{6i}*\text{lghwden}_i + \beta_{7i}*\text{year}_{2006i} + \beta_{8i} \\
&*\text{year}_{2011_i} + \beta_{9i}*\text{year}_{2016_i} + U_i + \varepsilon_i
\end{aligned} \tag{3}$$

Table 3. Description of the variables in the mixed-effect regression.

Dependent Variable	
lglandpct	Logarithm form of percentage urbanized land (%) in the county
lgimppct	Logarithm form of percentage weighted urbanized land (%) in the county
Independent Variable	
metro	=1 if the county is in a metropolitan area
lgpop	Logarithm form of the population in the county
lgjob	Logarithm form of jobs in the county
cpatent	=0, if the number of patents is 0 in the county; =1, if the number of patents is between 1 and 5 in the county; =2, if the number of patents is between 6 and 100 in the county; =3, if the number of patents is between 101 and 1000 in the county; =4, if the number of patents is above 1001 in the county
lggdp	Logarithm form GDP (in millions of dollars) in the county
year	2001, 2006, 2011, 2016
lnhwden	Logarithm of the length of the highway in the county (km)/Area of the county (km^2)

3. Results

3.1. Spatiotemporal Patterns of Urban Expansion in the Texas Triangle

Figure 3 shows the change in urbanized area from 2001 to 2016 in the Texas Triangle. The newly developed urban land is mainly concentrated in the periphery of the major metropolitan area evident in the figure. The newly developed urban area in other counties presents scattered patterns. Additional cluster analysis further confirms that there are no cluster effects in those counties.

Figure 3. Urbanized Area in the Texas Triangle.

Figures 4 and 5 show the spatial and temporal patterns of urban expansion in the Texas Triangle during the three time periods. The results, first, illustrate higher growth rates in metropolitan counties than other counties. From 2001 to 2016, the urbanized area has increased by 2887 km^2, while 95% of those expansions occurred in metropolitan areas. Moreover, Figure 5 shows the decreasing growth rate over time in the Texas Triangle.

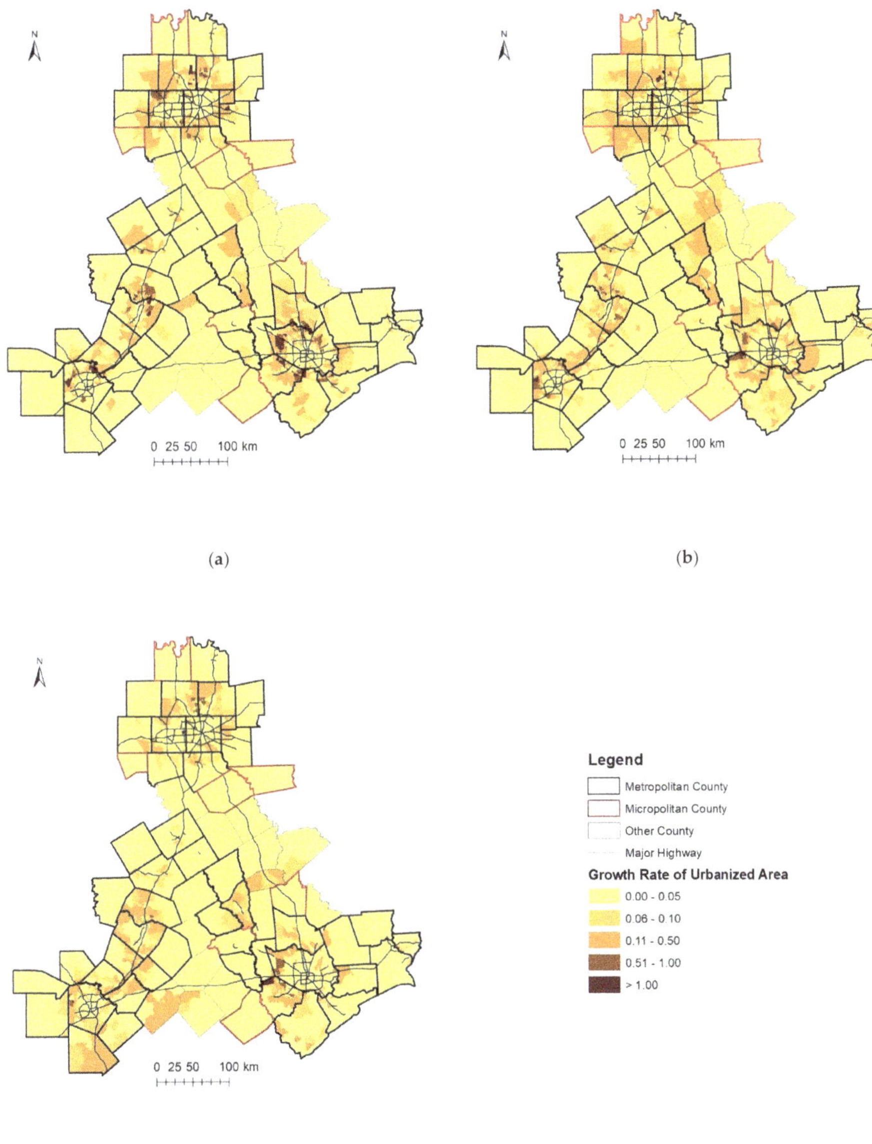

Figure 4. Maps of Urbanized Land Growth Rate in the Texas Triangle from 2001 to 2016: (**a**) 2001–2006; (**b**) 2006–2011; (**c**) 2011–2016.

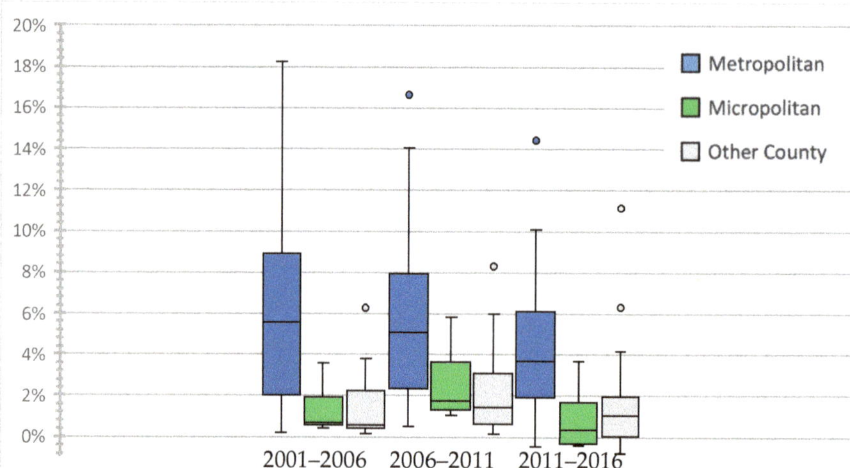

Figure 5. Urbanized land growth rate in different geographic area in the Texas Triangle.

Maps of growth spatial clusters and outliers in the megaregion (Figure 6) illustrate the spatial characteristics of urban expansion from 2001 to 2016 in the Texas Triangle. To start with, counties in four major metropolitan areas presented more high-high clusters of urban growth than micropolitan counties and other counties. The four major MSAs commonly exhibited patterns from principal cities outward: low-low cluster, low-high outlier, high-high cluster, and high-low outliers. This pattern was location-irrelevant in all periods. Specifically, central counties in major metropolitan areas, i.e., Travis county in Austin MSA, Dallas and Tarrant County in Dallas MSA, Bexar County in San Antonio MSA, and Harris County in Houston MSA, had a relatively low increase rate. Principal cities, Austin, San Antonio, Houston, Dallas, and Fort Worth, all presented significant low-low growth clusters, while the high-high cluster aggregated in slightly different localities at different time intervals. However, the urban expansion has an extraordinary intensity in the periphery area around the central county of metropolitan areas like Fort Bend County in Houston metropolitan, Rockwall County in Dallas metropolitan, and Williamson county in Austin metropolitan. Moreover, the urbanized land growth rates decreased over time, and the growth moves farther away from the principal cities. Interestingly, many census tracts with comparatively high growth rates are on the north side of metropolitan areas.

Besides the commonalities, four major metropolitan areas exhibited different urban expansion patterns. The Houston MSA shows relatively consistent and intensive expansion patterns in the outer ring on the north and east sides. This pattern might relate to its adjacency to the ocean to the south. Whereas the San Antonio MSA presents a relatively low expansion pace overall, and those expansions are concentrated on the north side. The Austin MSA has the fastest growth rate in the Texas Triangle. While development on the north Austin MSA dated back from 2001, the south side started to consume significantly more land as urbanized land from 2006. Lastly, the Dallas MSA has a high growth rate in the periphery places around Dallas and Fort Worth from 2001 to 2006. Then, from 2006 to 2011, the urbanized land grew primarily on the north and southwest in the metropolitan region. In the last time period, the urban land mainly expanded only on the north side.

In other smaller metropolitan areas, urban expansion patterns are slightly different from the major ones. Growth rates in those areas are generally slower, except for the College Station metropolitan. Besides, there are no prominent spatial clusters of high or low values in urbanized land growth. It is worth noting that, in the connecting MSA counties between four major MSAs, the urban expansion patterns are different over time. Specifically, counties between Dallas and Houston experienced a sizeable urban expansion in the first two periods, whereas more expansion was found between Austin and San Antonio in the third.

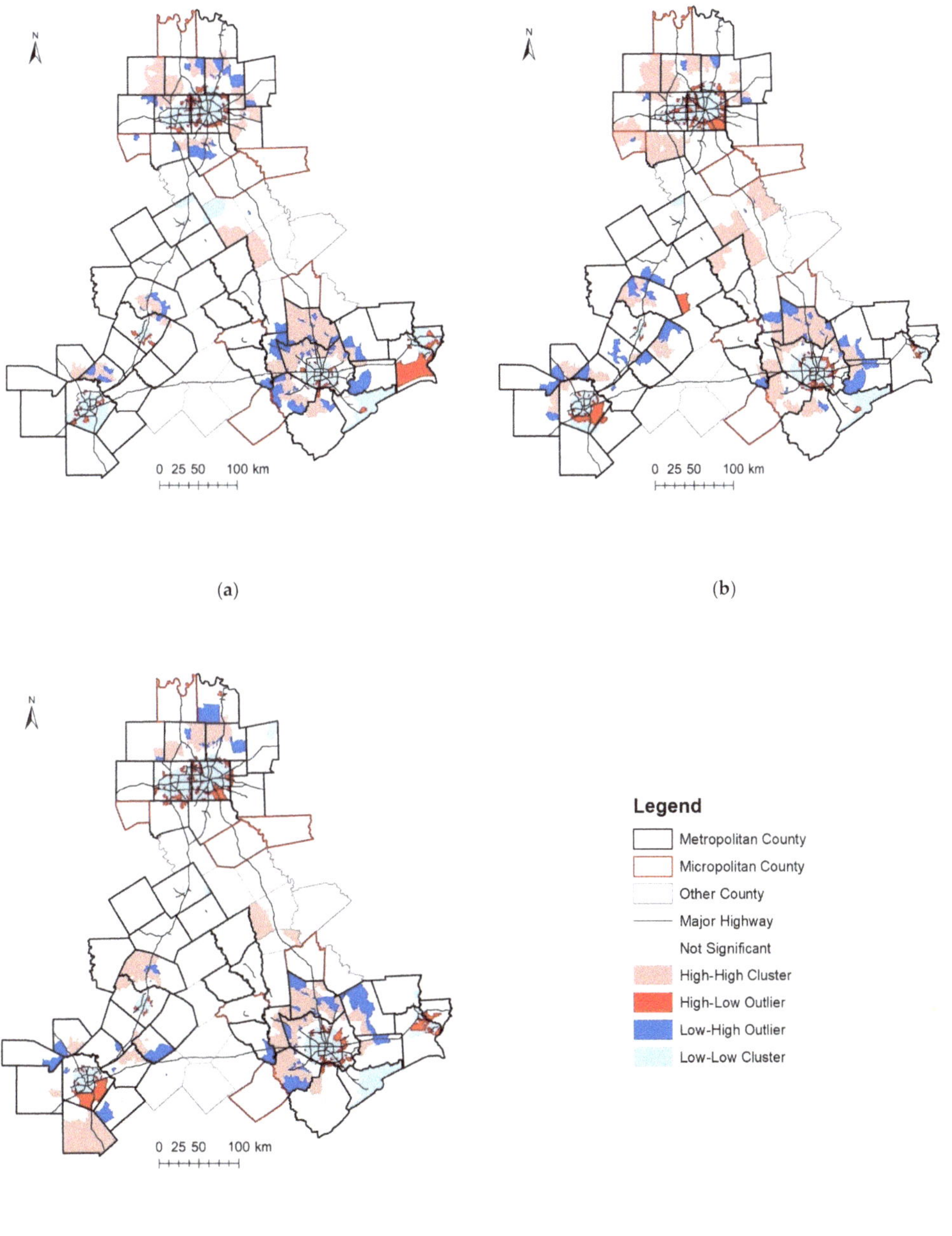

Figure 6. Spatial cluster and outliners of newly developed urbanized land (**a**) 2001–2006; (**b**) 2006–2011; (**c**) 2011–2016.

However, in non-metropolitan counties, the scale of urban expansion is relatively small. Moreover, from the cluster analysis, there are no significant urban expansion clusters in all periods, indicating a scattered expansion pattern in those counties.

To sum up, the newly developed urbanized land concentrates mainly on the periphery of core cities of major metropolitan areas. Overall, the urbanized land in the metropolitan area has expanded to suburbs, but the growth rate has declined over time. Even though more undeveloped land has changed to urbanized land during the whole period, the results above show that the expansion happened in a more aggregated manner instead of randomly sprawling in the metropolitan counties. Those patterns accord with the urban expansion patterns in major urban areas in the US Great Plains from 2000 to 2009 [45] with showing a compact development trend. Comparably, the growth in non-metropolitan counties is slower and more scattered.

3.2. Regression Results

Table 4 presents the result of the regression analysis. There are in total 262 observations, representing 66 counties in 4 years (the Milam County had no highway in 2001 and 2006 and therefore omitted in the regression models). The overall r^2 in the two models are 0.86 and 0.66, which explains most variations by the models.

Table 4. Results of the mixed-effect regression.

Dependent Variable	Lglandpct	Lgimppct
metro	0.0292	0.0325
	(0.39)	(0.49)
lgpop	0.3794 ***	0.143 **
	(8.25)	(2.38)
lgjob	0.0150	−0.0375
	(0.35)	(−0.66)
lggdp	0.0342 ***	0.0722 ***
	(3.43)	(5.16)
cpatent	0.0065	0.0117 *
	(1.54)	(1.95)
lghwden	0.0105 **	0.0215
	(2.68)	(1.51)
Year		
2001	0	0
	(.)	(.)
2006	−0.00047	0.0201 ***
	(0.2)	(3.01)
2011	0.0074	0.0442 ***
	(1.61)	(3.01)
2016	0.0041	0.0556 ***
	(0.72)	(6.36)
_cons	−2.724 ***	1.142 ***
	(−14.13)	(5.33)
σ_u	0.26542	0.20912
σ_e	0.02482	0.03268
ρ	0.99170	0.97614
overall r^2	0.8644	0.6553
N	262	262

Note: significance level: ***: $p < 0.001$; **: $p < 0.01$; *: $p < 0.05$.

First of all, the models coincide with previous literature and show that population growth statistically significantly leads to greater expansion. Surprisingly, albeit 95% of the expansion in the Texas Triangle happened in the metropolitan areas during the entire period, being a metropolitan county shows no statistically significant advantages than other counties in both models. The result implies that metropolitan setting does not explain the trend of urban expansion in the Texas Triangle megaregion.

In terms of the time variable, compared to 2001, while controlling for other factors, there is no significant growth of the urbanized area from 2001 to 2016. However, the weighted urbanized area shows totally different results, and all-time variables are highly positively correlated to weighted urban areas. Unlike economic development or GDP, where we assume there might be natural growth because of productivity or efficiency improvement, the total urbanized area shows no such natural growth in the Texas Triangle from 2001 to 2016 while controlling population, economic development, and other factors. Nevertheless, it is worth noting that the intensity of urban areas has such growth from 2001 to 2016 when controlling other variables. That is to say, from 2001 to 2016, the urban expansion process in the Texas Triangle is more compact rather than low-density development. From this aspect, it is possible to control the urban growth if policies are controlling for population and transportation infrastructure. Moreover, it shows the results of promoting compact development.

As for economic factors, the models reveal a complicated relationship to urban expansion. On the one hand, economic development requires land investment as space and capital. On the other hand, economic activity agglomeration is an important driving force to a greater urban expansion. The model results show that, while employment has no significant statistical relationship to urban expansion, GDP positively influences the urban expansion process in the Texas Triangle. In contrast, the patent variable as a measurement of technology innovation shows no significant relationship to urban expansion in the Texas Triangle.

As much concern to urban transportation planners, the result shows that the highway density is highly positively related to an urbanized area. However, interestingly, the highway density presents no relationship to a weighted urbanized area. That means the highway density might influence the urban in changing non-urban land to urban land but has little relationship to its intensity. This result provides information for planners to rethinking the use the transportation infrastructure to guide future urban growth. Planners should also consider their role in compact urban development.

4. Discussion and Conclusions

This study examines the urban expansion pattern influenced by six major driving factors in the Texas Triangle Megaregion from 2001 to 2016. We first conducted a spatial cluster analysis to explore where the expansion occurred and its magnitude during the period. We then employed a mixed-effect model at the county level to explain the relationships between urban expansion and the focal driving forces, including affiliation to a metropolitan area, population, employment, GDP, technology innovation, and transportation infrastructure. The cluster analysis results show that the urban expansion rate displayed a decreasing trend in the Texas Triangle between 2001 and 2016; 95 percent of new development occurred within the Triangle's metropolitan areas. While clustering patterns varied between the metropolitan areas, the expansion occurred largely in the periphery of central cities. Contrastingly, the urban expansion in non-metropolitan counties was rather scattered. The mixed-effect modeling shows that population, GDP, and highway density were significant predictors of urban expansion.

Between 2001 and 2016, metropolitan areas in the Texas Triangle displayed different patterns despite a shared experience of overall urban expansion. In particular, the Dallas-Houston corridor area showed clustered growth in the periods of 2001–2006 and 2006–2011. This clustered growth pattern, however, did not occur in 2011–2016. The San Antonio-Austin corridor presented clustered expansion throughout the study period of 2001–2016. The two metropolitan areas have now become contiguous, prompting Texas DOT to coordinate joint planning efforts by their respective metropolitan planning organizations [46]. The finding accords with previous studies by demonstrating the heterogeneous expansion patterns at the periphery of large cities [47]. Amid shifting expansion patterns in the areas between large cities and metro areas, state or joint state efforts are necessary to foster cooperation beyond the municipal or agency's jurisdictional boundaries.

Metropolitan planning agencies can play an important role in leveraging regional resource distribution to guide urban expansion [32,48], despite that municipalities make local land use decisions. Improved coordination and cooperation between local and regional entities could lead to desired development outcome. For example, transit-oriented development (TOD) has been widely considered as a tool to facilitate smart urban growth [49]. Since regional transit lines typically traverse multiple municipalities, coordination between MPOs, transit agencies, and local communities is essential to implement TOD strategy at the regional scale. A best-practice example from the Texas Triangle exists from the Dallas region where NCTCOG (North Central Texas Council of Governments), DART (Dallas Area Rapid Transit), City of Dallas as well as other communities along DART routes have coordinated joint efforts to practice TOD in the region [50]. City of Houston initiative Livable Places echoes H-GAC's (Houston-Galveston Area Council) Livable Centers program to promote walkable places and TOD [51,52].

The second regression model estimated in this study considered imperviousness or development intensity, that is, the model of weighted urban land expansion. Adding intensity information into the urban expansion modeling resulted in the loss of statistical significance for the variable highway density. However, the predictor number of patents which was statistically insignificant in the first model, turned to be significant ($p < 0.005$). The contrasting results between the two models suggest that capital investments in highways tend to drive urban expansion horizontally, whereas innovations and technological advances likely push urban expansion vertically towards increased land use efficiency. As concerns over climate change and sustainability grow, local and state governments need to rethink about the conventional strategy of investing in highways to accommodate population and economic growth.

This study has several limitations, suggesting directions for future research. First, this study utilized data on land cover from satellite images, which provide very limited information on land uses for various urban functions. In future research, detailed land use information may be incorporated to allow analyses on variations of urban expansion by different functional types of land uses. Second, this study used imperviousness information from NLCD as a proxy for development intensity; the information provides a rather coarse measure that cannot adequately capture the variation of vertical urban expansion across cities and regions. Lidar data could be used to enhance this study with detailed urban form and vertical development characteristics [53,54]. Lastly, considering the vast area of the Texas Triangle, this study selected county as the geographic unit of analysis. The study findings and discussions are thus limited to the county level. There exist significant within-county variations that this study did not capture. Hence, the study can be refined with use of finer-scale data, for instance, at the census track or block group level.

Despite these limitations, the study results suggest a megaregion-wide emerging trend deviating from the sprawling development course known in Texas' urban growth history. The changing trend can be attributed to the pro-sustainability initiatives taken by several anchor cities and metropolitan planning agencies in the Texas Triangle. Future planning and policy-making efforts should foster this trend toward a sustainable megaregion.

Author Contributions: Conceptualization, J.G. and M.Z.; data curation, J.G.; formal analysis, J.G.; funding acquisition, M.Z.; investigation, J.G.; methodology, J.G.; visualization, J.G.; writing—original draft, J.G.; writing—review and editing, M.Z. All authors have read and agreed to the published version of the manuscript.

Funding: This research was supported by USDOT UTC Cooperative Mobility for Competitive Megaregions (CM2) (Grant #: 69A3551747135) and the Snell Endowment Grant from the University of Texas at Austin Center for Sustainable Development (CSD).

Institutional Review Board Statement: Not applicable.

Informed Consent Statement: Not applicable.

Data Availability Statement: Data are retrieved from The U.S. Census Bureau (https://www.census.gov/data.html/ accessed on 9 November 2021), Texas Department of Transportation (https://www.txdot.gov/inside-txdot/division/transportation-planning/roadway-inventory.html/ accessed on 9 November 2021), National Landcover Database (https://www.mrlc.gov/data/nlcd-land-cover-conus-all-years/ accessed on 9 November 2021) and Bureau of Economic Analysis (https://www.bea.gov/data/gdp/ accessed on 9 November 2021), and U.S Patent and Trademark Office (https://developer.uspto.gov/data/ accessed on 9 November 2021).

Acknowledgments: Thanks to CM2 and CSD for providing funding support to this research. We also want to acknowledge all colleagues, especially Shunhua Bai, in the School of Architecture, University of Texas at Austin for their insightful advice.

Conflicts of Interest: The authors declare no conflict of interest. We certify that this submission is original work and is not under review at any other publication. The funders had no role in the design of the study; in the collection, analyses, or interpretation of data; in the writing of the manuscript, or in the decision to publish the results.

References

1. Bren d'Amour, C.; Reitsma, F.; Baiocchi, G.; Barthel, S.; Güneralp, B.; Erb, K.-H.; Haberl, H.; Creutzig, F.; Seto, K.C. Future Urban Land Expansion and Implications for Global Croplands. *Proc. Natl. Acad. Sci. USA* **2017**, *114*, 8939–8944. [CrossRef]
2. Georgescu, M. Challenges Associated with Adaptation to Future Urban Expansion. *J. Clim.* **2015**, *28*, 2544–2563. [CrossRef]
3. Hutyra, L.R.; Yoon, B.; Hepinstall-Cymerman, J.; Alberti, M. Carbon Consequences of Land Cover Change and Expansion of Urban Lands: A Case Study in the Seattle Metropolitan Region. *Landsc. Urban Plan.* **2011**, *103*, 83–93. [CrossRef]
4. Krayenhoff, E.S.; Moustaoui, M.; Broadbent, A.M.; Gupta, V.; Georgescu, M. Diurnal Interaction between Urban Expansion, Climate Change and Adaptation in US Cities. *Nat. Clim. Change* **2018**, *8*, 1097–1103. [CrossRef]
5. Dutt, A.K.; Noble, A.G.; Costa, F.J.; Thakur, R.R.; Thakur, S.K. (Eds.) *Spatial Diversity and Dynamics in Resources and Urban Development*; Springer: Dordrecht, The Netherlands, 2016. [CrossRef]
6. Brueckner, J.K.; Fansler, D.A. The Economics of Urban Sprawl: Theory and Evidence on the Spatial Sizes of Cities. *Rev. Econ. Stat.* **1983**, *65*, 479. [CrossRef]
7. Wei, Y.H.D. Restructuring for Growth in Urban China: Transitional Institutions, Urban Development, and Spatial Transformation. *Habitat Int.* **2012**, *36*, 396–405. [CrossRef]
8. McDonald, J.F. *Fundamentals of Urban Economics*; Prentice Hall: Upper Saddle River, NJ, USA, 1997.
9. Zeng, C.; Zhang, M.; Cui, J.; He, S. Monitoring and Modeling Urban Expansion—A Spatially Explicit and Multi-Scale Perspective. *Cities* **2015**, *43*, 92–103. [CrossRef]
10. Zambon, I.; Colantoni, A.; Salvati, L. Horizontal vs. Vertical Growth: Understanding Latent Patterns of Urban Expansion in Large Metropolitan Regions. *Sci. Total Environ.* **2019**, *654*, 778–785. [CrossRef] [PubMed]
11. Pham, H.M.; Yamaguchi, Y.; Bui, T.Q. A Case Study on the Relation between City Planning and Urban Growth Using Remote Sensing and Spatial Metrics. *Landsc. Urban Plan.* **2011**, *100*, 223–230. [CrossRef]
12. McHarg, I.L. *Design with Nature*, 25th ed.; John Wiley & Sons, Inc.: New York, NY, USA; Chichester, UK; Brisbane, Australia; Toronto, ON, Canada; Singapore, 1992.
13. Saiz, A. The Geographic Determinants of Housing Supply. *Q. J. Econ.* **2010**, *125*, 1253–1296. [CrossRef]
14. Jones, B.G.; Koné, S. An Exploration of Relationships between Urbanization and per Capita Income: United States and Countries of the World. *Pap. Reg. Sci.* **1996**, *75*, 135–153. [CrossRef]
15. Wang, K.; Zhou, W. Do Local Factors or Teleconnections Control Urbanization? The Shifting Balance in a Chinese Megaregion. *Landsc. Urban Plan.* **2018**, *180*, 179–186. [CrossRef]
16. del Mar López, T.; Aide, T.M.; Thomlinson, J.R. Urban Expansion and the Loss of Prime Agricultural Lands in Puerto Rico. *AMBIO J. Hum. Environ.* **2001**, *30*, 49–54. [CrossRef]
17. Martinuzzi, S.; Gould, W.A.; Ramos González, O.M. Land Development, Land Use, and Urban Sprawl in Puerto Rico Integrating Remote Sensing and Population Census Data. *Landsc. Urban Plan.* **2007**, *79*, 288–297. [CrossRef]
18. Song, Y.; Knaap, G.-J. Measuring Urban Form: Is Portland Winning the War on Sprawl? *J. Am. Plann. Assoc.* **2004**, *70*, 210–225. [CrossRef]
19. Haughton, G.; Counsell, D. Regions and Sustainable Development: Regional Planning Matters. *Geogr. J.* **2004**, *170*, 135–145. [CrossRef]
20. Camagni, R.; Gibelli, M.C.; Rigamonti, P. Urban Mobility and Urban Form: The Social and Environmental Costs of Different Patterns of Urban Expansion. *Ecol. Econ.* **2002**, *40*, 199–216. [CrossRef]
21. Kuang, W.; Chi, W.; Lu, D.; Dou, Y. A Comparative Analysis of Megacity Expansions in China and the U.S.: Patterns, Rates and Driving Forces. *Landsc. Urban Plan.* **2014**, *132*, 121–135. [CrossRef]
22. Campbell, S.D. The Planner's Triangle Revisited: Sustainability and the Evolution of a Planning Ideal That Can't Stand Still. *J. Am. Plann. Assoc.* **2016**, *82*, 388–397. [CrossRef]

23. Lea, B.D. Tax-Friendly Texas Poised for Flood of Corporate Relocations in 'Biggest Year Yet,' Expert Says. Available online: https://www.foxbusiness.com/economy/companies-relocate-tax-friendly-texas-2021 (accessed on 3 October 2021).

24. American Spatial Development and the New Megalopolis | Lincoln Institute of Land Policy. Available online: https://www.lincolninst.edu/publications/articles/american-spatial-development-new-megalopolis (accessed on 7 February 2021).

25. Bomar, G.W. *Texas Weather*; University of Texas Press: Austin, TX, USA, 1983.

26. Texas Water Development Board. GIS Data. Available online: https://www.twdb.texas.gov/mapping/gisdata.asp (accessed on 9 November 2021).

27. Butler, K.; Hammerschmidt, S.; Steiner, F.; Zhang, M. *Reinventing the Texas Triangle: Solutions for Growing Challenges*; Center for Sustainable Development, School of Architecture, The University of Texas at Austin: Austin, TX, USA, 2009; p. 31.

28. Zhang, M.; Steiner, F.; Butler, K. Connecting the Texas Triangle: Economic Integration and Transportation Coordination. The Healdsburg Research Seminar on Megaregions. 2007. Available online: https://www.researchgate.net/profile/Ming-Zhang-24/publication/228463644_Connecting_the_Texas_Triangle_Economic_Integration_and_Transportation_Coordination/links/56535f6408ae4988a7af89bc/Connecting-the-Texas-Triangle-Economic-Integration-and-Transportation-Coordination.pdf (accessed on 9 November 2021).

29. Bureau, U.C. Population. Available online: https://www.census.gov/topics/population.html (accessed on 3 October 2021).

30. Yang, L.; Jin, S.; Danielson, P.; Homer, C.; Gass, L.; Bender, S.M.; Case, A.; Costello, C.; Dewitz, J.; Fry, J.; et al. A New Generation of the United States National Land Cover Database: Requirements, Research Priorities, Design, and Implementation Strategies. *ISPRS J. Photogramm. Remote Sens.* **2018**, *146*, 108–123. [CrossRef]

31. Musakwa, W.; van Niekerk, A. Earth Observation for Sustainable Urban Planning in Developing Countries: Needs, Trends, and Future Directions. *J. Plan. Lit.* **2015**, *30*, 149–160. [CrossRef]

32. Rifat, S.A.A.; Liu, W. Quantifying Spatiotemporal Patterns and Major Explanatory Factors of Urban Expansion in Miami Metropolitan Area During 1992–2016. *Remote Sens.* **2019**, *11*, 2493. [CrossRef]

33. Terando, A.J.; Costanza, J.; Belyea, C.; Dunn, R.R.; McKerrow, A.; Collazo, J.A. The southern megalopolis: Using the past to predict the future of urban sprawl in the southeast, U.S. *PLoS ONE* **2014**, *9*, e102261. [CrossRef]

34. Zhang, S.; Fang, C.; Kuang, W.; Sun, F. Comparison of changes in urban land use/cover and efficiency of megaregions in China from 1980 to 2015. *Remote Sens.* **2019**, *11*, 1834. [CrossRef]

35. Anselin, L. Local Indicators of Spatial Association—LISA. *Geogr. Anal.* **1995**, *27*, 93–115. [CrossRef]

36. Pietrzak, M.B.; Wilk, J.; Kossowski, T.; Bivand, R. *The Identification of Spatial Dependence in the Analysis of Regional Economic Development—Join-Count Test Application*; Institute of Economic Research: Torun, Poland, 2013; Available online: https://www.econstor.eu/bitstream/10419/219556/1/ier-wp-2013-030.pdf (accessed on 9 November 2021).

37. Liévanos, R.S. Race, Deprivation, and Immigrant Isolation: The Spatial Demography of Air-Toxic Clusters in the Continental United States. *Soc. Sci. Res.* **2015**, *54*, 50–67. [CrossRef]

38. Hu, Z.; Rao, K.R. Particulate Air Pollution and Chronic Ischemic Heart Disease in the Eastern United States: A County Level Ecological Study Using Satellite Aerosol Data. *Environ. Health* **2009**, *8*, 26. [CrossRef] [PubMed]

39. Lin, Y.; Li, Y.; Ma, Z. Exploring the Interactive Development between Population Urbanization and Land Urbanization: Evidence from Chongqing, China (1998–2016). *Sustainability* **2018**, *10*, 1741. [CrossRef]

40. Montgomery, M.R. The Urban Transformation of the Developing World. *Science* **2008**, *319*, 761–764. [CrossRef]

41. Florida, R.; Gulden, T.; Mellander, C. The Rise of the Mega-Region. *Camb. J. Reg. Econ. Soc.* **2008**, *1*, 459–476. [CrossRef]

42. Friedman, T.L. *The World Is Flat: A Brief History of the Twenty-First Century*, 1st ed.; Farrar, Straus and Giroux: New York, NY, USA, 2007.

43. Alonso, W. *Location and Land Use: Toward a General Theory of Land Rent*; Harvard University Press: Cambridge, MA, USA, 1964.

44. Adams, J.S. Residential Structure of Midwestern Cities. *Ann. Assoc. Am. Geogr.* **1970**, *60*, 37–62. [CrossRef]

45. Nguyen, L.H.; Nghiem, S.V.; Henebry, G.M. Expansion of Major Urban Areas in the US Great Plains from 2000 to 2009 Using Satellite Scatterometer Data. *Remote Sens. Environ.* **2018**, *204*, 524–533. [CrossRef]

46. TxDOT. Transportation Planning & Programming Division, TxDOT. In *Capital-Alamo Connections Study*; 2019. Available online: https://www.txdot.gov/inside-txdot/projects/studies/statewide/capital-alamo-connections.html (accessed on 9 November 2021).

47. Yang, C.; Li, Q.; Hu, Z.; Chen, J.; Shi, T.; Ding, K.; Wu, G. Spatiotemporal Evolution of Urban Agglomerations in Four Major Bay Areas of US, China and Japan from 1987 to 2017: Evidence from Remote Sensing Images. *Sci. Total Environ.* **2019**, *671*, 232–247. [CrossRef]

48. Oden, M.; Sciara, G.C. The Salience of Megaregional Geographies for Inter-Metropolitan Transportation Planning and Policy Making. *Transp. Res. Part Transp. Environ.* **2020**, *80*, 102262. [CrossRef]

49. Knowles, R.D.; Ferbrache, F.; Nikitas, A. Transport's Historical, Contemporary and Future Role in Shaping Urban Development: Re-Evaluating Transit Oriented Development. *Cities* **2020**, *99*, 102607. [CrossRef]

50. North Central Texas Council of Governments. TOD Data Products. Available online: https://www.nctcog.org/trans/plan/land-use/tod/tod-data-products (accessed on 9 November 2021).

51. City of Houston. Livable Places Action Committee. Available online: https://houstontx.gov/planning/livableplaces.html (accessed on 9 November 2021).

52. Houston-Galveston Area Council (H-GAC). Livable Centers Planning Studies. Available online: https://www.h-gac.com/livable-centers/planning-studies (accessed on 9 November 2021).
53. Hoalst-Pullen, N.; Patterson, M.W. Applications and Trends of Remote Sensing in Professional Urban Planning: Remote Sensing in Professional Urban Planning. *Geogr. Compass* **2011**, *5*, 249–261. [CrossRef]
54. Zhao, C.; Weng, Q.; Hersperger, A.M. Characterizing the 3-D Urban Morphology Transformation to Understand Urban-Form Dynamics: A Case Study of Austin, Texas, USA. *Landsc. Urban Plan.* **2020**, *203*, 103881. [CrossRef]

 *land*

Article

Urban Innovation Efficiency Improvement in the Guangdong–Hong Kong–Macao Greater Bay Area from the Perspective of Innovation Chains

Wenzhong Ye [1], Yaping Hu [1] and Lingming Chen [1,2,*]

[1]   Department of Economics, School of Business, Hunan University of Science and Technology (HNUST), Xiangtan 411201, China; 1300030@hnust.edu.cn (W.Y.); yaping1225@mail.hnust.edu.cn (Y.H.)
[2]   Department of Economics & Statistics, School of Economics and Management, Xinyu University (XYU), Xinyu 338004, China
*   Correspondence: lingming1016@mail.hnust.edu.cn; Tel.: +86-158-7009-2285

Abstract: Against the background of globalization and informatization, innovation is the primary driving force for regional economic and social development. Urban agglomerations are the main body of regional participation in global competition, and promoting the construction of the Guangdong–Hong Kong–Macao Greater Bay Area is an important strategy for China's regional economic development. Aimed at the differences in location advantages among cities in the Guangdong–Hong Kong–Macao Greater Bay Area, based on the theory of innovation chain, we developed a three-stage model of "knowledge innovation-scientific research innovation-product innovation". A three-stage DEA model was used to measure the innovation efficiency of cities in the Greater Bay Area at different stages, and two progressive two-dimensional matrices are constructed to locate the innovation development of cities according to the efficiency value. The results show the following: ① The overall innovation efficiency of the Greater Bay Area urban agglomerations gradually decreased in the process from knowledge innovation and scientific research innovation to product innovation, and the innovation efficiency among cities was unbalanced. ② Shenzhen, Guangzhou, and Hong Kong all performed well in the whole innovation stage, while other cities in the Greater Bay Area showed weakness in innovation at different stages. Based on this, this paper puts forward relevant countermeasures and suggestions for promoting and optimizing collaborative innovation in the Greater Bay Area taking into account factor flow, industrial structure, and innovation network of urban agglomerations.

Keywords: innovation value chain; Guangdong–Hong Kong–Macao Greater Bay Area (GBA); innovation efficiency; urban agglomerations

check for updates

Citation: Ye, W.; Hu, Y.; Chen, L. Urban Innovation Efficiency Improvement in the Guangdong–Hong Kong–Macao Greater Bay Area from the Perspective of Innovation Chains. *Land* **2021**, *10*, 1164. https://doi.org/10.3390/land10111164

Academic Editors: Iwona Cieślak, Andrzej Biłozor and Luca Salvati

Received: 11 October 2021
Accepted: 29 October 2021
Published: 30 October 2021

1. Introduction

China's economy has moved from a stage of rapid growth to a new era of high-quality development, and innovation is a significant engine for high-quality regional development. Urban agglomerations are concentrated areas of economy, technology, workforce, information, and a leading area of national technological innovation. In recent years, China has attached great importance to the building of regional innovation centers, with key cities such as Beijing, Shanghai, Guangzhou and Shenzhen as the core, forming representative regional innovation urban agglomerations in the Beijing–Tianjin–Hebei Region, the Yangtze River Delta, the Pearl River Delta and the Guangdong–Hong Kong–Macao Greater Bay Area, highlighting these major cities' function as innovative hubs and driving the sustainable development of local economy [1]. The Guangdong–Hong Kong–Macao Greater Bay Area (hereinafter GBA) is one of the most open and economically active regions in China and plays an important strategic role in China's overall development and global economic development. The Outline Development Plan for the Guangdong–Hong

Kong–Macao Greater Bay Area issued by China at the beginning of 2019 sets out the strategic positioning of the GBA to become an international innovation center with global influence. It stresses the need to deepen innovation cooperation in the GBA and build an open regional collaborative innovation community for integrated development [2].

As the innovation center in China, the GBA is a crucial spatial carrier leading regional innovation and high-quality development. Its innovation intensity determines the level of regional economic high-quality development. With significant connotations, urban innovation efficiency has become a typical index to evaluate regional innovation and high-quality development. Thus, at this stage, has the GBA collaborative innovation community achieved coordinated progress? How can the innovation efficiency of the GBA be improved? Therefore, we adopted the DEA model based on the perspective of the innovation value chain to study the innovation efficiency of cities in the GBA at the stages of knowledge innovation, scientific research innovation, and product innovation, and we propose paths for improving the innovation efficiency of each city to provide a reference for decision-making to accelerate the innovation-driven strategy of the GBA and build a world-class innovation bay area.

The structure of this paper is as follows: The first section is the introduction, which introduces the research background. The second section is the literature review, which summarizes the research results of previous scholars and has a systematic and comprehensive cognition of the issues discussed in this paper. The third section provides the theoretical mechanism, discusses the connotation of the innovation value chain and the two-dimensional innovation efficiency matrix. The fourth section introduces the empirical model and discusses the data source, variable selection and data processing. The fifth section introduces the result analysis. According to the efficiency value measured by the three-stage DEA model, the positioning of each city in the two-dimensional matrix of innovation efficiency is analyzed. The sixth section is the research summary and recommendations.

2. Literature Review

As for the research on innovation efficiency, innovation efficiency is an important index reflecting the relationship between innovation input-output and the operation level of the regional innovation system [3,4]. In the efficiency measurement method, the existing research adopts the Stochastic Frontier Analysis (hereinafter SFA) based on parameters and the Data Envelopment Analysis (hereinafter DEA) based on non-parameters [5–9]. In terms of regional innovation efficiency, earlier research mostly focused on the research of innovation efficiency and its influencing factors at the national level [10,11]. Some scholars take provincial samples as research units and use the SFA analysis method to measure regional innovation efficiency [12–14], while others use the DEA method to research inter-provincial innovation efficiency [15–17]. The influencing factors of innovation efficiency mainly include human capital, economic development level and spillover effect [18,19], as well as the degree of marketization [20] and openness to the outside world [21].

The mode of the innovation value chain is an incremental process of multi-subject phased participation and continuous evolution of functional nodes. Its essence is the generation, transfer and diffusion of knowledge [22]. The core idea of the innovation value chain lies in the openness of innovation elements, the synergy of the whole operation and the added value [23]. Based on this idea, the innovation value chain can reflect the linkage process of innovation resources in an urban agglomeration to realize the added value of innovation value through the chain's activity structure. The interaction of different links in the innovation value chain affects the combination, flow and diffusion of innovation elements, and the coupling and synergistic development of each link contributes to the improvement of the overall innovation level and international competitiveness of urban agglomeration [24]. From the perspective of research, many scholars have also paid attention to the research on the phased efficiency of innovation, believing that the innovation process is divided into two stages: R&D of innovation results and application of innovation results [25,26], and analysis of the efficiency of the two stages, respectively. Based on the perspective of the innovation value chain, the innovation efficiency of knowledge innovation, R&D innovation and product innovation in China has been analyzed by using the three-stage DEA model and selecting the appropriate promotion path according to the advantages and disadvantages of the innovation efficiency in different regions at different stages [27].

For the research on the innovation efficiency of urban agglomerations, Zhao (2018) adopted the DEA method to study the innovation efficiency of cities in the GBA and compared the scale efficiency and technical efficiency [28]. Sheng et al. (2019) used the SFA method and spatial econometric model to study innovation efficiency and spatial spillover effect of urban agglomerations in Beijing–Tianjin–Hebei, the Yangtze River Delta and the Pearl River Delta in China [29]. Ye and Xu et al. (2021) conducted a comparative study on the innovation efficiency and influencing factors of the three major urban agglomerations in eastern China through the construction of an input-output index system [30,31]. The Bay Area economy, represented by the GBA, is an advanced form of regional economic development [32]. The collaborative innovation of urban agglomerations and their innovation efficiency play a driving role in developing a regional economy. Because of the spatial characteristics, collaborative innovation linkages and influencing factors of innovation efficiency within the GBA urban agglomeration, some Chinese scholars adopted a fixed-effect stochastic frontier model, the DEA–Malmquist index, the grey correlation analysis method, social network analysis and negative binomial regression model to conduct research and discussion [33–35]. Some scholars also studied the characteristics and influencing factors of innovation efficiency in the GBA from static and dynamic perspectives [36].

In the existing studies, many scholars have conducted in-depth analyses on the innovation efficiency of urban agglomerations from different perspectives and using different methods, which can provide important theoretical support for the development of this paper. At the same time, however, the current research findings still have the following shortcomings. Firstly, most of the literature uses provincial data to measure innovation efficiency, focusing on examining the mechanisms for improving innovation efficiency at the national level for the region as a whole. However, there is less literature that examines the optimization of innovation efficiency across cities in urban agglomerations. Secondly, the existing literature is relatively limited in its study of the innovation process in urban agglomerations and its efficiency at different stages. In fact, collaborative innovation between urban agglomerations occurs not only between different cities at the same stage of the value chain, but also at different innovation stages in the same city. Finally, most existing literature discusses the efficiency of scientific and technological innovation, while scientific and technological innovation is only a part of innovation activities, and the two cannot be generalized.

In the background of the new era, the study of innovation efficiency of urban agglomeration and its promotion path need to be further studied. The main marginal contribution and innovation of this paper mainly include the following aspects. First of all, from the perspective of the innovation value chain, a three-stage analysis framework of "knowledge innovation–scientific research innovation–product innovation" is constructed to clarify the main components and relations of each stage of innovation value chain. Secondly, the innovation efficiency of different stages of the GBA is measured by the broad innovation input–output index, and the improvement path of the innovation efficiency of each city is proposed, which provides decision-making reference for accelerating the innovation-driven strategy of the GBA and developing the world-class innovation Bay Area. Eventually, the research results of this paper will also provide development ideas and countermeasures for promoting the innovative development and sustainable development of urban agglomeration around the world.

3. Innovation Value Chain Theory and Innovation Path Setting

3.1. Innovation Process and Innovation Value Chain Theory

Previous research regards the innovation process as a "black box". Innovation is an output process that is input by multiple factors and through multiple innovation activities. Scholars at home and abroad have carried out much work on the research of the innovation process, and they believe that innovation is a whole that is interlinked by several innovation activities. At the same time, innovation activities are composed of interrelated multi-links such as design, research and development, production, and sales.

The concept of the innovation value chain was proposed by Hansen and Birkinshaw (2007), who explained that the generation, transformation and dissemination of ideas in innovation are multistage and interrelated [37]. Roperd et al. believe that it is the process of knowledge acquisition, transformation and utilization [38]. From the perspective of globalization, Kramer (2011) determined the contents of the innovation value chain from value sharing [39]. Lee, J. (2012) extended the perspective of the innovation value chain to developing countries, and further analyzed and supplemented its application value [40]. According to the relevant literature, the innovation chain is market-oriented and based on the innovation needs of enterprises, and takes innovation theory as the core element [41]. The innovation value chain is a new theoretical edge concept that combines technology innovation theory with value chain theory [42].

Based on the decomposition of technological innovation links from the perspective of production, the innovation value chain is recognized as a typical three-stage structure, which is a chain structure pattern formed by multiple innovation subjects based on ordinary interests around the whole process of knowledge generation, research and development, large-scale production and commercialization. Technological innovation is a multi–stage and multi–factor value chain transfer process from the input of innovative elements to the output of innovative products, including three stages from the input of innovation to the condensation of innovative knowledge and then to the realization of innovative results. This concept has a similar meaning to the three stages of basic research, applied research and experimental development in the process of innovation in China's statistics, which can match the research on regional economic reality and innovation efficiency well. The innovation process should include three closely linked stages: knowledge innovation, scientific research innovation and product innovation [27], which promote and link each other. The purpose of knowledge innovation and scientific research innovation is for product innovation, and knowledge innovation is the theoretical basis of scientific research innovation and product innovation. At the same time, they are independent of each other; each is a new stage of innovation, and the whole process is cyclic. Following this research idea and drawing on the model ideas of Yu et al. [27], we constructed the innovation value chain mechanism as shown in Figure 1. This paper analyzes and discusses the innovation efficiency of the Guangdong–Hong Kong–Macao Greater Bay Area from the three stages of knowledge innovation, scientific research innovation and product innovation in the innovation process, so as to further explore the status quo of the innovation and development of the GBA in detail.

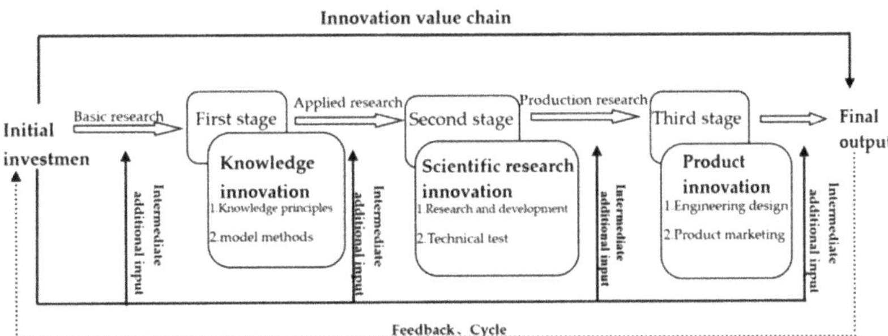

Figure 1. The three-stage operation mechanism of the innovation value chain in urban agglomerations.

3.2. The Theoretical Model of Innovation Path Setting

Cities in urban agglomerations show differences between the stages of efficiency of innovation and the three-phase contact operation mechanism of urban agglomeration of the innovation value chain. We constructed a two–dimensional innovation efficiency matrix of "knowledge–scientific research" innovation efficiency and "scientific research–product" innovation efficiency of the regional innovation urban agglomerations (as shown in the diagrams (a) and (b) in Figure 2) [27]. The average efficiency of cities at different stages can be divided into four efficiency combinations according to the value of the two efficiency dimensions. Different combination modes represent different efficiency states, thus providing a basic idea for the improvement of innovation efficiency in urban agglomeration. These four types of efficiency combination regions are, respectively, the combination of "one high and one low" represented by B and D, the combination of "double low" represented by C and the combination of "double high" represented by A. This division is convenient for cities in different quadrants when searching for appropriate theoretical reference paths to improve innovation efficiency according to their development conditions.

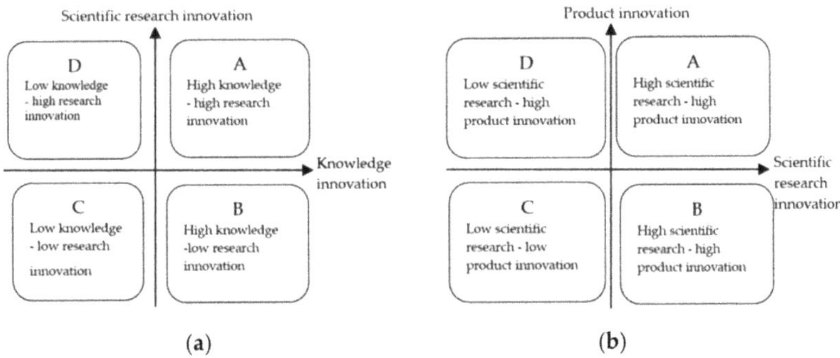

Figure 2. (**a**) Matrix of "knowledge–research" efficiency values; (**b**) matrix of "scientific research–product" efficiency values.

According to the diagrams (a) and (b) in Figure 2, this paper proposes the following three ways to improve the innovation efficiency: first, the breakthrough improvement path of unilateral supplement from B to A or D to A, which takes the link with low efficiency in one stage as the breakthrough point to improve the overall innovation efficiency on the premise of ensuring the high efficiency in the other stage. Second, C reaches the progressive path of improving the excellent and compensating the poor in region A through B or D; that is, the "double low" region can make up for the disadvantages while maintaining its own advantages step by step according to its situation, to achieve the double high goal. Thirdly, the breakthrough leapfrog promotion path from C to A, that is, the leapfrog of innovation efficiency, can be directly realized by simultaneously promoting the two stages.

4. Model Method and Variable Selection

4.1. Three-Stage DEA Model

In the measurement of innovation efficiency, most scholars focus on Data Envelopment Analysis (DEA) based on non-parametric techniques, which was proposed by Charnes, Coopor and Rhodes in 1978 [43]. The principle of this method is to determine the relatively effective production frontier by means of mathematical planning and statistical data by keeping the input or unchanged of Decision-Making Units (DMU), and projecting each DMU onto the production frontier of DEA. The relative effectiveness of DMU is evaluated by comparing its deviation from DEA frontier. The DEA method is an evaluation method based on the concept of relative efficiency using convex analysis and linear programming as tools, and using a data programming model to calculate the optimal input–output scheme of the comparison decision-making unit itself. The advantage of this model is that there is no need to consider the corresponding relationship between input and output, which can solve the linear corresponding relationship between input and output well, and there is no need to assign hypothetical weights to parameters and estimation indexes, which avoids the subjectivity of the research subject, and the measurement of relative efficiency value is more objective, real and reliable. Compared with the traditional efficiency evaluation model, DEA can better reflect the information and characteristics of the evaluation object when dealing with the efficiency evaluation of composite input and multi-type output system.

The classic DEA method has been optimized in the treatment of environmental factors, such as using the DEA-Tobit method. In the first stage, the method uses the classical DEA method to calculate the efficiency value of DMU, and then in the second stage, the efficiency value is taken as the dependent variable and environmental factors as the independent variable to establish a Tobit regression model to investigate the impact of environmental factors. This method can use regression technology to determine the intensity and direction of the impact of environmental factors on efficiency, but its effect is not to strip out environmental factors in the calculation of efficiency, so it does not change the efficiency value measured by the classical method. Due to random noise and environmental factors also having a certain degree of influence on the efficiency evaluation of decision-making units, Frided et al. (2002) [44] proposed a three-stage DEA method based on the classical DEA model that can separate environmental factors to make the calculated efficiency value more accurate. Among the cities in the GBA, there are differences in the level of economic development, the stage of industrial development and the development degree of science, technology and education, and the innovation environment facing each city is highly differentiated and unbalanced. Environmental factors should be taken into account in the innovation efficiency of different stages of innovation activities in different cities. Combined with the above characteristics, We adopted the three-stage DEA model to measure the efficiency of the GBA at each stage in the process of innovation value chain, and its theoretical model is as follows.

4.1.1. In the First Stage, the Initial Efficiency Is Evaluated by Using the Original Input–Output Data

The DEA model can choose different orientations according to the difference of analysis purpose and demand. The traditional DEA model has two forms: the CCR model and the BCC model. Among them, the CCR model assumes that DMU is in a fixed scale return situation, which is used to measure the total efficiency. The BCC model is used to measure pure technology and scale efficiency, assuming DMU is in a variable return to scale situation. In the applications discussed in existing literature, generally speaking, most three-stage DEA models will choose the input—oriented BCC (variable return to scale) model so as to analyze comprehensive technical efficiency from two aspects of pure technical efficiency and scale efficiency. For any decision unit, the dual-form BCC model under input guidance can be expressed as:

$$min\theta - \varepsilon(\hat{e}^T S^- + e^T S^+)$$

$$s.t. \begin{cases} \sum_{j=1}^{n} X_j \lambda_j + S^- = \theta X_0 \\ \sum_{j=1}^{n} Y_j \lambda_j - S^+ = Y_0 \\ \lambda_j \geq 0, S^-, S^+ \geq 0 \end{cases} \tag{1}$$

The formula $j = 1, 2, \cdots, n$ represents the decision unit, and X, Y are input and output vectors. The DEA model is essentially a linear programming problem.

If $\theta = 1, S^+ = S^- = 0$, then the decision-making unit DEA is valid;

If $\theta = 1, S^+ \neq 0$, or $S^- \neq 0$, then the weak DEA of the decision-making unit is valid;

If $\theta < 1$, then the decision-making unit is not valid by DEA.

The efficiency value calculated by the BCC model is the comprehensive technical efficiency (TE), which can be divided into scale efficiency (SE) and pure technical efficiency (PTE), that is, TE = SE × PTE. In the first stage, two values should be calculated: ① the initial DEA efficiency value without filtering environmental factors is used for comparison and analysis in subsequent links; ② the relaxation variable of input is the dependent variable of the second stage.

4.1.2. In the Second Stage, Environmental Factors and Statistical Noise Were Eliminated by SFA-Like Regression

The relaxation variable calculated in the first stage was taken as the opportunity cost of the decision unit, and the influence of environmental factors and random errors was considered. The relaxation variable $[x - X\lambda]$ can reflect the initial inefficiency, composed of environmental factors, management inefficiency, and statistical noise. The SFA model was used to decompose the slack variables in the first stage into the above three effects. Then the input and output were corrected and readjusted, or only the input or output was adjusted. The following regression function similar to SFA is constructed based on input orientation.

$$S_{ni} = f(Z_i; \beta_n) + v_{ni} + \mu_{ni}; i = 1, 2, \cdots, I; n = 1, 2, \cdots, N \tag{2}$$

In the formula, S_{ni} is the relaxation value of the input of item n of the decision unit i, Z_i is the environment variable; β_n is the coefficient of the environment variable; $v_{ni} + \mu_{ni}$ is the mixed error term; v_{ni} represents random interference, and μ_{ni} represents inefficiency of management. In the formula, $v \sim N(0, \sigma_v^2)$ is the random error term, representing the influence of random disturbance factors on input relaxation variables; μ refers to management inefficiency and represents the influence of management factors on input slack variables. It is assumed that it follows a normal distribution truncated at zero points, $\mu \sim N^+(0, \sigma_\mu^2)$.

4.1.3. The Third Stage: DEA Efficiency Analysis of the Adjusted Input-Output Variables

The adjusted input-output variables are used to calculate the efficiency of each decision unit again. At this point, the efficiency has removed the influence of environmental factors and random factors, which is relatively authentic and accurate.

4.2. Indicator Selection and Data Source

From the perspective of the innovation value chain, input–output variables are not the same in different stages according to the different focuses of research in the three stages of innovation activities. In the knowledge innovation stage, since higher education personnel are the main contributors to the output of scientific and technical papers, this paper identifies the number of higher education personnel and the cost of higher education expenditure as input indicators. In terms of output indicators, referring to the practice of Yu (2014), this paper chooses the number of published scientific papers and the number of scientific monographs [27]. In the stage of scientific research innovation, the number of R&D personnel and R&D investment are firstly considered based on the previous research ideas of scholars [45]. Among them, it is the essence of regional innovation that R&D personnel carry out scientific and technological research, which reflects the regional ability to attract talents. The total amount and intensity of R&D expenditure represent the overall scientific and technological R&D capability of the region as well as the degree of investment in innovation activities. In addition to the two input indicators mentioned above, we argue that the input indicators at the research and innovation stage should also include the outputs at the knowledge innovation stage, the number of scientific and technical papers, and scientific and technical monographs. The output index of scientific research innovation stage is the number of patent applications and the number of patent grants [45]. Similarly, when determining the input indicators for the product innovation stage in this paper, in addition to considering the number of patents granted in the previous stage, the indicator of new product development expenditure should also be included. With reference to the relevant literature [27], the output indicators for this phase were selected as two of the industrial enterprises' new product output and export value.

Combined with existing studies, the environmental variables selected in this paper include higher education investment level [46], government support level, economic development level, informatization level, fixed asset investment [47], scientific research foundation, industrial structure [48] and regional openness [49]. The level of investment in higher education is expressed by the proportion of higher education funding to GDP, the level of government support is expressed by the proportion of government funding in science and technology funding, and the level of economic development is expressed by the per capita GDP of the region. Informatization level refers to innovative information resources, mainly including the number of mobile phone users at the end of the year and the number of broadband Internet users, which can reflect the development level of regional information resources. Investment in fixed assets is a significant component of innovation material resources and the material basis for regional innovation activities. It reflects the degree of enrichment of innovation resources. The scientific research base is expressed as the number of enterprises with R&D activities in the region. The industrial structure is expressed by the ratio of the tertiary industry to the GDP of the region, and the proportion of the total import and export of the region to GDP is used to express the openness of the region. The specific input-output indicators and environmental variables of the three stages of innovation are shown in Table 1:

The index data are from China City Statistical Yearbook 2010–2020, Guangdong Statistical Yearbook and Guangdong Science and Technology Yearbook. For Hong Kong, the data are from Hong Kong Statistical Yearbook 2020, Statistics on Innovation Activities in Hong Kong 2019 and Statistics of Hong Kong Trade Development Council 2019, respectively, of the Census and Statistics Department of the Hong Kong Special Administrative Region. The data for Macau are derived from the Macao Statistical Yearbook 2020, Macao Economic Development Report for the fourth quarter of 2019, and the time-series database of The Macao Statistical Survey Bureau. The amount of Hong Kong dollars and Macau patacas in each indicator is converted into people's value by the exchange rate of Hong Kong dollars and Macau patacas against RMB in the current year.

Table 1. Three-stage input-output index system of innovation.

Stage	Input Variables	Output Variable	Environment Variable (Non-Dimensional)
Knowledge innovation	Number of higher education personnel (per year) Higher Education Expenditure (100 million yuan)	Number of scientific papers published (Piece) Number of scientific and technological monographs (pieces)	Level of investment in higher education Level of government support
Scientific research innovation	Number of R&D personnel (per year) R&D Expenditure (100 million yuan) Number of scientific papers published (Piece) Number of scientific and technological monographs (pieces)	Number of Patent Applications (Piece) Number of Patents Granted (Piece)	Level of economic development Level of informatization Investment in fixed assets Scientific research base
Product innovation	Number of Patents Granted (Piece) New product development expenses (100 million yuan)	The output value of new products of industrial enterprises (100 million yuan) Export of new industrial products (100 million yuan)	The industrial structure Regional openness

5. Empirical Results and Analysis

5.1. Efficiency Measurement Analysis of Innovation Chain Three-Stage DEA Model

According to the three stages of knowledge innovation, scientific research innovation and product innovation in the innovation value chain, the indicators are different. According to the analysis and research of Bruce [50] (1993), innovation output results have a certain lag. We assumed that innovation activities have a lag period of one year from input to output, that is, when the input index of year T is selected, the corresponding output index data should be selected in year T + 1. Therefore, when measuring the innovation efficiency of the GBA in this paper, the index data of the innovation input were selected from 2010 to 2018, and the variable data of innovation output were selected from 2011 to 2019. We followed the three-stage DEA analysis steps. In the first stage, DEAP2.1 software was used to make a preliminary efficiency calculation for input-output indicators. The results are shown in Table 2.

Table 2. Traditional DEA innovation efficiency value of GBA (average value of 2010 to 2018).

City	Knowledge Innovation				Scientific Research Innovation				Product Innovation			
	TE	PTE	SE		TE	PET	SE		TE	PST	SE	
Guangzhou	0.824	1	0.824	drs	0.525	0.543	0.968	irs	0.325	0.343	0.948	drs
Shenzhen	0.698	0.7	0.997	irs	0.941	1	0.941	drs	0.825	1	0.825	drs
Zhuhai	0.744	0.792	0.939	irs	0.637	0.781	0.815	irs	0.578	0.73	0.792	irs
Foshan	0.757	0.838	0.903	drs	0.858	0.889	0.965	irs	0.482	0.492	0.979	irs
Huizhou	0.741	0.772	0.96	irs	0.712	0.775	0.919	irs	0.677	0.747	0.907	irs
Dongguan	0.718	0.729	0.985	irs	0.916	0.964	0.95	irs	0.675	0.836	0.808	irs
Zhongshan	0.741	0.78	0.951	irs	1	1	1	-	0.331	0.635	0.52	irs
Jiangmen	0.816	0.839	0.973	irs	0.55	0.921	0.598	irs	0.298	0.841	0.354	irs
Zhaoqing	0.755	0.756	0.998	drs	0.367	0.859	0.427	irs	0.431	0.851	0.507	irs
Hong Kong	0.668	0.675	0.991	irs	0.999	1	0.999	drs	1	1	1	-
Macao	0.861	0.919	0.937	irs	0.728	0.925	0.788	irs	0.782	0.919	0.851	irs
Mean	0.757	0.8	0.951	-	0.748	0.878	0.852	-	0.582	0.763	0.772	-

In the second stage, Frontier4.0 software was used to establish the SFA regression model with the input relaxation variables calculated in the first stage as the dependent variables and the environmental variables in each stage as the explanatory variables. The results are shown in Table 3. In the knowledge innovation stage, higher education investment level and the number of higher education personnel were positively correlated. However, there was a negative correlation with higher education expenses and description of higher education, and there is a certain redundancy of financial investment; similarly, the government support and knowledge innovation phase 2 class negatively correlated to the inputs, which is also due to the government and the phenomenon of the lack of output conversion on spending too much. Among the four kinds of input resources in the stage of scientific research innovation, the number of scientific papers published has no apparent correlation with other influencing factors except the level of economic development. Economic development was positively correlated with the publication of scientific papers and monographs but negatively correlated with R&D personnel and R&D expenditure. The informatization level and fixed asset investment was beneficial to the number of R&D personnel but hurt R&D expenditure. The research basis had a positive effect on the other three input factors except for the number of scientific papers published. In the stage of product innovation, industrial structure and regional openness had a negative correlation with the number of patents granted. In contrast, new product development funds had a positive correlation with industrial structure and a negative correlation with regional openness.

Table 3. Results of stage 2 estimation based on SFA.

	Number of Higher Education Personnel	Expenses for Higher Education	Number of R&D Personnel	R&D Expenditure	Number of Scientific Papers Published	Number of Scientific and Technological Monographs	Number of Patents Granted	New Product Development Funds
Constant term	−9.61 * (0.67)	−5.32 (1.22)	10.17 ** (0.95)	−0.63 (1.04)	−7.95 * (1.28)	−2.27 (0.33)	3.12 * (0.26)	0.92 * (0.77)
Level of investment in higher education	6.97 * (2.19)	−0.73 ** (1.01)	-	-	-	-	-	-
Level of government support	−3.78 * (4.25)	−0.42 * (2.57)	-	-	-	-	-	-
Level of economic development	-	-	−1.21 * (1.27)	−5.69 ** (3.25)	4.52 ** (6.78)	8.47 * (0.17)	-	-
Level of informatization	-	-	1.33 *** (5.62)	−0.83 * (1.26)	−3.96 (0.86)	4.72 * (0.22)	-	-
Investment in fixed assets	-	-	5.22 ** (0.67)	−0.39 * (1.26)	−6.21 (5.94)	−0.87 * (7.41)	-	-
Scientific research base	-	-	7.25 * (0.19)	3.16 ** (4.66)	−5.44 (0.21)	1.99 * (2.64)	-	-
The industrial structure	-	-	-	-	-	-	−8.21 * (0.96)	10.21 * (0.77)
Regional openness	-	-	-	-	-	-	−2.85 (1.36)	−0.63 ** (5.64)
sigma-squared	0.00012	0.000027	0.00009	0.00014	0.000081	0.00011	0.000075	0.00018
gamma	0.751	0.699	0.801	0.774	0.725	0.623	0.822	0.739
log likelihood	20.31	100.3	5.62	42.36	56.23	7.86	10.91	26.77

The data in the table are the correlation coefficient, and the data in brackets are the standard deviation. ***, **, and * mean significant at the level of 1%, 5%, and 10%, respectively.

In the third stage, DEAP2.1 software was used to adjust the input variables according to the SFA regression results, and the adjusted data were substituted into the BCC model again for efficiency evaluation. The specific innovation efficiency measured by the adjusted three-stage DEA model is shown in Table 4.

Table 4. Three-stage DEA innovation efficiency of GBA after adjustment (average value from 2010 to 2018).

City	Knowledge Innovation				Scientific Research Innovation				Product Innovation			
	TE	PTE	SE		TE	PET	SE		TE	PST	SE	
Guangzhou	0.916	1	0.916	drs	0.96	1	0.96	drs	0.639	0.641	0.996	drs
Shenzhen	0.983	0.989	0.994	drs	1	1	1	irs	0.715	0.896	0.798	drs
Zhuhai	0.882	0.902	0.978	irs	0.899	1	0.899	irs	0.43	0.669	0.643	irs
Foshan	0.888	0.891	0.997	drs	0.858	0.889	0.965	drs	0.497	0.499	0.996	drs
Huizhou	0.819	0.828	0.989	irs	0.778	0.781	0.996	irs	0.587	0.625	0.939	irs
Dongguan	0.887	0.924	0.96	irs	0.761	0.855	0.89	irs	0.675	0.836	0.808	irs
Zhongshan	0.869	0.958	0.907	irs	0.767	1	0.767	irs	0.342	0.543	0.63	irs
Jiangmen	0.703	0.835	0.842	irs	0.558	0.815	0.684	irs	0.32	0.561	0.57	ins
Zhaoqing	0.684	0.718	0.952	irs	0.484	0.853	0.568	irs	0.232	0.594	0.391	ins
Hong Kong	0.948	0.951	0.996	irs	1	1	1	-	0.967	0.967	1	-
Macao	0.789	0.885	0.892	irs	0.598	0.81	0.738	irs	0.558	0.723	0.773	irs
Mean	0.852	0.898	0.948	-	0.788	0.909	0.861	-	0.542	0.687	0.777	-

According to the comparison between Tables 2 and 4, after eliminating environmental factors and random interference, the average comprehensive efficiency of the GBA in the knowledge innovation stage increased from 0.757 to 0.852, in which the pure technical efficiency also increased while the scale efficiency decreased slightly. The mean value of comprehensive efficiency increased from 0.748 to 0.788 in the stage of scientific research innovation, and both the pure technical efficiency and scale efficiency increased to a certain extent. In the product innovation stage, the average comprehensive efficiency decreased from 0.582 to 0.542, but both the pure efficiency and scale efficiency increased slightly.

As can be seen from Table 4, compared with the latter two stages, the efficiency of the knowledge innovation stage was higher, with an average value of 0.852. Among them, the efficiency of knowledge innovation in Guangzhou, Shenzhen, and Hong Kong was more than 0.9, ranking in the first echelon of innovation. The efficiency values of Zhuhai, Foshan, Huizhou, Dongguan and Zhongshan were between 0.8 and 0.9, which belong to the second echelon. The efficiency value of Jiangmen, Zhaoqing and Macao was below 0.8, the third echelon. In the scientific research innovation stage, the average efficiency was 0.788, which was lower than that in the knowledge innovation stage. Guangzhou, Shenzhen, and Hong Kong had higher efficiency values, which were all higher than 0.9. Among them, the efficiency values of Shenzhen and Hong Kong reached the allocation effective level of 1. In the second tier, including Zhuhai, Foshan, Huizhou, Dongguan and Zhongshan, their efficiency values were between 0.6 and 0.9. The efficiency value of Jiangmen, Zhaoqing, and Macao was below 0.6, among which the efficiency value of Zhaoqing was the lowest, only 0.484. In the product innovation stage, its average efficiency was low on the whole, only 0.542, which is far lower than the previous two innovation stages. Among them, Hong Kong had the highest product innovation efficiency (0.967), followed by Shenzhen, Dongguan and Guangzhou (0.6 to 0.8). The product innovation efficiency value of Macao, Foshan, Huizhou, and Zhuhai was between 0.4 and 0.6, which belongs to the second tier. In the third stage of the innovation chain, the efficiency of Zhongshan, Jiangmen, and Zhaoqing was low, with a value below 0.4. Among them, Zhaoqing was the lowest, with only 0.232. Based on the calculated efficiency value, it can be preliminarily analyzed that the reason why Guangzhou, Shenzhen, and Hong Kong achieved efficient innovation lies in their strong foundation conditions and distinct advantages in innovation platforms [51]. Compared with the three leading innovation cities, the overall innovation efficiency of the other cities was average, and there was a certain gap with the first echelon, which indicates the imbalance of innovation capacity within the GBA.

By comparing the efficiency values of the three stages in the innovation process (as shown in Figure 3), it is not difficult to find that with the progress of innovation, the innovation efficiency becomes increasingly lower in the process from knowledge innovation to scientific research innovation and finally to product innovation. The innovation efficiency decreased from 0.852 in the knowledge innovation stage to 0.788 in the scientific research innovation stage and was only 0.542 in the product innovation stage. From the composition of the main bodies of innovation activities, the main bodies of innovation in the GBA are the government, enterprises, universities, research institutes and science and technology intermediaries. Knowledge innovation mainly relies on universities. There are many universities (162 in total) in the GBA [52], so the GBA has high efficiency in the stage of knowledge innovation. Scientific research innovation mainly relies on scientific research institutes and scientific research intermediaries, and the number of patent applications is mainly taken into account when testing the results in this stage. The overall efficiency of the scientific research innovation stage was lower than that of the first stage, indicating that the innovation efficiency of the transfer from knowledge innovation to scientific research innovation in the innovation process needs to be enhanced. The relatively low efficiency in the stage of product innovation reflects that the transformation of innovative product results is not ideal. The reason may be that the technical maturity of the scientific and technological achievements of some universities is low, and they lack accurate docking with the industry and market demand [53]. At the same time, the evaluation and pricing mechanism of scientific research results is not perfect; the steps are tedious, time–consuming, and lengthy, which affects the transformation time and leads to the low efficiency of overall product innovation in the GBA. At present, there is a lack of interaction and communication among innovation players in the GBA, and the coordination ability is weak. The synergistic effect between enterprises and other innovation players such as universities and research institutes has not been played. The knowledge spillover benefit and technology diffusion effect produced by innovation activities are non-significant [54]; the synergistic value-added effect of the value chain innovation between enterprises and the system innovation effect between enterprises and the government did not lead to an effective value [55].

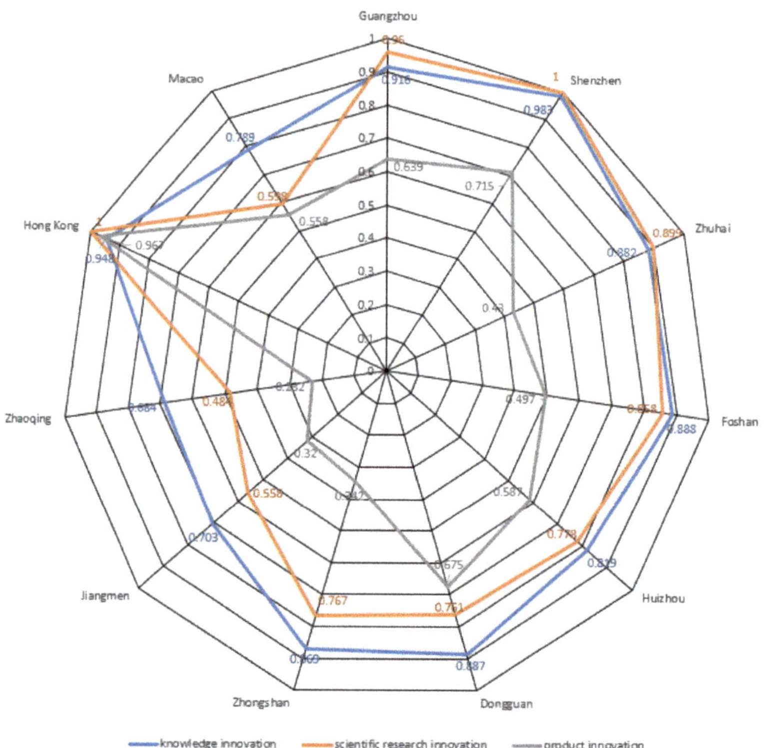

Figure 3. Comparison of innovation efficiency values in the three stages of innovation in the GBA.

5.2. The Positioning Analysis of the Two-Dimensional Matrix of Innovation Efficiency of Each City

Based on the empirical results in Table 4 and the previous research ideas, we put the innovation efficiency of each stage in a two-dimensional distribution map (using the mean value of innovation efficiency of each stage as the division standard). The details are shown in Figures 4 and 5.

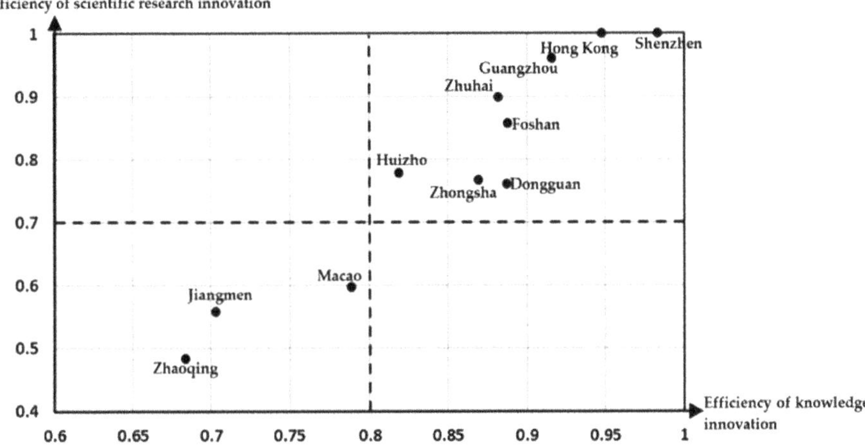

Figure 4. "Knowledge-scientific research" innovation efficiency matrix.

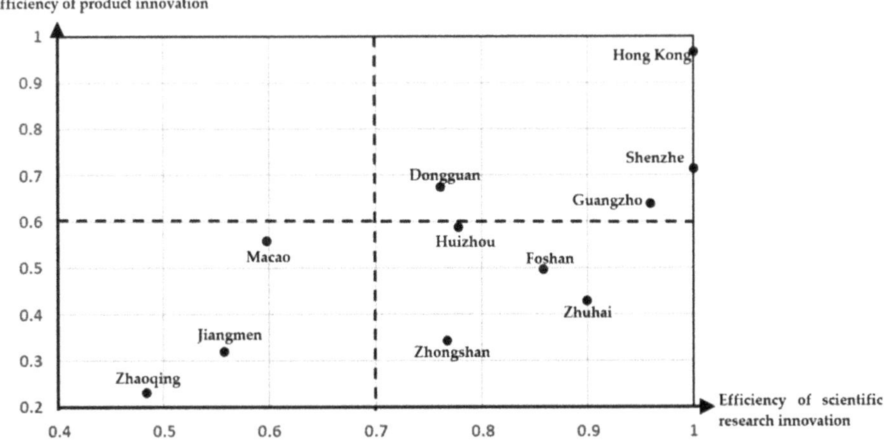

Figure 5. "Scientific research–product" innovation efficiency matrix.

According to the analysis of the "knowledge–scientific research" innovation efficiency matrix (Figure 4), cities in the GBA are distributed in regions A and C, namely, intensive and efficient knowledge–scientific research innovation and extensive and inefficient knowledge–scientific research innovation. The cities in Zone A include eight cities: Guangzhou, Shenzhen, Hong Kong, Zhuhai, Foshan, Huizhou, Dongguan and Zhongshan. Higher education and basic scientific and technological innovation facilities in these cities are relatively developed, laying a talent base and material guarantee for knowledge innovation and scientific research innovation. The innovation-leading development cities represented by Guangzhou, Shenzhen and Hong Kong have a high level of economic development and a strong "siphon effect" to attract resources from surrounding cities. Therefore, a large number of innovative elements are gathered here and their innovation ability is relatively prominent. In recent years, Zhuhai has been gradually developing as an innovation growth pole [56], while Foshan and Dongguan, as extensions of the GBA's "Guangzhou-Shenzhen-Hong Kong-Macao" innovation corridor, have also been steadily improving their innovation capabilities. By connecting with Shenzhen and Guangzhou, Huizhou and Zhongshan will further cluster in the innovation corridor and develop their innovative advantages.

Macao, Jiangmen and Zhaoqing are the two cities with low efficiency in knowledge innovation and scientific research innovation. Although these three cities are all located in region C, the causes of double inefficiency are different. Macao has a relatively high level of economic development. The low efficiency in the knowledge stage may be caused by excessive input of human and material resources in innovation and limited innovation output in this link. In contrast, the low efficiency in the scientific research stage reflects the lack of ability to integrate knowledge innovation into scientific research innovation. Jiangmen and Zhaoqing are cities with relatively low economic strength in the GBA, so their cities have limited innovation resources and lack the momentum of innovation development.

For the three cities with low efficiency of knowledge and scientific research innovation, the optimization path includes the C $\rightarrow$ A skipping of the development path, and the C $\rightarrow$ B $\rightarrow$ A and C $\rightarrow$ D $\rightarrow$ A progressive improvement path. The leap-forward development path requires a solid economic foundation and reserves of resource elements, and Macao is suitable for the leap-forward development path. The gradual promotion path is to give full play to the existing advantages and make up the disadvantages on this basis, that is, through the transition of B or D region, and finally towards A region. The low efficiency of Jiangmen and Zhaoqing in the stage of knowledge innovation is mainly due to the lack of input resources provided by the cities, which makes the improvement of efficiency in the stage of knowledge innovation difficult. Therefore, when choosing the path of improvement, the two cities should choose the C $\rightarrow$ D $\rightarrow$ A progressive innovation development path that first improves the efficiency of scientific research innovation while maintaining the level of knowledge innovation efficiency.

Let us look at the "scientific research–product" innovation efficiency matrix (Figure 5). In the process of linking scientific research and product innovation, cities in the GBA are distributed in regions A, B, and C, namely, intensive and efficient scientific research–product innovation, high scientific research innovation efficiency and low product innovation efficiency, and extensive and low-efficiency scientific research–product innovation. The cities in the double high efficiency A region, including Hong Kong, Guangzhou, Shenzhen and Dongguan, also performed well in the previous phase of knowledge innovation. Because they had a higher level of economic development of the city itself, the innovation efficiency of the transformation provided the economic foundation, which was also due to the development of the related policy funds, talents, and supporting environment brought by various advantages, making scientific research and product innovation efficiency higher; these areas also reflect the scientific research and innovation of the elements of the transition to product innovation conversion efficiency better.

The cities of high-efficiency scientific research innovation and low-efficiency product innovation are located in region B are Huizhou, Foshan, Zhuhai and Zhongshan. The "scientific research–product" innovation of these four cities also proves that regional scientific research innovation is not able to transform into product innovation. Therefore, cities in Region B should focus on enhancing the efficiency of product innovation while not ignoring the efficiency of scientific research innovation. Located in the geometric center of the GBA, Zhongshan plays a crucial role in "connecting the east with the west". As a start-up resource city, Huizhou can connect with Shenzhen, learn from its research results and complete the transformation suitable for its development. Zhuhai can take advantage of its proximity to Macao to explore new paths of innovative development. Foshan should further upgrade its existing leading industries. At the same time, we should increase the investment in scientific research and cooperate with Guangzhou to complete the integration of traditional industries and emerging industries [57], develop complementary industrial systems, and increase cooperation with cities on the east coast. The four cities of Huizhou, Foshan, Zhuhai and Zhongshan should strengthen the efficiency of product innovation according to their regional development characteristics and strive to solve the problem of low efficiency of product innovation.

Macao, Jiangmen and Zhaoqing are the two cities with low efficiency in scientific research innovation and product innovation. The efficiency of these three cities in the stage of knowledge innovation is relatively lower than that of other cities. The fundamental reason is the imbalance of input and output of innovation factors, and the transformation of innovative product results is not ideal. These three cities need to strengthen scientific research innovation and product innovation. They can maximize their advantages with the help of functional division of urban agglomeration, and then focus on improving their weaknesses. At the same time, the advantage of the overall coordinated development of the GBA urban agglomerations can be utilized to make full use of the technology spillover effect, absorb the scientific research achievements of other regions and complete the industrialization transformation suitable for local development, to improve the product innovation efficiency of the region.

6. Conclusions and Recommendations for Countermeasures

6.1. Conclusions

Based on the three-stage innovation value chain model of "knowledge innovation–scientific research innovation–product innovation", we constructed a generalized innovation input-output index system and use the three–stage DEA model that examines environmental factors to measure the innovation efficiency in each stage of innovation activities in the GBA from 2010 to 2018. The results show the following: ① On the whole, the innovation efficiency of the Greater Bay Area decreased with the advancement of innovation activities. The average innovation efficiency of knowledge innovation, scientific research innovation, and product innovation was 0.852, 0.788, and 0.542, respectively. ② From the perspective of different stages in the innovation process of each city, Shenzhen, Guangzhou, and Hong Kong all performed well in the whole stage of innovation, while other cities in the GBA had innovation shortcomings in different stages. Based on the above research results, we constructed the "knowledge–scientific research" innovation efficiency matrix and the "scientific research–product" innovation efficiency matrix using a two-dimensional matrix, and divided the innovation efficiency at each stage into four types. From the matrix analysis, the current state of innovation development in each city was located, and the path to improve its innovation efficiency was analyzed according to the characteristics of each city.

6.2. Recommendations for Countermeasures

Innovation is an important starting point and strategic path for promoting the development of the GBA. To promote the construction of a national innovation system and strengthen strategic scientific and technological forces, it is necessary to construct and perfect the new mechanism of regional collaborative innovation and improve the efficiency of regional innovation. The GBA is an important bridge linking the internal and external circulation of China's economy, and will become a pioneer and demonstration area for China to build a new development pattern. In recent years, China has issued a series of policy documents on collaborative innovation around the construction of an international science and technology innovation center in the GBA and carried out a series of fruitful explorations of institutional mechanisms to promote the upgrading of cooperation models in collaborative innovation in the GBA [58]. However, due to the lack of resource transformation in the innovation process, the innovation efficiency gradually declines along with the innovation chain, and the imbalance of innovation development among cities in the GBA shows that there is still considerable room for improvement in collaborative innovation in the GBA. The mechanisms and systems of collaborative innovation in the GBA are not yet perfect, and the scientific collaborative innovation system has not been completely established. The existing government management system has not adapted to the coordinated governance of metropolitan groups. As an important territory to promote innovation in China, the GBA still needs to improve its collaborative innovation mechanism and establish a collaborative system among innovation bodies such as cities, enterprises and R&D institutions. In addition, the cooperation mechanism, communication mechanism and guarantee mechanism based on collaborative innovation in the Greater Bay Area need to be further deepened. Based on this, relevant countermeasures and suggestions for optimizing collaborative innovation in the GBA are put forward from the aspects of factor flow, industrial structure, innovation networks.

6.2.1. Promoting the Flow of Innovation Factors and Building a Circular and Smooth Bay Area

At present, the GBA has problems such as uneven distribution of innovation factors and uncoordinated ability to allocate resources. At the same time, administrative barriers caused by policy and economic factors have placed very significant constraints on the convenient and efficient flow of innovation resources within the GBA, which is not conducive to the formation of technology diffusion effects and hinders the development of collaborative regional innovation. The key to enhancing innovation effectiveness in the GBA city cluster lies in the free flow of innovation factors. Therefore, the Greater Bay Area should improve the phenomenon of excessive siphoning from the central cities, strengthen the diffusion effect of innovation technologies, and accelerate the introduction and market-based allocation of innovation factor resources. These improvements would attract more high-quality elements to the GBA, optimally fill in the missing links of the innovation chain in cities with low and medium levels of coordination, promote positive interaction among the links and enhance the level of chain coordination. At the same time, the mechanism channel for the flow of innovation factors will be unblocked, and the cities in the GBA will be promoted to jointly enhance the level of coupling and coordination of science and technology innovation, so as to achieve the construction goal of smooth circulation of innovation factors in the GBA.

6.2.2. Improving Innovative Cooperation Mechanisms to Build a Mutually Beneficial Bay Area

There is currently a lack of interaction and communication between innovation agents in the Greater Bay Area, and coordination is weak. Universities and research institutes with their enterprises and other innovation subjects are many but not strong, the interaction and coordination between the subjects is insufficient, and the sharing platform of science and technology innovation elements is relatively lacking. The existing mechanism does not play a synergistic role between enterprises and other innovation bodies such as universities and research institutes, thus resulting in problems such as low value-added effect of scientific and technological innovation results and low conversion rate. Therefore, creating a collaborative and mutually supportive partnership among multiple players and improving the innovation cooperation mechanism are the backbone of the GBA's innovation-driven strategy. The GBA should further promote the interconnection of the city's industrial, value and innovation chains, attach importance to the investment and accumulation of talent capital, and cultivate highly qualified talents with core competitiveness. Improving the institutional mechanisms of Guangdong, Hong Kong, and Macao in areas such as mutual recognition of talent qualifications, the use of science and technology funds and customs clearance facilitation would achieve the goal of building a mutually beneficial Bay Area.

6.2.3. Optimizing the Industrial Structure and Building a Bay Area with a Full Industrial Chain

In terms of industrial structure, the GBA has uneven development. Hong Kong and Macao have entered the post-industrialization stage, while the Pearl River Delta is between the post-industrialization stage and post-industrialization stage, and the non-Pearl River Delta region is still in the mid-industrialization stage. In the face of the lack of synergistic added value and technology diffusion effects in the industrial chain, the Greater Bay Area needs to optimize its industrial structure in order to strengthen innovation cooperation, achieve complementary advantages and create an "innovation ecosystem". It is necessary to build the Guangzhou–Shenzhen–Hong Kong and Guangzhou-Shenzhen-Australia Science and Technology Corridors to a high level and to co-ordinate the use of quality science and technology innovation resources from Hong Kong and Macao to create a secure bay area for industrial chains. The development of high–tech industries; the integration of the existing advantageous industries of Hong Kong, Macao; and the leading innovation cities in the Pearl River Delta (Shenzhen, Guangzhou, etc.) form a collaborative division of labor patterns among the cities around the dual-chain integration model of innovation and industrial chains and create an innovation economic belt suitable for the development of the GBA. The optimization of the industrial structure of the GBA and the improvement of the construction of the whole industrial chain are important engines for the construction of a world-class innovation-leading Bay Area posture.

6.2.4. Establishing a Collaborative Innovation Network and Building a Synergistic Development Bay Area

The building of collaborative innovation networks in the Bay Area city clusters is a great boost to the overall flow of innovation resources in the region. Strengthening the construction of collaborative innovation networks in the GBA and neighboring cities and accelerating the construction of infrastructure networks with complementary functions to create a network system with an adequate flow of resource elements would make the spatial connection of the city cluster as a whole even closer. It is necessary to improve the mechanism for sharing innovation platform resources and open services, promote a market-based model for efficient sharing of science and technology innovation platform resources and industrial cooperation resources within the Greater Bay Area, and establish a unified project pool and talent pool. At the same time, the existing basic innovation resources of the GBA will be brought into full play, and global innovation resources will be introduced through the International Innovation Demonstration Zone to build a world-class innovative Bay Area, enhance the overall innovation strength of the region and raise the status of the Greater Bay Area in global innovation.

Author Contributions: Conceptualization, W.Y. and Y.H.; methodology, W.Y. and Y.H.; software, Y.H. and L.C.; formal analysis, Y.H. and L.C.; resources, W.Y. and Y.H.; data curation, Y.H. and L.C.; writing—original draft preparation, Y.H. and W.Y.; writing—review and editing, Y.H. and L.C.; supervision, Y.H. and W.Y.; project administration, Y.H. and W.Y.; funding acquisition, W.Y. All authors have read and agreed to the published version of the manuscript.

Funding: This work was supported by The National Social Science Fund of China [grant number 20BGL299].

Institutional Review Board Statement: Not applicable.

Informed Consent Statement: Not applicable.

Data Availability Statement: The data supporting the findings of the article is available in the China City Statistical Yearbook 2010–2020, Guangdong Statistical Yearbook and Guangdong Science and Technology Yearbook. For Hong Kong, the data are from Hong Kong Statistical Yearbook 2020, Statistics on Innovation Activities in Hong Kong 2019 and Statistics of Hong Kong Trade Development Council 2019, respectively, of the Census and Statistics Department of the Hong Kong Special Administrative Region. The data for Macau are derived from the Macao Statistical Yearbook 2020, Macao Economic Development Report for the fourth quarter of 2019, and the time-series database of The Macao Statistical Survey Bureau.

Acknowledgments: We would like to express our gratitude to all those who helped us during the writing of this article. Our deepest gratitude goes first and foremost to Lu Wang, who is from the Hunan University of Science and Technology in China, for his constant encouragement and guidance.

Conflicts of Interest: The authors declare no conflict of interest.

References

1. Xu, Y.Q.; Zeng, G.; Wang, Q.Y. The evolution and optimization strategy of collaborative innovation network pattern in Yangtze River Delta urban agglomerations. *Econ. Geography.* **2018**, *38*, 133–140.
2. CPC Central Committee, State Council. Outline of the Development Plan for the Guangdong–Hong Kong–Macao Greater Bay Area [EB/OL]. 18 February 2019. Available online: http://www.gov.cn/gongbao/content/2019/content_5370836.htm (accessed on 25 May 2021).
3. Sharma, S.; Thomas, V.J. Inter-country R&D efficiency analysis: An application of data envelopment analysis. *Scientometrics* **2008**, *76*, 483–501.
4. Fritsch, M.; Slavtchev, V. Determinants of the efficiency of regional innovation systems. *Reg. Stud.* **2011**, *45*, 905–918. [CrossRef]
5. Batz, F.J.; Janssen, W.; Peters, K.J. Predicting technology adoption to improve research priority–Setting. *Agric. Econ.* **2003**, *28*, 151–164.
6. Rametsteiner, E.; Hansen, E.; Niskanen, A. Introduction to the special issue on innovation and entrepreneurship in the forest sector. *For. Policy Econ.* **2006**, *8*, 669–673. [CrossRef]
7. Comin, D.; Hobijn, B.; Rovito, E. A new approach to measuring technology with an application to the shape of the diffusion curves. *Technol. Transf.* **2008**, *33*, 187–207. [CrossRef]
8. Ghebremichael, A.; Potter-Witter, K. Effects of tax incentives on long-run capital formation and total factor productivity growth in the Canadian sawmilling industry. *For. Policy Econ.* **2009**, *11*, 85–94. [CrossRef]
9. Mardani, A.; Zavadskas, E.K.; Streimikiene, D.; Jusoh, A.; Khoshnoudi, M. A comprehensive review of data envelopment analysis (DEA) approach in energy efficiency. *Renew. Sustain. Energy Rev.* **2017**, *70*, 1298–1322. [CrossRef]
10. Nasierowski, W.; Arcelus, F.J. On the efficiency of national innovation systems. *Socio-Econ. Plan. Sci.* **2003**, *37*, 215–234. [CrossRef]
11. Guo, D.B.; Lei, J.X.; Zhang, J.F.; Peng, B. National Innovation System efficiency and its influencing factors: A two-step analysis based on DEA-Tobit. *J. Tsinghua Univ. (Philos. Soc. Sci.)* **2012**, *27*, 142–150.
12. Shi, X.S.; Zhao, S.D.; Wu, F.X. Research on regional innovation efficiency and its spatial difference in China. *J. Quant. Tech. Econo.* **2009**, *26*, 45–55.
13. Lan, Q.X.; Han, J.; Li, M.P. Innovation efficiency of Chinese manufacturing industry from 2003 to 2008–Based on SFA method. In Proceedings of the 2010 IEEE 17th International Conference on Industrial Engineering and Engineering Management, Xiamen, China, 29–31 October 2010; Volume 1, pp. 524–528.
14. Wang, R.Q.; Peng, L.T.; Jiang, N. Regional innovation efficiency and its influencing factors based on SFA and Malmquist Method. *Sci. Sci. Manag. S. T.* **2010**, *31*, 121–128.
15. Cullmann, A.; Schmidt-Ehmcke, J.; Zloczysti, P. Innovation, R&D Efficiency and the Impact of the Regulatory Environment: A Two-Stage Semi-Parametric DEA Approach. *Discuss. Pap. DIW Berl.* **2009**, *8*, 36–42.
16. Sun, H.B.; Xiang, G. Innovation efficiency evaluation analysis of urban innovation system based on DEA. *Sci. Technol. Prog. Countermeas.* **2011**, *28*, 130–135.
17. Guo, L.; Liu, Z.Y.; Zhou, Z.X. Research on regional technological innovation efficiency evaluation based on DEA cross efficiency model. *Sci. Sci. Manag. Res.* **2011**, *32*, 138–143.
18. Maudos, J.; Pastor, J.M.; Serrano, L. Total factor productivity measurement and human capital in OECD countries. *Econ. Lett.* **1999**, *63*, 39–44. [CrossRef]
19. Gui, H.B. Spatial econometric analysis of innovation efficiency and its influencing factors of high-tech industries in China. *Econ. Geogr.* **2014**, *34*, 100–107.
20. Fang, J.W.; Chiu, Y.H. Research on Innovation Efficiency and Technology Gap in China Economic Development. *Asia-Pac. J. Oper. Res.* **2017**, *34*, 1750005. [CrossRef]
21. Fan, H.; Zhou, D.Q. The evolution and influencing factors of scientific and technological innovation efficiency in Chinese provinces. *Sci. Res. Manag.* **2012**, *33*, 10–18.
22. Sen, N. Innovation chain and CSIR. *Curr. Sci.* **2003**, *85*, 570–574.
23. Dai, M.; Liang, Y.M.; Dai, Y. Research on innovation chain Deconstruction. *Sci. Technol. Prog. Countermeas.* **2009**, *26*, 157–160.
24. Ge, P.F.; Han, Y.N.; Wu, X.X. The measurement and evaluation of the coupling coordination between innovation and economic development in China. *J. Quant. Tech. Econ.* **2020**, *37*, 101–117.
25. Peng, R.Z.; Li, P. China's industrial innovation: Process, efficiency and model: Based on the data of large and medium-sized industrial enterprises from 2001 to 2008. *Ind. Econ. Res.* **2011**, *2*, 1–8.
26. Yu, Y.Z. Technology Innovation efficiency and its influencing factors: A two-stage analysis based on the perspective of the value chain. *Econ. Sci.* **2009**, *4*, 62–74.
27. Yu, Y.Z.; Liu, D.Y. Research on the improvement path of regional innovation efficiency in China from the perspective of the innovation value chain. *Sci. Res. Manag.* **2014**, *35*, 27–37.
28. Zhao, S. Analysis of Urban Innovation Efficiency in Guangdong-Hong Kong-Macao Greater Bay Area. *Open J. Bus. Manag.* **2018**, *6*, 539–550. [CrossRef]
29. Sheng, Y.; Zhao, J.; Zhang, X.; Song, J.; Miao, Y. Innovation efficiency and spatial spillover in urban agglomerations: A case of the Beijing-Tianjin-Hebei, the Yangtze River Delta, and the Pearl River Delta. *Growth Chang.* **2019**, *50*, 1280–1310. [CrossRef]

30. Ye, T.L.; Li, L.; Wang, X.Y. A comparative study on innovation efficiency and influencing factors of three urban agglomerations in eastern China. *Sci. Technol. Prog. Countermeas.* **2021**, *27*, 1–10.
31. Xu, L. Evaluation of innovation efficiency in Yangtze River Delta urban agglomerations. *Stat. Decis.* **2021**, *37*, 84–87.
32. Tan, H.; Xie, L. Guangdong-Hong Kong-Macao Greater Bay Area: Shaping the International Technology and Innovation Center. *China Open. J.* **2019**, *2*, 61–66.
33. Chen, Y.; Lin, Z.H. Grey correlation analysis and innovation of coordination mechanism of inter-city industrial coordination in Guangdong-Hong Kong-Macao Greater Bay Area. *J. Guangdong Univ. Financ. Econ.* **2018**, *33*, 89–97.
34. Tan, C.L.; Huang, L.J. Collaborative innovation links and influencing factors among cities in the Guangdong-Hong Kong-Macao Greater Bay Area. *J. Beijing Univ. Technol. (Soc. Sci.)* **2020**, *20*, 56–65.
35. Han, Z.Z.; Zhu, F.Y.; Li, Z.Y. Research on innovation efficiency in the Guangdong-Hong Kong-Macao greater bay area. *Stat. Decis.* **2020**, *36*, 80–85.
36. Zhang, P.; Li, L.X.; Zeng, Y.Q. Research on the efficiency evaluation of science and technology innovation in Guangdong-Hong Kong-Macao Greater Bay Area-based on DEA-Malmquist Index. *J. Ind. Tech. Econ.* **2021**, *40*, 12–17.
37. Hansen, M.T.; Birkinshaw, J. The Innovation Value Chain. *Harv. Bus. Rev.* **2007**, *85*, 121–135.
38. Roper, S.; Du, J.; Love, J.H. Modelling the innovation value chain. *Res. Policy* **2008**, *37*, 961–977. [CrossRef]
39. Porter, M.E.; Kramer, M.R. Creating Shared Value. *Harv. Bus. Rev.* **2011**, *89*, 62–77.
40. Lee, J.; Gereffi, G.; Beauvais, J. Global value chains and agrifood standards: Challenges and possibilities for smallholders in developing countries. *Proc. Natl. Acad. Sci. USA* **2012**, *109*, 12326–12331. [CrossRef]
41. Lin, M. Research on knowledge transfer, Innovation Chain and innovation policy. *Beijing Econ. Sci. Press.* **2018**, 121–122.
42. Wang, W.G.; Zhang, Z.Y.; Hou, J.L. Innovation value chain and its structure: A theoretical framework. *Sci. Technol. Prog. Strategy* **2019**, *36*, 36–43.
43. Charnes, A.; Cooper, W.W.; Rhodes, E. Measuring the efficiency of decision-making units. *Eur. J. Oper. Res.* **1978**, *2*, 429–444. [CrossRef]
44. Freid, H.O.; Lovell, C.A.K.; Schmidt, S.S.; Yaisawarng, S. Accounting for Environmental Effects and Statistical Noise in Data Envelopment Analysis. *J. Product. Anal.* **2002**, *17*, 157–174. [CrossRef]
45. Bai, J.H.; Li, J. Government R&D funding and firm innovation: An empirical analysis from the perspective of efficiency. *Financ. Res.* **2011**, *6*, 181–193.
46. Li, X.B. An empirical analysis of regional innovation capacity change in China: Based on the innovation system perspective. *Manag. World* **2007**, *12*, 18–30.
47. Furman, J.L.; Porter, M.E.; Stern, S. The determinants of national innovative capacity. *Res. Policy* **2002**, *31*, 899–933. [CrossRef]
48. Arrow, K. Economic Welfare and the Allocation of Resources for Invention. *Nber Chapters* **1972**, *12*, 609–626.
49. Niu, Z.D.; Zhang, Q.X.; Wang, W. Empirical analysis of the impact of FDI on the independent innovation capability of China's high-tech industry. *Sci. Technol. Prog. Countermeas.* **2011**, *28*, 51–55.
50. Bruce, A.K. The influence of R&D expend it ureson new firm formation and economic growth. *Small Bus. Econ.* **1993**, *13*, 275–317.
51. Zhang, Z. Blockchain + Guangdong-Hong Kong-Macao greater bay area collaborative innovation community construction analysis. *Acad. Forum* **2020**, *43*, 42–49.
52. Ning, J.; Liu, Y.; Chen, Y. "Two-chain integration" and innovation system optimization in Guangdong-Hong Kong-Macao Greater Bay Area. *Reg. Econ. Rev.* **2021**, *1*, 97–104.
53. Fan, X.; Liu, W. Regional innovation collaborative governance based on innovation chain: A case study of Guangdong-Hong Kong-Macao Greater Bay Area. *Contemp. Econ. Manag.* **2020**, *42*, 54–60.
54. Tan, Y.H.; Cao, X.Y. Research on collaborative innovation of science and technology in the Guangdong-Hong Kong-Macao Greater Bay Area. *J. Cent. China Norm. Univ. (Nat. Sci. Ed.)* **2019**, *53*, 255–262.
55. Wang, Y.; Yang, Y.; Liu, Y. Global vision and theoretical model of Guangdong-Hong Kong-Macao Greater Bay Area building international Science and Technology Innovation Center. *Geogr. Res.* **2020**, *39*, 1958–1971.
56. Ye, Y.Y.; Wang, J.S.; Wu, K.M.; Tu, Z.W.; Wang, Y.; He, S.Y.; Liu, Z.Q. Strategic thinking on the construction of Guangdong-Hong Kong-Macao greater bay area as an international science and technology innovation center. *Trop. Geogr.* **2020**, *40*, 27–39.
57. Ye, L.; Song, X.Z. Regional collaborative innovation system of Guangdong-Hong Kong-Macao greater bay area: Based on the perspective of planning outline. *Exec. Forum* **2019**, *26*, 87–94.
58. Chen, X.Q.; Zhao, D.X.; Lian, X.S. Research on the development of collaborative innovation of science and Technology in Guangdong-Hong Kong-Macao Greater Bay Area: Based on the perspective of factor collaboration. *Sci. Technol. Manag. Res.* **2020**, *40*, 36–42.

 *land*

Article

An Analysis of an Area's Vulnerability to the Emergence of Land-Use Conflicts

Iwona Cieślak and Andrzej Biłozor *

Department of Socio-Economic Geography, Faculty of Geoengineering, University of Warmia and Mazury in Olsztyn, Prawocheńskiego 15, 10-720 Olsztyn, Poland; isidor@uwm.edu.pl
* Correspondence: abilozor@uwm.edu.pl

Abstract: The optimization of space is the priority goal of spatial planning. Spatial planning policies have numerous objectives, including the prevention of land-use conflicts. Conflicts arise whenever two entities have contradictory expectations regarding the surrounding space. In the process of spatial development, humans impart new characteristics to space, which, under specific circumstances, can give rise to land-use conflict. The elements of space that are particularly vulnerable to conflict include boundary points, property boundaries, density of development, or the shared use of infrastructure. The main aim of this study was to develop a procedure for evaluating the risk of land-use conflict based on the characteristic attributes of space. The proposed procedure for assessing the accumulation of conflict-generating traits in space was developed with the use of databases, GIS tools, and statistical data processing methods.

Keywords: land management; land-use conflicts; components of space; spatial analysis; GIS tools

check for updates

Citation: Cieślak, I.; Biłozor, A. An Analysis of an Area's Vulnerability to the Emergence of Land-Use Conflicts. *Land* **2021**, *10*, 1173. https://doi.org/10.3390/land10111173

Academic Editor: Robert Gilmore Pontius

Received: 21 October 2021
Accepted: 28 October 2021
Published: 1 November 2021

Publisher's Note: MDPI stays neutral with regard to jurisdictional claims in published maps and institutional affiliations.

1. Introduction

Conflict is a widespread phenomenon that always involves at least two parties, where one party attempts to derive personal gain at the other party's expense and without its explicit consent [1]. Conflict is a concept that can be examined in various contexts, including as a social (social approach) or an economic (functional approach) process, where one party attempts to maximize its own gain by eliminating or dominating over the other party to the conflict [2].

The potential sources of conflict have to be determined to identify the existing or potential conflicts. The main categories of conflict have been systematized by Bogetoft and Pruzan [3] who posited that conflicts can arise from one or several different factors. These factors can be divided into four main groups:

1. Value system factors:

    1.1. The parties have different values and aims;
    1.2. The parties take different measures to represent their aims;
    1.3. The parties have different preference hierarchies (for example, by applying different weights for different solutions);

2. Effect distribution factors:

    2.1. The parties can incur different costs and derive various indirect benefits associated with the anticipated effects of their actions;
    2.2. The breakdown of costs and benefits is perceived as uneven and unfair;

3. Uncertainty factors:

    3.1. The parties cannot reach agreement on the probable effects of actions;
    3.2. The parties are not certain of the effects of their actions;
    3.3. The evidence and the rationale behind the effects of their actions may be insufficient or incomprehensible;

3.4.   The parties may have doubts about the associations between the effects of an action and other actions;

4.   Process factors:

4.1.   The parties find it difficult to exchange information about their values, aims, criteria, preferences, and expectations.

All of the above factors have a common denominator: the actions undertaken by the parties are contradictory, and each action is localized in space. A big proportion of conflicts are directly or indirectly associated with space. Conflicts that are directly related to space are referred to as spatial conflicts [4].

Spatial conflict is an inseparable element of spatial development and human existence [5]. The significance and progression of land-use conflicts are irreversibly linked to the use of space, and the quality of life is influenced by the quality of the surrounding space [6–8]. Profit maximization has been the driving force behind land-use conflict ever since humans discovered that land can be a critical source of capital and wealth [9,10]. The struggle for land took on various forms and progressed on a different scale, and due to the nature and the object of that struggle, it became known as spatial conflict or land-use conflict [11,12].

Spatial conflicts are widely encountered, and they can occur at any time and in any location. Changes in the hierarchy of needs and needs satisfaction, as well as individual traits (greed, envy, etc.) can promote the belief that the existing space is insufficient. This conviction increases the value of space, and it is the main cause of spatial conflict.

Most spatial conflicts have highly negative consequences for the economy, society, space, and the environment [13]. The above applies particularly to developing and transition countries, where the real estate market is weakly developed and where illegal practices involving land create an opportunity for rapid wealth accumulation [14]. In this case, land ceases to be a public good, and land ownership is often transferred to a small and wealthy social group. However, the extent to which the parties to a conflict manifest their views is correlated with the level of democracy in a given country [15]. The causes, course, and consequences of land-use conflict should be analyzed to develop effective spatial planning tools [16].

In theory, land-use conflict does not differ from other types of conflict, but space is always the subject matter of such disputes [17]. Land-use planning shapes our living and working environment; the integrity and sophistication of the overall planning process govern the sustainability of proposed spatial allocations in urban and rural areas [18]. Spatial conflict is usually driven by various land-use options, contradictory interests, and goals, including those pertaining to the use of natural resources [19–23].

Spatial conflicts can have different causes, and they are rarely homogeneous. In other words, spatial conflict is not driven by a single factor, and it is a complex and multidimensional phenomenon. This complexity can be attributed not only to the physical characteristics of space, such as size, diversity, and form, but also to changes in space and subjective perceptions of space [24]. Humans have different needs and expectations that are satisfied in space [25]. Land is a common good that is essential for human activity. Due to the diversity of human needs and the limited supply of land, land has been the source of conflict throughout human history [26]. Owing to the complexity of space, various spatial relationships can be identified, but not all of them can be described in detail. However, they can be generalized [27]. The main types of conflict relationships are associated with environmental features, economic activities, and social expectations [28]. The conflict relationships based on these three main pillars of human existence are presented in Figure 1.

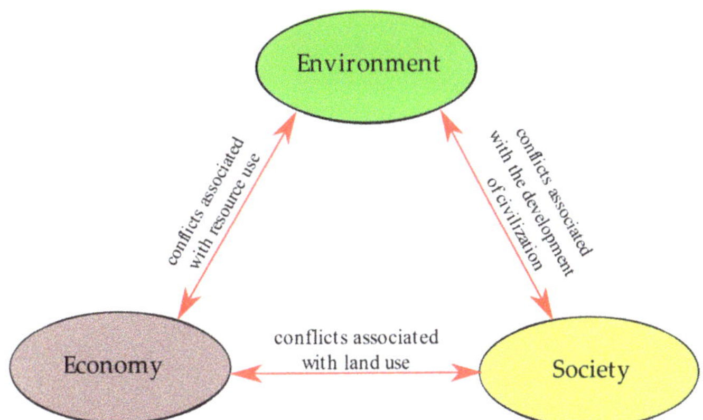

Figure 1. Conflicts arising from relationships in space. Source: Own compilation based on [29].

Conflicts between the economy and the environment relate to the use of natural resources and the extent to which the environment is transformed by human activity [30]. Mineral resource mining and the acquisition of land for development projects are prime examples of the above. The expectations associated with environmental protection stand in opposition to the maximization of profit derived from resource use.

Conflicts between the society and the environment have a similar context, but they focus on the quality of human life. The expected comfort and convenience of the modern lifestyle, such as access to public infrastructure or pest-free households, run contrary to environmental protection.

Conflicts between the economy and the society generally result from contradictory expectations regarding access to public infrastructure and natural resources between the consumers, investors, and public administration bodies that analyze the economic consequences of various uses of space. Disputes surrounding the location of land development projects are an example of such conflicts [31]. These types of conflict gave rise to NIMBY (not in my backyard) syndrome, where residents generally agree that a facility producing goods or services is needed, but oppose its construction in their neighborhood [32]. Conflicts can also erupt in a reverse scenario, when local communities advocate for the construction of certain types of infrastructure, but are faced with opposition from supralocal groups that derive benefits from the use of space in its existing form [33].

Conflicts waged on a wider scale are referred to as functional-economic conflicts [34], where factors leading to the aggregation and dispersion of various land-use functions occur simultaneously and undermine the welfare of local communities [6,35]. These factors are usually associated with considerable disproportions in the development of adjacent areas. Selected functions are excessively concentrated in some areas (such as cities), but they are lacking in peripheral and underdeveloped areas. Globalization contributes to the emergence of spatial conflict in the social-economic dimension. Urban sprawl increases the prices of agricultural land, leads to conflict over farmland protection and, consequently, drives conflict over land ownership [5].

The presented classification of spatial conflicts is not exhaustive or sufficiently detailed. Many conflicts combine the features of all of the above categories, where contradictory expectations regarding land use are influenced by the object of the dispute and the involved parties [36].

Spatial conflicts can emerge over land ownership and land-use rights, but also over land itself, the neighborhood, or the use of private or public land [37]. Regardless of the above, space is always the essence of conflict. The object of conflict can also differ in scale or magnitude, and it can include buildings, undeveloped land, or property boundaries. Entire regions and countries can be embroiled in spatial conflict.

In view of the above, the object of spatial conflict can be classified based on its location, scope, and specific claims. When these criteria are taken into consideration, the objects of spatial conflicts can be divided into the following groups:

- space in the geodetic sense:

    cadastral plot;
    building;
    property boundaries.

- land-use rights:

    ownership, perpetual usufruct, co-ownership;
    other rights to derive income from land: easement, lease, tenure, etc.;
    specific land-use (land management) rights.

- space in the geographic sense:

    immediate space (own—private);
    local space (neighborhood, municipality);
    regional and ethnic space;
    national space;
    global space.

The object of spatial conflict should be classified to identify the scope of the dispute and the involved parties. The resulting classification also contributes to the selection of the optimal conflict resolution tools [38].

Many land-use conflicts are caused by different expectations regarding the attributes of space [39]. The above implies that space cannot be evaluated unambiguously [40]. However, fragments of space with a high accumulation of potentially conflict-generating attributes can be identified [41]. Most spatial conflicts are caused by social factors as conflicts are driven by humans and do not exist without humans [42]. Conflicts are often fueled by personal traits which are difficult to assess and are rarely analyzed in studies on spatial conflicts. However, the risk of social conflict can also be exacerbated by the characteristic traits of entire social groups [43]. These include:

- high population density;
- major cultural and political differences among local community members;
- number of households;
- differences in educational attainment.

The above factors strongly influence the quality of life and individual expectations towards space. In less affluent societies, land-use strategies are implemented to generate economic benefits and improve the standard of living, and these goals are accomplished at the expense of other attributes, such as the quality of the natural environment [44]. Social attributes that contribute to spatial conflict can also influence economic factors, including:

- land prices;
- unemployment;
- differences in economic status;
- sources of income (different types of employment).

These differences are often responsible for various expectations towards land use and land management. In many cases, different expectations can be the secondary cause of local and personal land-use conflict [45].

The risk of spatial conflict increases in communities and regions with diverse social characteristics and contradictory expectations towards land management. These differences are manifested by:

- real estate prices;
- fragmentation of ownership (fragmentation of cadastral plots);
- development density;
- building height;

- length of property boundaries;
- diverse land-use types in the immediate neighborhood;
- large area of ecologically valuable land.

The specificity of land-use conflict has to be explored to expand our understanding of the scope and potential consequences of such conflicts, which could play a key role in rational and sustainable management of space [46–48].

The attributes of space where land-use conflicts occur usually have a local character. Land division, changes in land-use, and urbanization take place locally, and spatial conflicts are generally intensified by anthropogenic factors [49]. Some of these attributes can be identified and evaluated intuitively, whereas others become apparent only when a conflict erupts or when the causes of the conflict are analyzed in detail [50].

One of the goals of spatial planning is to limit the negative impact of conflict-generating attributes on human existence [51]. Such attributes have to be identified, eliminated, or minimized [52]. However, conflict-generating factors can be reliably assessed only if they can be unambiguously identified and described in the entire area under analysis. The above can be achieved with the use of GIS databases which are characterized by extensive thematic content and broad spatial coverage [53,54].

The characteristic attributes of space that incite spatial conflict usually evolve at the local level [55]. Land division, changes in land use, and urbanization are processes that occur locally, and spatial conflicts are fueled mostly by anthropogenic factors.

As previously mentioned, space has many attributes that can generate land-use conflict. Some of these attributes are recognized and evaluated intuitively, whereas others are identified only when a conflict erupts or after its causes have been thoroughly analyzed [49]. The authorities responsible for land management have to limit the negative effects of these factors on human life [56]. Therefore, conflict-generating factors need to be identified, eliminated, or minimized. However, such factors can be determined, and their conflict-generating potential can be evaluated only when these factors are unambiguously identified and described during the entire research process. Not all factors that increase the risk of spatial conflict are described; therefore, they are not listed in the databases that are accessible to analysts [57,58]. Undoubtedly, such factors can be described in individual cases for the needs of specific analyses [59]. However, this is a highly laborious process that is not always cost-effective. The identified factors also tend to have limited spatial coverage. Therefore, the use of the existing databases appears to be a much simpler and sensible solution [60].

Local space is usually described in considerable detail. Legal regulations and, increasingly often, market processes necessitate the creation of geospatial databases with different content and significance. Geospatial databases differ in the homogeneity, validity, and reliability of the accumulated information [61]. Databases developed by centers for geodetic and cartographic documentation appear to be the most robust sources of homogeneous, valid, and reliable data.

In view of the above, the main aim of this study was to develop a procedure for evaluating an area's vulnerability to land-use conflict based on the spatial attributes described by geodetic and cartographic databases and with the use of GIS tools.

2. Materials and Methods

2.1. Study Area and Data

For the needs of this study, the attributes of space that can potentially contribute to the risk of land-use conflict were identified based on an analysis of the existing databases and the relevant literature. Spatial attributes were selected for the study based on the extent to which they can be identified in space. The studied area was divided into comparative units for the needs of delimitation based on the results of the conducted analysis. Comparative units were created within the cadastral districts of the examined municipalities. This approach was adopted to ensure the cohesiveness of research stages, to minimize susceptibility to spatial conflict, and, potentially, to integrate the results with the findings of social studies [62]. Cadastral districts are often associated with settlement units (in particular in rural areas) or city districts, which implies that they can be linked with local communities. Therefore, the extent to which the characteristic traits of local communities contribute to the risk of spatial conflict can be analyzed in greater detail.

Spatial features that can contribute to the intensification of land-use conflicts were identified with the use of GIS data and based on a review of the literature [63,64]. The analyzed traits were selected based on the extent to which they can be identified and based on the results of a survey involving fifteen teams of real estate and spatial planning experts, including researchers and practitioners such as land surveyors and real estate agents. In each of the fifteen teams, the experts and the moderator (leading researcher) discussed the extent to which the identified attributes contribute to spatial conflict. The results of these discussions were used to prepare 15 questionnaires and determine the conflict-generating potential of each attribute. A questionnaire survey was conducted between 20 May and 25 September 2020 in organizations employing real estate and spatial planning experts. This approach was adopted due to the complex nature of the undertaken research. Direct communication with the experts generated valuable insights about the studied topic. The two-tiered survey procedure also enhanced the objectivity of the results.

The study area was the rural municipality of Purda in the Olsztyn county, Region of Warmia and Mazury in Poland. The Region of Warmia and Mazury has a rather unique settlement structure. The region abounds in natural resources, and agriculture has long been the main source of income for the local population [65]. The rapid development of Olsztyn county drives growth in the surrounding municipalities, including Purda. The Purda municipality has an area of more than 300 km^2, with a prevalence of agricultural land and forests. Due to local specificity and the proximity of a large urban center, Purda is susceptible to spatial conflict in both the social (progressive exploitation of natural resources resulting from local population growth) and the economic dimension (growing number of land development projects). The influx of new residents whose land-use preferences differ from those of the local population also drives conflict. These problems are typically encountered at the rural–urban interface. The location of Purda municipality is presented in Figure 2.

Purda is undergoing rapid urbanization, and it is an important source of land reserves for the spatial development of Olsztyn, the capital city of the Region of Warmia and Mazury. To facilitate a cohesive evaluation and classification of results, the municipality was divided into territorial units corresponding to cadastral districts. The availability of cohesive information regarding the intensity of the evaluated attributes in the analyzed units and the calculation of indices relating to every attribute played an important role in the selection process.

The analyzed municipality was divided into comparative units within the cadastral districts of Purda municipality. The aim of this procedure was to guarantee the consistency of the adopted measures for investigating possible triggers of land-use conflict and to consolidate our findings with research into social causes of conflict [66]. Cadastral districts are often equated with settlement units (in particular in rural areas) or city districts, which implies that they are closely linked to local communities [67]. The above assumption can be taken into account in spatial analyses to determine the extent to which the characteristic features of local communities can incite or exacerbate land-use conflict. The attributes were selected for analysis based on two considerations. The first consideration was the list of attributes that were identified by the surveyed experts. Each of the fifteen surveyed expert teams developed a list of attributes that could potentially contribute to spatial conflict. The second consideration was the extent to which the identified attributes could be measured with the use of the existing geospatial databases. Not all attributes can be evaluated in this approach. Databases characterizing local communities are not available. A total of 12 features that are described in the geospatial database were identified and assessed in 33 districts of Purda municipality.

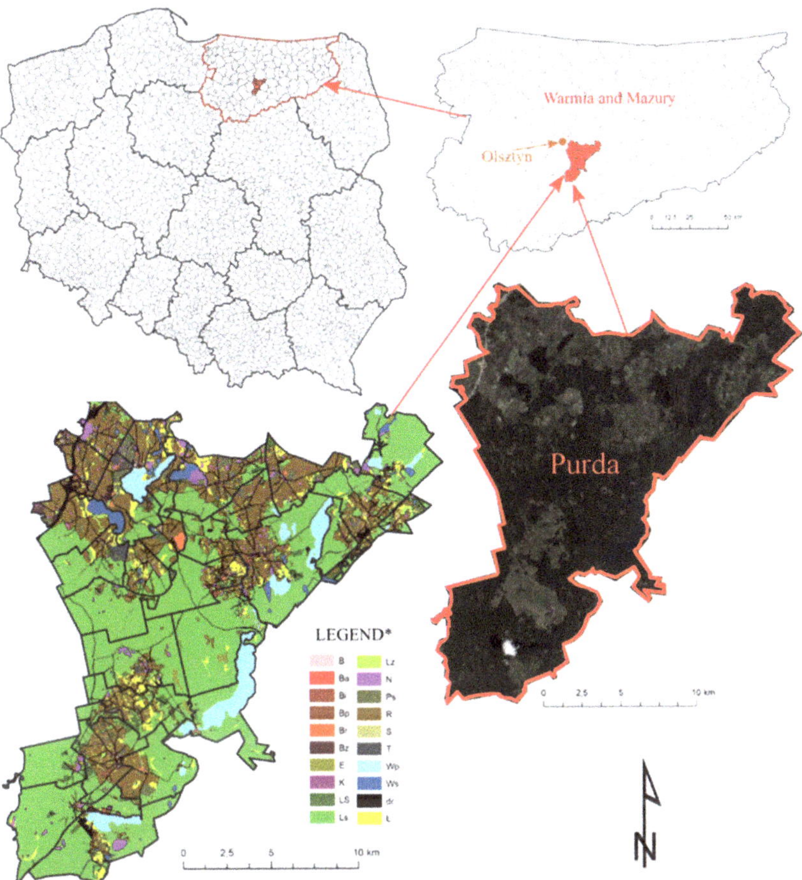

Figure 2. Location of Purda municipality and the local land-use pattern. * B—residential areas; Ba—industrial areas; Bi—other built-up areas; Bp—urbanized areas that are not built-up or are under construction; Br—developed agricultural land; Bz—recreational areas; E—ecological sites; K—mining sites; Ls—forests; Lz—areas with tree and shrub cover; N—fallow land; Ps—permanent pastures; R—arable land; S—orchards; T—transport routes; Wp—flowing waters; Ws—standing waters; dr—roads; and Ł—permanent meadows.

Most conflict-generating features have anthropogenic origin. Land-use conflicts can also be driven by factors that are not always associated with human activity, including topography and various types of land cover such as water bodies and forests. However, human activities exert a considerable influence on these attributes of space [68,69].

The significance of the identified spatial features was determined. The extent to which the analyzed attributes contribute to land-use conflict was described with the use of dedicated measures or indices [70]. These indices and the relevant measurement methods are described below:

Length of cadastral plot boundaries (**A.1**)—expressed by the ratio of the total length of all cadastral plot boundaries to the area of the cadastral district.

Complex boundaries (**A.2**)—expressed by the ratio of the number of boundary points in cadastral districts to the area of the cadastral district.

Area of cadastral plots (**A.3**)—expressed by the average area of a cadastral plot in a cadastral district.

Density of cadastral plots (**A.4**)—expressed by the ratio of the number of cadastral plots to the area of the cadastral district.

Development density (**A.5**)—expressed by the ratio of built-up areas in a cadastral district to the area of that district.

Land-use homogeneity (**A.6**)—expressed by the land-use homogeneity index SJ_i [71]:

$$SJ_i = \sum_{z=1}^{u} J_{iz}^2 \tag{1}$$

where: J_{iz} is the proportion of a given land-use type (z) in the total area of cadastral district i and u is the number of land-use types in cadastral district.

Technical infrastructure—the availability of technical infrastructure was evaluated in three categories: roads (**A.7**), power grid (**A.8**), and the water supply network (**A.9**). All networks and grids were evaluated with the same key. Cadastral districts were assessed based on the proportion of plots without access to each of the three types of infrastructure in the total number of plots in that district. The availability of roads was defined as the presence of a road in the plot's immediate vicinity, whereas the availability of the remaining infrastructure was defined as the presence of the relevant utilities at a distance of up to 100 m from the plot. The studied municipality was inventoried based on the above principles, and the results are presented in Figure 3.

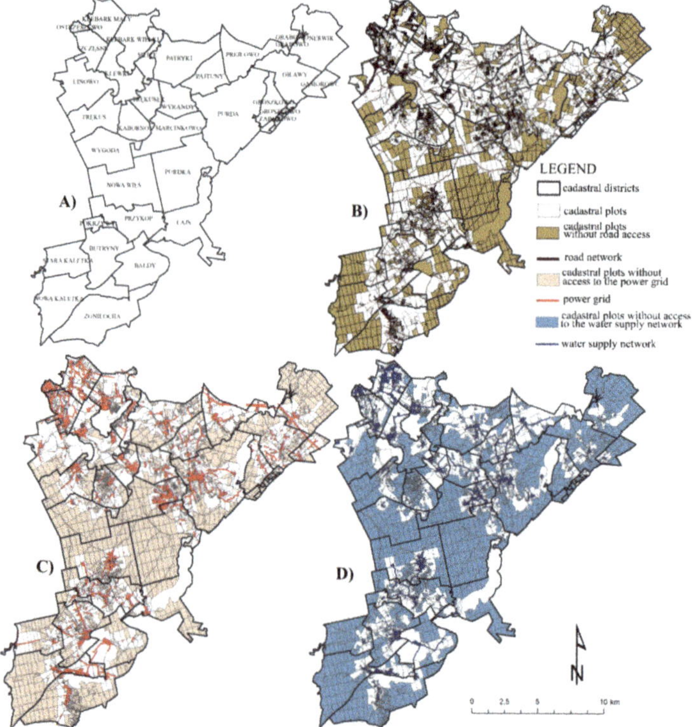

Figure 3. Division of Purda municipality into cadastral districts (**A**); Cadastral plots without access to: roads (**B**), power grid (**C**), and water supply network (**D**). Source: own elaboration based on data from the County Centre for Geodetic and Cartographic Documentation (PODGiK) data.

Topography (**E.1**)—topographic features were determined based on the maximum terrain curvature in the cadastral district. Terrain curvature was calculated based digital elevation model data from the Municipal Center for Geodetic and Cartographic Documentation. Digital elevation model data were interpolated (Figure 4A), and terrain curvature was determined at the nodes of a regular grid with a side length of 100 m. The results were used to calculate the range of curvature values in area K in the analyzed cadastral districts. The calculated values were used to evaluate conflict-generating attributes in these districts (Figure 4B).

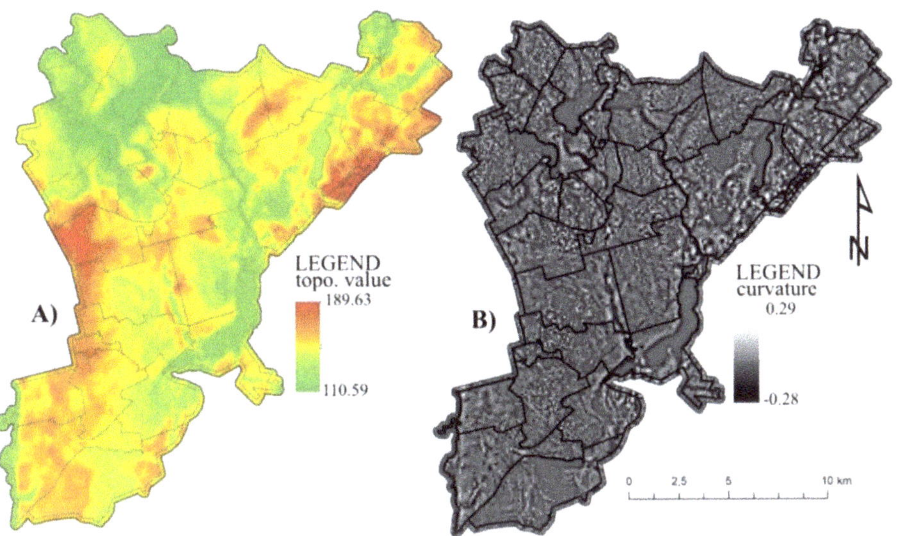

Figure 4. (**A**) Topography of Purda municipality; (**B**) Terrain curvature in Purda municipality. Source: own elaboration based on data from the Municipal Centre for Geodetic and Cartographic Documentation (GODGiK).

Boundaries of areas with natural land cover—this analysis included forests (**E.2**) and surface water bodies (**E.3**). This indicator was expressed by the ratio of the length of boundaries of forests and surface water bodies to the area of forests and water bodies in the analyzed cadastral district.

The results of the evaluation of conflict-generating attributes in the cadastral districts of Purda municipality are presented in Table 1.

2.2. Methodology

The identified attributes had a varied contribution to the risk of land-use conflict. These attributes were weighed to assess their importance. Shannon's entropy method [72] is widely used to determine the weights of various evaluation criteria. The concept of Shannon's entropy plays an important role in information theory, and it is regarded as a general measure of uncertainty. In transportation models, entropy is a measure of the dispersal of trips between the point of origin and the destination [73]. In physics, entropy represents the state of disorder of a system [74]. Entropy associated with an event is also a measure of the event's degree of randomness. The concept of entropy can also be used to measure fuzziness [75]. This method is relatively objective, and it can be effectively deployed in the decision-making process [76]. The entropy weight method is also widely used to determine the weights of criteria and attributes [77–80]. In the discussed method, relative importance is assigned to the analyzed attributes by measuring the existing differences between sets of data [81].

Table 1. Conflict-generating attributes in the cadastral districts of Purda municipality—matrix **M**.

| | | FEATURES * | | | | | | | | | | | |
| | District | A.1 | A.2 | A.3 | A.4 | A.5 | A.6 | A.7 | A.8. | A.9 | E.1 | E.2 | E.3 |
j \ i		1	2	3	4	5	6	7	8.	9	10	11	12
1	Bałdy	63.11	0.73	8.21	0.12	13.89	0.99	0.31	0.34	0.31	0.35	21.80	41.78
2	Butryny	146.50	2.42	1.60	0.63	29.79	0.96	0.20	0.24	0.31	0.29	14.68	58.07
3	Gąsiorowo	161.52	2.46	1.69	0.59	22.18	0.99	0.34	0.23	0.58	0.38	29.47	123.32
4	Giławy	135.24	1.98	2.86	0.35	14.83	0.98	0.32	0.31	0.63	0.49	25.91	82.14
5	Grabowo	3362.40	12.23	0.27	1.88	0.00	1.00	0.17	1.00	1.00	0.05	912.47	0.00
6	Groszkowo	96.01	2.18	3.06	0.33	11.02	0.99	0.21	0.24	0.30	0.53	14.13	132.96
7	Kaborno	191.10	3.71	1.12	0.89	19.74	0.99	0.43	0.43	0.68	0.41	62.98	61.67
8	Klebark Mały	217.56	4.97	0.74	1.35	44.43	0.97	0.54	0.05	0.08	0.26	16.98	55.40
9	Klebark W.	157.61	2.80	1.23	0.81	32.52	0.37	0.43	0.19	0.28	0.31	41.09	7.81
10	Klewki	130.99	3.37	1.15	0.87	46.49	0.87	0.15	0.12	0.25	0.35	38.58	60.22
11	Łajs	67.10	0.99	6.52	0.15	5.22	0.83	0.75	0.36	0.40	0.48	19.10	37.07
12	Linowo	71.23	1.16	5.28	0.19	7.51	0.43	0.17	0.19	0.35	0.35	17.98	40.25
13	Marcinkowo	146.40	3.04	1.25	0.81	33.79	0.95	0.38	0.19	0.20	0.49	23.08	46.17
14	Nerwik	91.04	1.66	4.85	0.21	4.24	0.95	0.45	0.50	0.69	0.40	35.27	118.69
15	Nowa Kaletka	91.33	1.43	2.22	0.45	17.59	0.94	0.23	0.16	0.19	0.33	4.83	56.69
16	Nowa Wieś	117.55	1.78	2.35	0.43	22.06	0.88	0.32	0.30	0.34	0.36	16.83	53.10
17	Ostrzeszewo	508.80	14.56	0.22	4.57	270.18	0.96	0.40	0.02	0.02	0.18	33.02	3.96
18	Pajtuny	172.19	2.95	1.58	0.63	18.53	0.99	0.35	0.37	0.46	0.28	27.35	57.37
19	Patryki	146.83	2.56	1.85	0.54	19.67	0.92	0.33	0.29	0.37	0.37	37.51	55.72
20	Pokrzywy	135.96	1.66	3.07	0.33	21.29	1.00	0.18	0.20	0.23	0.23	0.00	81.95
21	Prejłowo	106.38	1.72	2.10	0.47	27.24	0.94	0.27	0.17	0.21	0.27	23.40	51.56
22	Przykop	141.55	2.38	1.73	0.58	19.08	0.93	0.27	0.20	0.32	0.31	23.78	79.78
23	Purda	121.11	2.52	1.86	0.54	21.25	0.85	0.30	0.19	0.30	0.56	27.82	87.01
24	Purdka	72.54	0.89	6.24	0.16	5.72	0.83	0.44	0.53	0.68	0.46	9.85	46.60
25	Silice	145.20	3.16	1.11	0.90	24.93	0.99	0.34	0.42	0.54	0.32	26.92	38.26
26	Stara Kaletka	91.56	1.13	4.55	0.22	7.63	1.00	0.32	0.34	0.51	0.27	0.79	60.25
27	Szczęsne	224.58	5.77	0.61	1.65	77.83	0.97	0.37	0.08	0.19	0.27	30.64	10.01
28	Tnękus	135.05	2.04	2.18	0.46	16.09	0.87	0.43	0.30	0.44	0.34	15.59	60.23
29	Tnękusek	163.54	2.82	1.22	0.82	124.92	0.38	0.42	0.23	0.28	0.34	14.83	50.59
30	Wygoda	75.35	0.86	6.60	0.15	5.40	0.99	0.25	0.53	0.67	0.40	1.08	48.84
31	Wyrandy	126.34	2.03	2.89	0.35	27.69	1.00	0.30	0.26	0.37	0.29	19.59	107.36
32	Zaborowo	203.20	2.75	2.53	0.40	12.87	1.00	0.13	0.34	0.40	0.40	33.13	74.21
33	Zgniłocha	110.37	1.51	2.50	0.40	9.68	0.95	0.31	0.49	0.53	0.32	37.06	45.52

* all attributes were measured with the use of the described procedure in the analyzed units.

The identified conflict-generating attributes were used to build decision matrix **M** (Table 1), which was normalized with the following equation:

$$n_{ij} = \frac{m_{ij}}{\sum_{i=1}^{k} m_{ij}} \tag{2}$$

where m_{ij} is the results of the evaluation of conflict-generating attributes; i is the successive districts in the total number of districts k; and j is the successive attributes in the total number of attributes z.

The normalized matrix **N** with elements n_{ij} was used to assign weights to conflict-generating attributes (Table 2).

Table 2. Elements of matrix **N** in the analyzed districts.

| | DISTRICT | A.1 | A.2 | A.3 | A.4 | A.5 | A.6 | A.7 | A.8. | A.9 | E.1 | E.2 | E.3 |
j \ i		1	2	3	4	5	6	7	8.	9	10	11	12
1	Bałdy	0.01	0.01	0.00	0.01	0.01	0.03	0.03	0.03	0.02	0.03	0.01	0.02
2	Butryny	0.02	0.02	0.02	0.03	0.03	0.03	0.02	0.02	0.02	0.03	0.01	0.03
3	Gąsiorowo	0.02	0.03	0.02	0.03	0.02	0.03	0.03	0.02	0.04	0.03	0.02	0.06
4	Giławy	0.02	0.02	0.01	0.02	0.01	0.03	0.03	0.03	0.05	0.04	0.02	0.04
5	Grabowo	0.42	0.12	0.15	0.08	0.00	0.03	0.02	0.10	0.08	0.00	0.55	0.00
6	Groszkowo	0.01	0.02	0.01	0.01	0.01	0.03	0.02	0.02	0.02	0.05	0.01	0.07
7	Kaborno	0.02	0.04	0.04	0.04	0.02	0.03	0.04	0.04	0.05	0.04	0.04	0.03

Table 2. *Cont.*

| | DISTRICT | A.1 | A.2 | A.3 | A.4 | A.5 | A.6 | A.7 | A.8. | A.9 | E.1 | E.2 | E.3 |
		i 1	2	3	4	5	6	7	8.	9	10	11	12
8	Klebark Mały	0.03	0.05	0.05	0.06	0.04	0.03	0.05	0.00	0.01	0.02	0.01	0.03
9	Klebark W.	0.02	0.03	0.03	0.03	0.03	0.07	0.04	0.02	0.02	0.03	0.02	0.00
10	Klewki	0.02	0.03	0.03	0.04	0.04	0.03	0.01	0.01	0.02	0.03	0.02	0.03
11	Łajs	0.01	0.01	0.01	0.01	0.01	0.03	0.07	0.04	0.03	0.04	0.01	0.02
12	Linowo	0.01	0.01	0.01	0.01	0.01	0.06	0.02	0.02	0.03	0.03	0.01	0.02
13	Marcinkowo	0.02	0.03	0.03	0.03	0.03	0.03	0.04	0.02	0.02	0.04	0.01	0.02
14	Nerwik	0.01	0.02	0.01	0.01	0.00	0.03	0.04	0.05	0.05	0.03	0.02	0.06
15	Nowa Kaletka	0.01	0.01	0.02	0.02	0.02	0.03	0.02	0.02	0.01	0.03	0.00	0.03
16	Nowa Wieś	0.01	0.02	0.02	0.02	0.02	0.03	0.03	0.03	0.03	0.03	0.01	0.03
17	Ostrzeszewo	0.06	0.15	0.18	0.20	0.26	0.03	0.04	0.00	0.00	0.02	0.02	0.00
18	Pajtuny	0.02	0.03	0.03	0.03	0.02	0.03	0.03	0.04	0.04	0.02	0.02	0.03
19	Patryki	0.02	0.03	0.02	0.02	0.02	0.03	0.03	0.03	0.03	0.03	0.02	0.03
20	Pokrzywy	0.02	0.02	0.01	0.01	0.02	0.03	0.02	0.02	0.02	0.02	0.00	0.04
21	Prejłowo	0.01	0.02	0.02	0.02	0.03	0.03	0.03	0.02	0.02	0.02	0.01	0.03
22	Przykop	0.02	0.02	0.02	0.02	0.02	0.03	0.03	0.02	0.02	0.03	0.01	0.04
23	Purda	0.02	0.03	0.02	0.02	0.02	0.03	0.03	0.02	0.02	0.05	0.02	0.04
24	Purdka	0.01	0.01	0.01	0.01	0.01	0.03	0.04	0.05	0.05	0.04	0.01	0.02
25	Silice	0.02	0.03	0.04	0.04	0.02	0.03	0.03	0.04	0.04	0.03	0.02	0.02
26	Stara Kaletka	0.01	0.01	0.01	0.01	0.01	0.03	0.03	0.03	0.04	0.02	0.00	0.03
27	Szczęsne	0.03	0.06	0.07	0.07	0.08	0.03	0.03	0.01	0.01	0.02	0.02	0.01
28	Trękus	0.02	0.02	0.02	0.02	0.02	0.03	0.04	0.03	0.03	0.03	0.01	0.03
29	Trękusek	0.02	0.03	0.03	0.04	0.12	0.07	0.04	0.02	0.02	0.03	0.01	0.03
30	Wygoda	0.01	0.01	0.01	0.01	0.01	0.03	0.02	0.05	0.05	0.03	0.00	0.03
31	Wyrandy	0.02	0.02	0.01	0.01	0.03	0.03	0.03	0.03	0.03	0.03	0.01	0.06
32	Zaborowo	0.03	0.03	0.02	0.02	0.01	0.03	0.01	0.03	0.03	0.04	0.02	0.04
33	Zgniłocha	0.01	0.02	0.02	0.02	0.01	0.03	0.03	0.05	0.04	0.03	0.02	0.02

The information content of matrix **N** was determined by calculating the entropy value E_j of every conflict-generating attribute with the use of the below formula [82]:

$$E_j = -K \sum_{i=1}^{k} n_{ij} \ln n_{ij} \tag{3}$$

where i is the successive districts in the total number of districts k; j is the successive attributes in the total number of attributes r; and $K = 1/\ln k$ (for $k = 33$ $K = 0.29$) is the constant value which guarantees that the value of E falls within the range of [0;1] (Table 3).

Table 3. Value of E in the analyzed districts.

| | DISTRICT | A.1 | A.2 | A.3 | A.4 | A.5 | A.6 | A.7 | A.8. | A.9 | E.1 | E.2 | E.3 |
| | *j* | 1 | 2 | 3 | 4 | 5 | 6 | 7 | 8 | 9 | 10 | 11 | 12 |
i							$n_{ij} \ln n_{ij}$						
1	Bałdy	−0.04	−0.04	−0.03	−0.03	−0.06	−0.09	−0.10	−0.12	−0.09	−0.11	−0.06	−0.08
2	Butryny	−0.07	−0.09	−0.09	−0.10	−0.10	−0.10	−0.07	−0.09	−0.09	−0.09	−0.04	−0.11
3	Gąsiorowo	−0.08	−0.09	−0.09	−0.09	−0.08	−0.09	−0.11	−0.09	−0.14	−0.11	−0.07	−0.18
4	Giławy	−0.07	−0.08	−0.06	−0.06	−0.06	−0.09	−0.10	−0.11	−0.15	−0.14	−0.07	−0.13
5	Grabowo	−0.36	−0.26	−0.28	−0.20	0.00	−0.09	−0.06	−0.23	−0.20	−0.02	−0.33	0.00
6	Groszkowo	−0.05	−0.08	−0.06	−0.06	−0.05	−0.09	−0.08	−0.09	−0.09	−0.14	−0.04	−0.18
7	Kaborno	−0.09	−0.12	−0.12	−0.13	−0.08	−0.09	−0.13	−0.14	−0.15	−0.12	−0.12	−0.11
8	Klebark Mały	−0.10	−0.15	−0.16	−0.17	−0.14	−0.10	−0.15	−0.03	−0.03	−0.09	−0.05	−0.10
9	Klebark W.	−0.08	−0.10	−0.11	−0.12	−0.11	−0.18	−0.13	−0.08	−0.08	−0.10	−0.09	−0.02

Table 3. Cont.

i	DISTRICT	A.1	A.2	A.3	A.4	A.5	A.6	A.7	A.8.	A.9	E.1	E.2	E.3
	j	1	2	3	4	5	6	7	8	9	10	11	12
							$n_{ij}\ \ln n_{ij}$						
10	Klewki	−0.07	−0.12	−0.12	−0.12	−0.14	−0.10	−0.06	−0.05	−0.07	−0.11	−0.09	−0.11
11	Łajs	−0.04	−0.05	−0.03	−0.03	−0.03	−0.11	−0.18	−0.12	−0.11	−0.13	−0.05	−0.08
12	Linowo	−0.04	−0.05	−0.04	−0.04	−0.04	−0.17	−0.07	−0.08	−0.10	−0.11	−0.05	−0.08
13	Marcinkowo	−0.07	−0.11	−0.11	−0.12	−0.11	−0.10	−0.12	−0.08	−0.06	−0.14	−0.06	−0.09
14	Nerwik	−0.05	−0.07	−0.04	−0.04	−0.02	−0.10	−0.13	−0.15	−0.16	−0.12	−0.08	−0.17
15	Nowa Kaletka	−0.05	−0.06	−0.07	−0.08	−0.07	−0.10	−0.08	−0.07	−0.06	−0.10	−0.02	−0.10
16	Nowa Wieś	−0.06	−0.07	−0.07	−0.07	−0.08	−0.10	−0.10	−0.11	−0.09	−0.11	−0.05	−0.10
17	Ostrzeszewo	−0.18	−0.28	−0.31	−0.32	−0.35	−0.10	−0.12	−0.01	−0.01	−0.06	−0.08	−0.01
18	Pajtuny	−0.08	−0.11	−0.09	−0.10	−0.07	−0.09	−0.11	−0.12	−0.12	−0.09	−0.07	−0.10
19	Patryki	−0.07	−0.10	−0.08	−0.09	−0.08	−0.10	−0.11	−0.10	−0.10	−0.11	−0.09	−0.10
20	Pokrzywy	−0.07	−0.07	−0.06	−0.06	−0.08	−0.09	−0.07	−0.08	−0.07	−0.08	0.00	−0.13
21	Prejłowo	−0.06	−0.07	−0.08	−0.08	−0.10	−0.10	−0.09	−0.07	−0.07	−0.09	−0.06	−0.10
22	Przykop	−0.07	−0.09	−0.09	−0.09	−0.07	−0.10	−0.09	−0.08	−0.09	−0.10	−0.06	−0.13
23	Purda	−0.06	−0.09	−0.08	−0.09	−0.08	−0.10	−0.10	−0.08	−0.09	−0.15	−0.07	−0.14
24	Purdka	−0.04	−0.04	−0.03	−0.03	−0.03	−0.11	−0.13	−0.16	−0.15	−0.13	−0.03	−0.09
25	Silice	−0.07	−0.11	−0.12	−0.13	−0.09	−0.09	−0.11	−0.14	−0.13	−0.10	−0.07	−0.08
26	Stara Kaletka	−0.05	−0.05	−0.04	−0.04	−0.04	−0.09	−0.10	−0.12	−0.13	−0.09	0.00	−0.11
27	Szczęsne	−0.10	−0.17	−0.18	−0.19	−0.19	−0.10	−0.12	−0.04	−0.06	−0.09	−0.07	−0.03
28	Trękus	−0.07	−0.08	−0.07	−0.08	−0.06	−0.10	−0.13	−0.11	−0.11	−0.10	−0.04	−0.11
29	Trękusek	−0.08	−0.10	−0.11	−0.12	−0.26	−0.18	−0.13	−0.09	−0.08	−0.10	−0.04	−0.10
30	Wygoda	−0.04	−0.04	−0.03	−0.03	−0.03	−0.09	−0.09	−0.16	−0.15	−0.12	0.00	−0.09
31	Wyrandy	−0.07	−0.08	−0.06	−0.06	−0.10	−0.09	−0.10	−0.10	−0.10	−0.09	−0.05	−0.16
32	Zaborowo	−0.09	−0.10	−0.07	−0.07	−0.05	−0.09	−0.05	−0.12	−0.11	−0.12	−0.08	−0.13
33	Zgniłocha	−0.06	−0.06	−0.07	−0.07	−0.04	−0.10	−0.10	−0.15	−0.13	−0.10	−0.08	−0.09
	SUM	−2.61	−3.19	−3.03	−3.10	−2.88	−3.44	−3.43	−3.33	−3.36	−3.45	−2.16	−3.34
	E_j	0.75	0.91	0.87	0.89	0.82	0.98	0.98	0.95	0.96	0.99	0.62	0.95

The differences in the value of n_{ij} relative to successive attributes were calculated with the following formula:

$$d_j = 1 - E_j \tag{4}$$

where j is the successive attributes in the total number of attributes r.

The result is the attribute vector d_j.

$$d_j = [0.25\ \ 0.09\ \ 0.13\ \ 0.11\ \ 0.18\ \ 0.02\ \ 0.02\ \ 0.05\ \ 0.04\ \ 0.01\ \ 0.38\ \ 0.05] \tag{5}$$

The value of d_j was used to determine the weights w_j of conflict-generating attributes with the use of the below formula:

$$w_j = \frac{d_j}{\sum_{j=1}^{r} d_j} \tag{6}$$

The result is the vector of attribute weights $\mathbf{W}$.

$$\mathbf{W} = [0.19\ \ 0.07\ \ 0.10\ \ 0.08\ \ 0.13\ \ 0.01\ \ 0.01\ \ 0.04\ \ 0.03\ \ 0.01\ \ 0.29\ \ 0.03] \tag{7}$$

The normalized matrix $\mathbf{N}$ was multiplied by vector $\mathbf{W}$ to produce matrix $\mathbf{D}$. The risk of land-use conflict in the analyzed districts was expressed by the sum of the rows of matrix $\mathbf{D}$. The results were represented by indicator V_i for improved readability.

$$V_i = 100 \times \sum_{j=1}^{r} (n_{ij} w_j) \tag{8}$$

where i is the successive districts in the total number of districts k.

The value of indicator V_i can range from 0 to 100. Districts with higher values of V_i are characterized by a greater accumulation of attributes that can give rise to spatial conflict.

3. Results and Discussion

Indicator V_i was calculated for the analyzed districts, and the results are presented in Table 4. In Grabowo district, V_i considerably exceeded the average value for the examined districts (27.60). The above result could be attributed to the very small area of Grabowo relative to other districts, as well as the fact that other attributes in Grabowo also substantially exceeded the average value. Grabowo was thus eliminated from classification. Only districts where the value of V_i ranged from 0.00 to 10.00 were included in the classification.

Table 4. Value of V_i in the analyzed districts.

	DISTRICT	A.1	A.2	A.3	A.4	A.5	A.6	A.7	A.8.	A.9	E.1	E.2	E.3	SUM	V_i
	j	1	2	3	4	5	6	7	8.	9	10	11	12	$r = 12$	
i								$n_{ij}\,w_j$							
1	Bałdy	0.0015	0.0005	0.0005	0.0004	0.0018	0.0003	0.0004	0.0013	0.0007	0.0003	0.0038	0.0008	0.0123	1.23
2	Butryny	0.0035	0.0016	0.0025	0.0023	0.0039	0.0003	0.0003	0.0009	0.0007	0.0003	0.0026	0.0010	0.0198	1.98
3	Gąsiorowo	0.0039	0.0017	0.0024	0.0022	0.0029	0.0003	0.0004	0.0009	0.0013	0.0003	0.0051	0.0022	0.0235	2.35
4	Giławy	0.0033	0.0013	0.0014	0.0013	0.0019	0.0003	0.0004	0.0011	0.0014	0.0004	0.0045	0.0015	0.0189	1.89
5	Grabowo	0.0811	0.0082	0.0147	0.0069	0.0000	0.0003	0.0002	0.0037	0.0022	0.0000	0.1587	0.0000	0.2760	27.60
6	Groszkowo	0.0023	0.0015	0.0013	0.0012	0.0014	0.0003	0.0003	0.0009	0.0007	0.0005	0.0025	0.0024	0.0152	1.52
7	Kaborno	0.0046	0.0025	0.0036	0.0033	0.0026	0.0003	0.0006	0.0016	0.0015	0.0004	0.0110	0.0011	0.0329	3.29
8	Klebark M.	0.0052	0.0033	0.0054	0.0049	0.0058	0.0003	0.0007	0.0002	0.0002	0.0002	0.0030	0.0010	0.0302	3.02
9	Klebark W.	0.0038	0.0019	0.0032	0.0030	0.0042	0.0008	0.0006	0.0007	0.0006	0.0003	0.0071	0.0001	0.0263	2.63
10	Klewki	0.0032	0.0023	0.0035	0.0032	0.0060	0.0003	0.0002	0.0004	0.0005	0.0003	0.0067	0.0011	0.0277	2.77
11	Łajs	0.0016	0.0007	0.0006	0.0006	0.0007	0.0003	0.0010	0.0013	0.0009	0.0004	0.0033	0.0007	0.0121	1.21
12	Linowo	0.0017	0.0008	0.0008	0.0007	0.0010	0.0007	0.0002	0.0007	0.0008	0.0003	0.0031	0.0007	0.0114	1.14
13	Marcinkowo	0.0035	0.0020	0.0032	0.0030	0.0044	0.0003	0.0005	0.0007	0.0004	0.0004	0.0040	0.0008	0.0233	2.33
14	Nerwik	0.0022	0.0011	0.0008	0.0008	0.0005	0.0003	0.0006	0.0018	0.0015	0.0004	0.0061	0.0021	0.0183	1.83
15	Nowa Kal.	0.0022	0.0010	0.0018	0.0016	0.0023	0.0003	0.0003	0.0006	0.0004	0.0003	0.0008	0.0010	0.0127	1.27
16	Nowa Wieś	0.0028	0.0012	0.0017	0.0016	0.0029	0.0003	0.0004	0.0011	0.0007	0.0003	0.0029	0.0010	0.0169	1.69
17	Ostrzeszewo	0.0123	0.0098	0.0184	0.0167	0.0350	0.0003	0.0005	0.0001	0.0000	0.0002	0.0057	0.0001	0.0990	9.90
18	Pajtuny	0.0042	0.0020	0.0025	0.0023	0.0024	0.0003	0.0005	0.0014	0.0010	0.0003	0.0048	0.0010	0.0226	2.26
19	Patryki	0.0035	0.0017	0.0022	0.0020	0.0025	0.0003	0.0004	0.0011	0.0008	0.0003	0.0065	0.0010	0.0224	2.24
20	Pokrzywy	0.0033	0.0011	0.0013	0.0012	0.0028	0.0003	0.0002	0.0007	0.0005	0.0002	0.0000	0.0015	0.0131	1.31
21	Prejłowo	0.0026	0.0012	0.0019	0.0017	0.0035	0.0003	0.0004	0.0006	0.0005	0.0003	0.0041	0.0009	0.0179	1.79
22	Przykop	0.0034	0.0016	0.0023	0.0021	0.0025	0.0003	0.0004	0.0007	0.0007	0.0003	0.0041	0.0014	0.0198	1.98
23	Purda	0.0029	0.0017	0.0021	0.0020	0.0028	0.0003	0.0004	0.0007	0.0007	0.0005	0.0048	0.0016	0.0205	2.05
24	Purdka	0.0017	0.0006	0.0006	0.0006	0.0007	0.0003	0.0006	0.0019	0.0015	0.0004	0.0017	0.0008	0.0117	1.17
25	Silice	0.0035	0.0021	0.0036	0.0033	0.0032	0.0003	0.0005	0.0016	0.0012	0.0003	0.0047	0.0007	0.0249	2.49
26	Stara Kal..	0.0022	0.0008	0.0009	0.0008	0.0010	0.0003	0.0004	0.0013	0.0011	0.0003	0.0001	0.0011	0.0102	1.02
27	Szczęsne	0.0054	0.0039	0.0066	0.0060	0.0101	0.0003	0.0005	0.0003	0.0004	0.0002	0.0053	0.0002	0.0392	3.92
28	Trękus	0.0033	0.0014	0.0018	0.0017	0.0021	0.0003	0.0006	0.0011	0.0010	0.0003	0.0027	0.0011	0.0173	1.73
29	Trękusek	0.0039	0.0019	0.0033	0.0030	0.0162	0.0008	0.0006	0.0009	0.0006	0.0003	0.0026	0.0009	0.0349	3.49
30	Wygoda	0.0018	0.0006	0.0006	0.0006	0.0007	0.0003	0.0003	0.0020	0.0015	0.0004	0.0002	0.0009	0.0097	0.97
31	Wyrandy	0.0030	0.0014	0.0014	0.0013	0.0036	0.0003	0.0004	0.0010	0.0008	0.0003	0.0034	0.0019	0.0187	1.87
32	Zaborowo	0.0049	0.0018	0.0016	0.0015	0.0017	0.0003	0.0002	0.0013	0.0009	0.0004	0.0058	0.0013	0.0215	2.15
33	Zgniłocha	0.0027	0.0010	0.0016	0.0015	0.0013	0.0003	0.0004	0.0018	0.0011	0.0003	0.0064	0.0008	0.0192	1.92

Indicator V_i was used to determine the risk of land-use conflict in the analyzed cadastral districts. Classification results are presented in Figure 5.

The analyzed districts were divided into five classes with the use of the geometric interval classification method. Class 1, with the highest accumulation of conflict-generating attributes, comprised only one district, Ostrzeszewo, where V_i reached 9.90. Class 2, with a high accumulation of the analyzed attributes, contained districts subjected to considerable urbanization and human pressure. Classes 4 and 5, with the lowest accumulation of conflict-generating attributes, were composed of ecological sites (forests and water bodies) in their entirety or in most part. The analysis of land-use types in the Purda municipality indicates that the proposed method generates reliable results. The value of indicator V_i was markedly higher in areas subjected to considerable anthropogenic pressure.

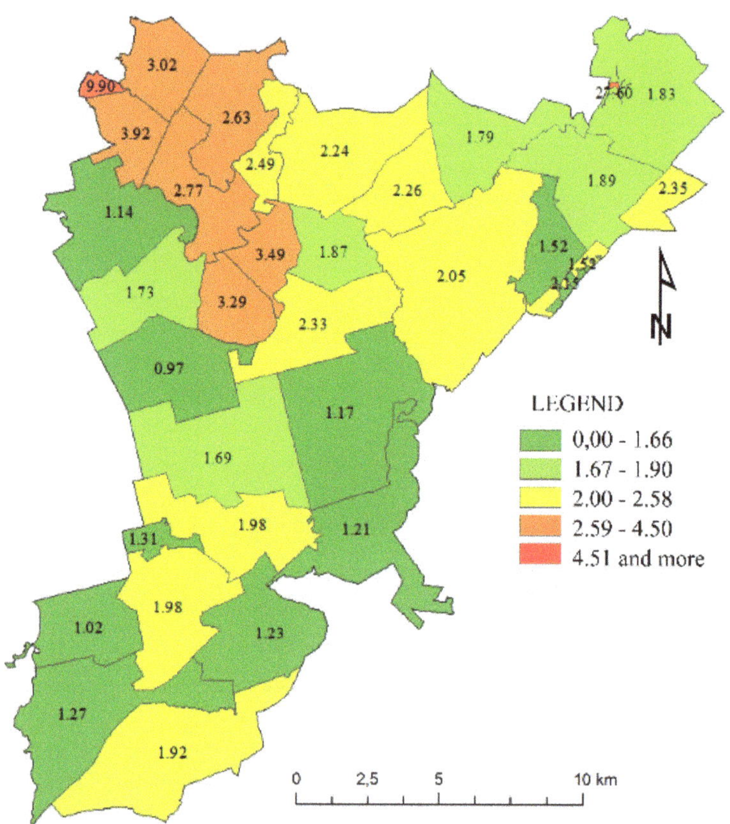

Figure 5. Values of indicator V_j in the cadastral districts of Purda municipality.

4. Conclusions

One of the greatest challenges in land management is to minimize the "pretexts" for spatial conflict. Conflict-generating factors should be eliminated or minimized in all stages of the process, including land-use planning, land protection, and land management. The aim of all land management procedures and decisions should be to improve land-use comfort. Therefore, the identification of potential hotbeds of conflict is a very important consideration that tends to be overlooked in research.

In this study, attempts were made to fill in this knowledge gap. The present research is by no means exhaustive, but the results indicate that geospatial databases can be used effectively to reduce the risk of spatial conflict.

This study proposes a procedure for evaluating the risk of land-use conflict based on the attributes that increase the probability of such risk. Conflict-generating attributes were calculated based on data that are available in municipal centers for geodetic and cartographic documentation. The analysis was conducted with the use of GIS tools in ArcGIS 10.4.1 software.

The proposed procedure was tested in a rural municipality with a varied spatial structure in the Region of Warmia and Mazury. The physical and geodetic attributes of space were identified in considerable detail at the local level. A procedure for weighing and measuring the intensity of attributes that can compromise spatial harmony and contribute to spatial conflict was proposed. The importance of conflict-generating attributes was assessed with Shannon's entropy method, and indicator V_i was calculated as the sum of attribute weights. The procedure of calculating indicator V_i involved well-known metrics as well as the measurement methods proposed by the authors. The results of the analysis indicate that anthropogenic factors are the main drivers of land-use conflict. Therefore, the developed procedure can be used to identify areas with a higher risk of land-use conflict. Conflict-prone areas should be identified based on analyses of cadastral data, local zoning plans and social participation to optimize spatial planning and improve the quality of life in local communities.

The results of the study demonstrated that the risk of land-use conflict is particularly high in the immediate vicinity of areas that are subjected to considerable anthropogenic pressure. Areas where human activities generate social controversy should be identified to promote sustainable land-use planning and prevent conflict.

The proposed procedure can be modified to account for specific research objectives, the characteristic features of the studied area, and access to spatial data. The developed methodology constitutes a valuable tool which supports planning and monitoring of sustainable land-use practices. The results of the analysis can be used to plan future actions with the aim of enhancing harmonious spatial development and minimizing the risk of land-use conflict.

Author Contributions: Conceptualization, I.C. and A.B.; methodology, I.C. and A.B.; software, I.C. and A.B.; validation, I.C. and A.B.; formal analysis I.C. and A.B.; investigation I.C. and A.B.; resources I.C. and A.B.; data curation, I.C. and A.B.; writing—original draft preparation, I.C.; writing—review and editing, A.B.; visualization, I.C.; supervision, I.C. and A.B.; project administration, I.C. and A.B.; and funding acquisition I.C. and A.B. All authors have read and agreed to the published version of the manuscript.

Funding: This research received no external funding.

References

1. Mucha, J. *Konflikt I Społeczeństwo: Z Problematyki Konfliktu Społecznego We Współczesnych Teoriach Zachodnich*; PWN: Warszawa, Poland, 1978.
2. Pietrzak, H. *Agresja, Konflikt, Społeczeństwo*; College of Socio-Economics: Tyczyn, Poland, 2000.
3. Bogetoft, P.; Pruzan, P. *Planning with Multiple Criteria: Investigation, Communication and Choice*; Copenhagen Business School Press: Copenhagen, Denmark, 1997.
4. Cieślak, I. Spatial conflicts: Analyzing a burden created by differing land use. *Acta Geogr. Slov.* **2019**, *59*, 43–57. [CrossRef]
5. Hite, J. *Land Use Conflicts on the Urban Fringe: Causes and Potential Resolution*; Strom Thurmond Institute, Clemson University: Clemson, SC, USA, 1998.
6. De Groot, R. Function-analysis and valuation as a tool to assess land use conflicts in planning for sustainable, multi-functional landscapes. *Landsc. Urban Plan.* **2006**, *75*, 175–186. [CrossRef]
7. Pawlewicz, K. Differences in development levels of urban gminas in the Warmińsko-Mazurskie voivodship in view of the main components of sustainable development. *Bull. Geogr. Socio-Econ. Ser.* **2015**, *29*, 93–102. [CrossRef]
8. Renigier-Bilozor, M.; Biłozor, A.; Wisniewski, R. Rating engineering of real estate markets as the condition of urban areas assessment. *Land Use Policy* **2017**, *61*, 511–525. [CrossRef]
9. Cieślak, I.; Pawlewicz, K.; Pawlewicz, A. Sustainable Development in Polish Regions: A Shift-Share Analysis. *Pol. J. Environ. Stud.* **2018**, *28*, 565–575. [CrossRef]
10. Wehrmann, B. *Land Conflicts: A Practical Guide to Dealing with Land Disputes*; Deutsche Gesellschaft für Technische Zusammenarbeit: Eschborn, Germany, 2008.
11. Imbusch, P.; Zoll, R. (Eds.) *Friedens- Und Konfliktforschung*; Springer: Berlin, Germany, 2006.
12. Renigier-Biłozor, M.; Biłozor, A. Optimization of the variables selection in the process of real estate markets rating. *Oeconomia Copernic.* **2015**, *6*, 139. [CrossRef]
13. Von der Dunk, A.; Grêt-Regamey, A.; Dalang, T.; Hersperger, A. Defining a typology of peri-urban land-use conflicts—A case study from Switzerland. *Landsc. Urban Plan.* **2011**, *101*, 149–156. [CrossRef]
14. Lehtovuori, P. *Experience and Conflict: The Production of Urban Space*; Routledge: London, UK, 2016.
15. Most, B.A.; Starr, H. Diffusion, Reinforcement, Geopolitics, and the Spread of War. *Am. Polit. Sci. Rev.* **1980**, *74*, 932–946. [CrossRef]
16. Flyvbjerg, B.; Richardson, T. Planning and Foucault: In search of the dark side of planning theory. In *Planning Futures: New Directions for Planning Theory*; Allmendinger, P., Tewdwr-Jones, M., Eds.; Routledge: London, UK, 2002; pp. 44–62.

17. Nolon, S.; Ferguson, O.; Field, P. *Land in Conflict: Managing and Resolving Land Use Disputes*; Lincoln Institute of Land Policy: Cambridge, UK, 2013.
18. Karakostas, S.M. Land-use planning via enhanced multi-objective evolutionary algorithms: Optimizing the land value of major Greenfield initiatives. *J. Land Use Sci.* **2016**, *11*, 595–617. [CrossRef]
19. Arndt, C. *Information Measures: Information and Its Description in Science and Engineering*; Springer Science & Business Media: Berlin, Germany, 2004. [CrossRef]
20. Gray, R.M. *Entropy and Information*; Springer: New York, NY, USA, 1990; pp. 21–55.
21. O'Lear, S.; Gray, A. Asking the right questions: Environmental conflict in the case of Azerbaijan. *Area* **2006**, *38*, 390–401. [CrossRef]
22. Santorineou, A.; Hatzopoulos, J.; Siakavara, K.; Davos, C. Spatial Conflict Management in Urban Planning. In Proceedings of the International Conference on Studying, Modeling & Sense Making of Planet Earth, Lesvos, Greece, 1–6 June 2008.
23. Hersperger, A.M.; Oliveira, E.; Pagliarin, S.; Palka, G.; Verburg, P.; Bolliger, J.; Grădinaru, S.R. Urban land-use change: The role of strategic spatial planning. *Glob. Environ. Chang.* **2018**, *51*, 32–42. [CrossRef]
24. Stojanovic, D. Space, Territory and Sovereignty: Critical Analysis of Concepts. *Nagoya Univ. J. Law Politics* **2017**, *275*, 111–185.
25. Cherry, K. Hierarchy of Needs. Retrieved 16 August 2014. Available online: http://blogs.jefftwp.org/wordpress/rzegas/files/2014/02/Maslow-Reading.pdf (accessed on 30 August 2019).
26. Brennan, R.E. The conservation "myths" we live by: Reimagining human–nature relationships within the Scottish marine policy context. *Area* **2018**, *50*, 159–168. [CrossRef]
27. O'Lear, S. Resource concerns for territorial conflict. *GeoJournal* **2005**, *64*, 297–306. [CrossRef]
28. Basiago, A.D. Economic, social, and environmental sustainability in development theory and urban planning practice. *Environmentalist* **1998**, *19*, 145–161. [CrossRef]
29. Campbell, S. Green Cities, Growing Cities, just Cities? Urban Planning and the Contradictions of Sustainable Development. *Class. Read. Urban Plan.* **2018**, *62*, 308–326. [CrossRef]
30. Reuveny, R. Economic Growth, Environmental Scarcity, and Conflict. *Glob. Environ. Politics* **2002**, *2*, 83–110. [CrossRef]
31. Dmochowska-Dudek, K. Konflikty społeczno-przestrzenne związane z rozwojem infrastruktury technicznej na obszarach wiejskich w ujęciu funkcjonalnym i społecznym. *Studia Obsz. Wiej.* **2014**, *35*, 109–120.
32. Borell, K. Westermark, Åsa Siting of human services facilities and the not in my back yard phenomenon: A critical research review. *Community Dev. J.* **2016**, *53*, 246–262. [CrossRef]
33. Giddings, B.; Hopwood, B.; O'Brien, G. Environment, economy and society: Fitting them together into sustainable development. *Sustain. Dev.* **2002**, *10*, 187–196. [CrossRef]
34. Chmielewski, J.M. *Teoria Urbanistyki W Projektowaniu I Planowaniu Miast*; Oficyna Wydawnicza Politechniki Warszawskiej: Warszawa, Poland, 2005.
35. Glasson, J.; Marshall, T. *Regional Planning*; Routledge: London, UK, 2007.
36. Ene, C.M.; Gheorghiu, A.; Burghelea, C.; Gheorghiu, A. The conflict between economic development and planetary ecosystem in the context of sustainable development. *arXiv* **2011**, arXiv:1102.5747.
37. Goetz, S.J.; Shortle, J.S.; Bergstrom, J.C. (Eds.) *Land Use Problems and Conflicts: Causes, Consequences and Solutions*; Routledge: London, UK, 2005.
38. Cieślak, I. *Wieloaspektowa Analiza Konfliktów Przestrzennych*; Wydawnictwo Uniwersytetu Warmińsko-Mazurskiego: Olsztyn, Poland, 2018.
39. Zhou, H.; Chen, Y.; Tian, R. Land-Use Conflict Identification from the Perspective of Construction Space Expansion: An Evaluation Method Based on 'Likelihood-Exposure-Consequence'. *ISPRS Int. J. Geo-Inf.* **2021**, *10*, 433. [CrossRef]
40. Norcliffe, G.B. *Inferential Statistics for Geographers*; John Wiley & Sons: New York, NY, USA, 1977.
41. Ye, Q.; Wei, R.; Zhang, P. A Conflict Identification Method of Urban, Agricultural and Ecological Spaces Based on the Space Conversion Matrix. *Sustainability* **2018**, *10*, 3502. [CrossRef]
42. Dietz, K.; Engels, B. Analysing land conflicts in times of global crises. *Geoforum* **2020**, *111*, 208–217. [CrossRef]
43. Rawal, N. Social Inclusion and Exclusion: A Review. *Dhaulagiri J. Sociol. Anthr.* **2008**, *2*, 161–180. [CrossRef]
44. Waller-Hunter, J.; Jones, T. Globalisation and sustainable development. *Int. Rev. Environ. Strateg.* **2002**, *3*, 53–62.
45. Wehrmann, B. *Understanding, Preventing and Solving Land Conflicts. A Practical Guide and Toolbox*; Schleunungdruck GmbH: Eltertstr, Germany, 2017; Volume 27, p. 97828.
46. Cieślak, I.; Szuniewicz, K. Analysis of the investment potential of location using the AHP method. *Géod. Vestnik* **2018**, *62*, 279–292. [CrossRef]
47. Biłozor, A.; Cieślak, I.; Czyża, S. An Analysis of Urbanisation Dynamics with the Use of the Fuzzy Set Theory—A Case Study of the City of Olsztyn. *Remote Sens.* **2020**, *12*, 1784. [CrossRef]
48. Cieślak, I.; Biłozor, A.; Szuniewicz, K. The Use of the CORINE Land Cover (CLC) Database for Analyzing Urban Sprawl. *Remote Sens.* **2020**, *12*, 282. [CrossRef]
49. Graham, S.; Healey, P. Relational concepts of space and place: Issues for planning theory and practice. *Eur. Plan. Stud.* **1999**, *7*, 623–646. [CrossRef]
50. Frey, W.H.; Zimmer, Z. Defining the City. In *Handbook of Urban Studies*; SAGE Publications Ltd.: London, UK, 2012; pp. 14–35.
51. Cieślak, I. Identification of areas exposed to land use conflict with the use of multiple-criteria decision-making methods. *Land Use Policy* **2019**, *89*, 104225. [CrossRef]

52. Strumiłło-Rembowska, D.; Bednarczyk, M.; Cieślak, I. Data generation of vector maps using a hybrid method of analysis and selection of geodata necessary to optimize the process of spatial planning. In Proceedings of the 9th International Conference Environmental Engineering, Vilnius, Lithuania, 22–23 May 2014; Volume 9. [CrossRef]

53. Szuniewicz, K.; Cieślak, I.; Czyża, S. The Use of Webgis Services in Public Administration in Poland. In Proceedings of the 15th International Multidisciplinary Scientific GeoConference SGEM2015, Ecology, Economics, Education And Legislation, Albena, Bulgaria, 18–24 June 2015; Volume 2, pp. 891–898. [CrossRef]

54. Renigier-Bilozor, M.; Janowski, A.; Walacik, M. Geoscience Methods in Real Estate Market Analyses Subjectivity Decrease. *Geosciences* **2019**, *9*, 130. [CrossRef]

55. Pasto, I.D. *Spatial Planning as Co-Evolution: Linking Expectations, Uncertainties and Conflicts*; Wageningen University: Wageningen, The Netherlands, 2009.

56. Zou, L.; Liu, Y.; Wang, J.; Yang, Y.; Wang, Y. Land use conflict identification and sustainable development scenario simulation on China's southeast coast. *J. Clean. Prod.* **2019**, *238*, 238. [CrossRef]

57. Kim, I.; Arnhold, S. Mapping environmental land use conflict potentials and ecosystem services in agricultural watersheds. *Sci. Total Environ.* **2018**, *630*, 827–838. [CrossRef]

58. Brown, G.; Raymond, C. Methods for identifying land use conflict potential using participatory mapping. *Landsc. Urban Plan.* **2014**, *122*, 196–208. [CrossRef]

59. Hersperger, A.M.; Ioja, C.; Steiner, F.; Tudor, C.A. Comprehensive consideration of conflicts in the land-use planning process: A conceptual contribution. *Carpathian J. Earth Environ. Sci.* **2015**, *10*, 5–13.

60. Biłozor, A.; Kowalczyk, A.M.; Bajerowski, T. Theory of Scale-Free Networks as a New Tool in Researching the Structure and Optimization of Spatial Planning. *J. Urban Plan. Dev.* **2018**, *144*, 04018005. [CrossRef]

61. McLeod, P.; Martin, A.; Crompvoets, J. Spatial Data Infrastructure (SDI) Manual for the Americas. In *Global Spatial Data Infrastructures Association*; PC-IDEA SDI Manual for the Americas: Addis Ababa, Ethiopia, 2013.

62. Rudnicki, R.; Męczekalski, M.; Dubownik, A. *Obręb Geodezyjny jako Jednostka Przestrzenna Badań Geograficznoturystycznych–Przykład Powiatu Suwalskiego In Uwarunkowania I Plany Rozwoju Turystyki*; Młynarczyk, Z., Zajadacz, A., Eds.; Bogucki Wydawnictwo Naukowe: Poznań, Poland, 2015; Volume 15, pp. 33–52.

63. Available online: Codgik.gov.pl/index.php/darmowe-dane/nmt-100.htmldatadostępu (accessed on 30 October 2017).

64. Cieślak, I.; Szuniewicz, K.; Gerus-Gościewska, M. Evaluation of the Natural Value of Land Before and after Planning Procedures. *Rural Dev.* **2013**, *6*, 3.

65. Cieślak, I.; Szuniewicz, K.; Czyża, S.; Ogrodniczak, M. Location of the forest complexes endangered by the vicinity to diverse form of land management. *Sylwan* **2019**, *163*, 300–310.

66. Cieślak, I.; Biłozor, A.; Źróbek-Sokolnik, A.; Zagroba, M. The Use of Geographic Databases for Analyzing Changes in Land Cover—A Case Study of the Region of Warmia and Mazury in Poland. *ISPRS Int. J. Geo-Inf.* **2020**, *9*, 358. [CrossRef]

67. Pawlewicz, K.; Olsztynie, U.W.-M. w relationships between social capital and socio-economic development based on rural communes in the warmińsko-mazurskie voivodeship. *J. Agribus. Rural. Dev.* **2016**, *10*, 373–381. [CrossRef]

68. Drobne, S.; Lisec, A. Multi-attribute decision analysis in GIS: Weighted linear combination and ordered weighted averaging. *Informatica* **2009**, *33*, 459–474.

69. Biłozor, A.; Renigier-Biłozor, M. Procedure of Assessing Usefulness of the Land in the Process of Optimal Investment Location for Multi-family Housing Function. *Proc. Eng.* **2016**, *161*, 1868–1873. [CrossRef]

70. Biłozor, A.; Renigier-Biłozor, M. Optimization and polioptimization in the management of land. In Proceedings of the SGEM2015 GeoConference on Informatics, Geoiformatics and Remote Sensing, Albena, Bulgaria, 18–24 June 2015; Volume 2, pp. 1011–1018, ISBN 978-619-7105-35-3.

71. Marzęcki, W. Ciągłość kulturowa w kształtowaniu przestrzeni miejskiej: Charakterystyka i metoda oceny jakości i zmienności tej przestrzeni [Cultural continuity in shaping urban space: Characteristics and method of assessing the quality and variability of this space]. *Inst. Archit. I Plan. Przestrz. Pr. Nauk. Politech. Szczecińskiej* **2000**, *564*, 5–214.

72. Shannon, C.E. A Mathematical Theory of Communication. *Bell Syst. Tech. J.* **1948**, *27*, 379–423. [CrossRef]

73. Ge, Q.; Fukuda, D. Updating origin–destination matrices with aggregated data of GPS traces. *Transp. Res. Part C Emerg. Technol.* **2016**, *69*, 291–312. [CrossRef]

74. Massoudi, M. A Possible Ethical Imperative Based on the Entropy Law. *Entropy* **2016**, *18*, 389. [CrossRef]

75. Al-Sharhan, S.; Karray, F.; Gueaieb, W.; Basir, O. Fuzzy entropy: A brief survey. In Proceedings of the 10th IEEE International Conference on Fuzzy Systems. (Cat. No.01CH37297), Melbourne, VIC, Australia, 2–5 December 2001; pp. 1135–1139. [CrossRef]

76. Tang, C.M.; Leung, A.Y.; Lam, K.C. Entropy Application to Improve Construction Finance Decisions. *J. Constr. Eng. Manag.* **2006**, *132*, 1099–1113. [CrossRef]

77. Chen, T.-Y.; Li, C.-H. Determining objective weights with intuitionistic fuzzy entropy measures: A comparative analysis. *Inf. Sci.* **2010**, *180*, 4207–4222. [CrossRef]

78. Li, D. TOPSIS Based Nonlinear Programming Methodology for Multiattribute Decision Making with Interval-Valued Intuitionistic Fuzzy Sets. *IEEE Trans. Fuzzy Syst.* **2010**, *18*, 299–311. [CrossRef]

79. Zhang, Y.; Li, P.; Wang, Y.; Ma, P.; Su, X. Multiattribute Decision Making Based on Entropy under Interval-Valued Intuitionistic Fuzzy Environment. *Math. Probl. Eng.* **2013**, *2013*, 1–8. [CrossRef]

80. Zhang, Y.; Wang, Y.; Wang, J. Objective Attributes Weights Determining Based on Shannon Information Entropy in Hesitant Fuzzy Multiple Attribute Decision Making. *Math. Probl. Eng.* **2014**, *2014*, 1–7. [CrossRef]
81. Jacquemin, A.P.; Berry, C.H. Entropy Measure of Diversification and Corporate Growth. *J. Ind. Econ.* **1979**, *27*, 359. [CrossRef]
82. Kobryń, A. *Wielokryterialne Wspomaganie Decyzji W Gospodarowaniu Przestrzenią [Multi-Criteria Decision Support in Space Management]*; Difin SA: Warsaw, Poland, 2014.

Article

Analysis and Evaluation of the Spatial Structure of Cittaslow Towns on the Example of Selected Regions in Central Italy and North-Eastern Poland

Marek Zagroba , Katarzyna Pawlewicz and Adam Senetra *

Department of Socio-Economic Geography, Faculty of Geoengineering, University of Warmia and Mazury in Olsztyn, Prawocheńskiego 15, 10-720 Olsztyn, Poland; mazag@uwm.edu.pl (M.Z.); katarzyna.pawlewicz@uwm.edu.pl (K.P.)
* Correspondence: adam.senetra@uwm.edu.pl; Tel.: +48-89-5234948

Abstract: Cittaslow International promotes harmonious development of small towns based on sustainable relationships between economic growth, protection of local traditions, cultural heritage and the environment, and an improvement in the quality of local life. The aim of this study was to analyze and evaluate the differences and similarities in the spatial structure of Cittaslow towns in the Italian regions of Tuscany and Umbria and the Polish region of Warmia and Mazury. The study examined historical towns which are situated in different parts of Europe and have evolved in different cultural and natural environments. The presented research attempts to determine whether the spatial structure of historical towns established in different European regions promotes the dissemination of the Cittaslow philosophy and the adoption of sustainable development principles. The urban design, architectural features and the composition of urban and architectural factors which are largely responsible for perceptions of multi-dimensional space were evaluated. These goals were achieved with the use of a self-designed research method which supported a subjective evaluation of spatial structure defined by historical urban planning and architectural solutions. The study demonstrated that Medieval urban layouts can be successfully incorporated into the modern urban fabric to promote sustainable development and slow living.

Keywords: Slow City; small towns; spatial structure; sustainable development; old market square; historical urban layout

check for updates

Citation: Zagroba, M.; Pawlewicz, K.; Senetra, A. Analysis and Evaluation of the Spatial Structure of Cittaslow Towns on the Example of Selected Regions in Central Italy and North-Eastern Poland. *Land* **2021**, *10*, 780. https://doi.org/10.3390/land10080780

Academic Editors: Luca Salvati, Iwona Cieślak and Andrzej Biłozor

Received: 23 June 2021
Accepted: 22 July 2021
Published: 25 July 2021

Publisher's Note: MDPI stays neutral with regard to jurisdictional claims in published maps and institutional affiliations.

1. Introduction

The objective of the Cittaslow International network of small towns is to promote harmonious development and slow living. This philosophy offers an alternative to globalization and the fast pace of life in large cities [1–3]. Cittaslow towns implement sustainable development policies by creating a healthy balance between economic growth, protection of local traditions, cultural heritage and the environment, and improving the quality of local life [4–6]. The members of the Cittaslow network promote local and specific values and products, and they initiate active measures to protect the natural environment and the historical urban fabric. These policies build a sense of solidarity among local community members who feel responsible for their place of residence, including its past and future [7,8]. The unique spatial structure of historical towns plays an important role in building harmonious relationships and adapting these towns to contemporary needs and aspirations [9,10].

Small towns are human settlements that occupy a relatively limited area and organize their living space in a similar manner regardless of their location. The scale of urban development, small population, absence of industry and low levels of technical infrastructure relative to large cities contribute to sustainable development and reluctance towards modern trends that rely heavily on globalization and a fast pace of life.

Considerable research has been dedicated to the development and future outlook of urban expansion. Zambon et al. [11] performed a quantitative analysis of urban plans to predict future urban growth patterns. The authors relied on an extensive database of social and economic indicators to examine the consequences of urban expansion in the Spanish region of Catalonia which was divided into local zones for the needs of the study. They found that urbanization had contributed little to the sustainable development of towns in recent decades. The authors concluded that future urban development in Catalonia will not follow a polycentric model, but will fuel the growth of small and medium-sized towns. The historical center of Santiago de Chile was modernized based its architectural and urban heritage. The elements of the city's original urban fabric were recreated and used to propose a new urban development concept. This approach was adopted to bridge the gap between modern requirements and the city's colonial and republican past [12]. In Senigallia, Italy, the open-design approach was implemented to regenerate the town's historical urban heritage through participatory design [13]. The town's historical district was redesigned with the use of Multicriteria Decision Analysis (MCDA). Historical districts pose a challenge for urban designers who have to reconcile conservation with modern welfare and safety standards. The MCDA approach was validated experimentally in Biella, Italy [14].

The aim of this study was to analyze the differences and similarities in the spatial structure of small towns belonging to the Cittaslow network. The evaluated towns are located in the Italian regions of Tuscany and Umbria and the Region of Warmia and Mazury in north-eastern Poland. Based on the aim of the study, the authors made an attempt to answer the following research questions:

1. Does the spatial structure of towns which are situated in different parts of Europe and have evolved under different historical, cultural and environmental conditions promote the achievement of Cittaslow goals in equal measure?
2. Which components of historical spatial structures (that developed under different functional conditions) foster the implementation of sustainable development policies and the slow living philosophy?

2. Cittaslow International

Current trends in urban and economic development favor measures that strengthen the growth potential of mainly large urban agglomerations. Rapid globalization has contributed to a faster pace of life. The citizens of highly developed countries show a growing appreciation for the slower approach to everyday life in smaller settlement units [15,16]. In small and medium-sized towns, most businesses operate only on the local market. These towns often collaborate and join various organizations to expand their economic potential. One of such organizations is the Cittaslow Movement, an international network of cities that was founded in 1999 to endorse the Slow City concept. Towns can apply for Cittaslow membership if their population does not exceed 50,000, and if they support and implement the organization's goals [17]. Candidates are evaluated in seven areas: energy and environmental policy, infrastructure policies, quality of urban life policies, agricultural, tourism and artisan policies, policies for hospitality, awareness and training, social cohesion and partnerships, based on 72 detailed criteria. Towns that meet at least 51% of these requirements are eligible for membership. They are re-evaluated (re-certified) based on the same criteria after five years [18].

The overreaching goal of the Cittaslow movement is to improve the quality of life, promote economic growth, protect the natural environment and cultural heritage, promote regional products, crafts, traditions and customs [19]. The movement advocates a slower pace of life as an alternative to the hustle and bustle of large cities and metropolises. The organization does not oppose technological progress, and its members are local economic hubs that place a high value on preserving their unique character [8]. Cittaslow members promote their towns as places with a high quality of life in the proximity of nature, high-quality products and strong local communities [20].

The goals of the Cittaslow movement coincide with the principles of sustainable development, with the main emphasis on increasing local incomes, improving local safety, fostering a friendly social environment, improving the quality of the natural environment, enhancing spatial order and the functionality of urban spatial structures [21]. Sustainable cities are characterized by a cohesive relationship between the urban and the natural environment. Social and environmental movements play an important role, and they focus on preserving the city's unique character, restoring historical architecture, promoting local products and cuisine. In cities that are popular tourist destinations, considerable emphasis is placed on the promotion of local hospitality traditions [16,22]. Globalization increases the risk of cultural uniformity by destroying local values. The Cittaslow concept stimulates local growth as the core element of sustainable development. Member towns strive to preserve their identity and protect local values, both in the Cittaslow network and in the global context [7]. Sustainable development encompasses the environmental, social and economic aspects of urban growth. Community participation plays an important role in decision-making, and participatory processes further the sustainable development of all Cittaslow members at the local level. Cittaslow towns initiate a variety of events and projects with the involvement of the local authorities and residents, which is why the Cittaslow network is also regarded as a social movement and a local governance model [23].

The spatial layout and the potential of small historical towns are well aligned with the main tenets of the Cittaslow movement which are strongly rooted in sustainability [24]. Towns that boast the snail logo strive to improve the quality of local life and cultivate the culture of good living. The small size of urban structures, the cozy ambience of public spaces (squares, streets, parks) and small population encourage the preservation of local traditions and the search for harmony between a slow-paced life and sustainable development [25,26]. Historical towns are characterized by strong place identity, which promotes social integration and fosters a sense of attachment to one's home [27,28]. In towns with historical urban design and architecture, the residents are more likely to identify with and feel responsible for their place of residence than large city dwellers. These factors contribute to the search for novel urban planning solutions that are consistent with the Cittaslow philosophy of a slow life and eco-friendly development [27,29]. Historical structures in small towns well serve the above goals.

The philosophy and ideological assumptions of Cittaslow movement also fit in the evolution of spatial and urban solutions. Changes in various fields of life (economy, culture, etc.) have influenced the changes in its style. Moreover, they have an impact on the physical side of the anthropogenic space [30]. Globalization and computerization have contributed to the modification of social relations and homogenization of space in all dimensions [31]. The dominant role is played by the flow of information, idea, capital, goods and also people [32]. The quality of life indices for the certification and monitoring of the member cities are the premises of the local councils' actions. The modern ideas of spatial planning allow claiming that the sustainable development and the idea of Slow City are based on similar, or even often the same, assumptions. Duplicating the same norms in sustainable planning and the idea of the movement has become the norm emphasized in the modern debates on cities development. Consequently, the social awareness increase, spatial structures are harmonized, the elements of the environment are respected and the results of intensive urbanization are balanced. As a result, all of the above-mentioned elements become the widely acknowledged rule of good practices in planning and spatial management [30]. Moreover, they have the chance to be more widely implicated in other, more complex and more urbanized spatial arrangements.

The integral element of city development is a constant process of changes in which one of the leading roles is assigned to the processes of revitalization. It is a process that includes comprehensive operations in old towns, old districts neglected in terms of the economy, aesthetics, space, infrastructure, communication and functionality. In numerous points, the process converges with the idea of Slow City and sustainable planning. It leads to the prevention of social exclusion, facilitates socio-economic development and it also leads to the improvement of the quality of life of the local society. The revitalization process must be conducted according to the principles of sustainable development with the integration of social, spatial, economic, cultural, construction, investment and ecological aspects within the determined aim [33,34]. The quoted scientific and technical discourse fully confirms the necessity for the integration of the idea of Slow City, the sustainable development and technical actions (f.eg. revitalization) in a widely-acknowledged planning and land use of Cittaslow towns.

3. Materials and Methods

3.1. Study Areas

The study areas were small towns, which are Cittaslow members, in the Italian regions of Tuscany and Umbria, and the Polish region of Warmia and Mazury (Figure 1).

Figure 1. The analyzed regions. Source: own elaboration.

The reason for choosing Cittaslow network for the study is the fact that it is a consistent network of small towns. Moreover, it gives the possibility of conducting the research on various urban organisms. While achieving the objectives, Cittaslow solicits raising the effectiveness of economic growth, it puts emphasis on a high level of education and helps in reducing poverty; it also develops newer and newer innovative economic projects. The objectives fit in the ideas of sustainable development. Among strong points of Cittaslow network, one can list [35]:

- possibility of uniting nationalities that exchange experiences and create the new quality of local life;
- creating common frames for development based on "the true-life" values;
- possibility of individualized towns development based on endogenous resources;
- creating a common brand that would assure the identification on the world scale as well as on the national (local) scale;
- building partnership with inhabitants for the need of taking developmental decisions;
- building the identity of the towns' inhabitants (Cittaslow Sunday/Week);
- exchanging good practices, including study visits;
- building common socio-economic undertakings.

Cittaslow network unites mostly historical towns which drew the attention of inter alia the scientists from Massachusetts Institute of Technology City Science (USA). They stated: "We went back in time and looked at towns in which we all would like to be—a lot of them belong to the historical European towns". They think that towns that work the best are towns that had been organized before the car was invented. The towns include districts with all essential everyday life facilities. It is the so-called [36].

The value of small towns is emphasized also in The European Commission studies on the basis of which one can state that [37]:

- the towns have influence on the well-being and the living not only of their inhabitants but also the surrounding rural population,
- they make public and private service centers and they collect knowledge, create innovations and infrastructure on the local and regional scale,
- they make the base of urban regions and they give the character of regional landscapes, and determine their distinctiveness,
- their structure of growth and development in Western Europe makes the best sustainable urban system in the world,
- generic features of small and medium-sized towns, especially the human dimension, living conditions, the friendly nature of the districts and geographic rooting and historical character, make the perfectly sustainable urbanization,
- they are the indispensable element to prevent depopulation in rural areas and migration to towns; they are also necessary to reach regional sustainable development, unity and sustainable Europe.

These regions were selected for the study due to similarities in their spatial layout. The analyzed regions are highly attractive and popular tourist destinations that are widely visited by domestic tourists as well as tourists from other countries in Europe and the world. The popularity of the studied regions can be attributed to their unique environment, landscape, topographic features and cultural heritage, including historical urban fabric and world-class monuments such as castles, churches and parks [38–41]. Moreover, Tuscany is the cradle of the Cittaslow movement, whereas Warmia and Mazury is a region with the highest number of Cittaslow towns in Poland. One should mention that Polish National Cittaslow Network is the second in the terms of the number of town members in the world [35].

Small towns are characterized by similarities and differences in their topographic features, historical factors which influenced the evolution of urban structures, culture, habits, building traditions, as well as climate which shaped the local architecture. The above factors are responsible for the differences and similarities in the spatial structure of small towns, but towns that are members of the Cittaslow network pursue similar goals and strive to fulfill modern needs and aspirations. The analyzed towns differ considerably in population which ranges from 505 to nearly 28,000 in the studied Italian regions, and from 1964 to more than 23,000 in the Polish region.

3.1.1. Italian Towns

Tuscan and Umbrian towns are much older than their Polish counterparts. Some Italian towns were established in Roman times, and they are regarded as the cradles of European civilization [42]. The founding dates of some towns cannot be determined accurately due to the turbulent history of the Italian peninsula. Small settlements and Roman military camps were transformed into towns that were destroyed and reconstructed on numerous occasions over the centuries. Other towns were built as part of the urbanization effort of the Roman Empire [43,44]. Topographic features and other environmental factors significantly influenced the spatial structure of Italian towns, most of which have a highly irregular street pattern [45]. However, these factors have little bearing on the preserved urban fabric which dates back to later periods in history. The present size and population of Italian towns is unrelated to topographic features [46]. The territorial reach of Italian towns continued to expand as their economies grew in successive centuries, but their original layout has been preserved to this day. The study covered 15 towns in Tuscany and 11 towns in Umbria:

1. Tuscany—Abetone Cutigliano, Anghiari, Barga, Bucine, Capalbio, Castelnuovo Berardenga, Civitella in val di Chiana, Cortona, Greve in Chianti, Marradi, Massa Marittima, Pratovecchio, San Miniato, San Vicenzo, Suvereto;
2. Umbria—Amelia, Citta della Pieve, Ficulle, Monte Castello di Vibio, Montefalco, Orvieto, Parrano, Preci, Todi, Torgiano, Trevi.

3.1.2. Polish Towns

The founding dates of towns in the Polish region of Warmia and Mazury are much easier to determine. Source documents citing charters that granted town rights to settlements in Warmia and Mazury have survived to this day. The relevant dates are also easier to establish because Polish towns were founded later than their Italian counterparts. Warmian and Masurian towns were granted charters because they were established on virgin territory. Their founding dates are indicated in historical documents [47].

The analyzed towns have evolved over the centuries. Their present size and population differ significantly from the original figures, but similarly to Italian towns, their historical layout can still be identified. The study analyzed 22 towns in the Polish Voivodeship of Warmia and Mazury: Barczewo, Bartoszyce, Biskupiec, Bisztynek, Braniewo, Dobre Miasto, Działdowo, Gołdap, Górowo Iławeckie, Jeziorany, Lidzbark Warmiński, Lidzbark, Lubawa, Nidzica, Nowe Miasto Lubawskie, Olsztynek, Orneta, Pasym, Reszel, Ryn, Sępopol, and Wydminy.

3.2. Methods

Multidimensional analyses and evaluations are required to assess the multiperspective functioning of historical towns and the ability of their spatial structures to meet the needs and aspirations of modern communities. The evaluated towns are characterized by similarities as well as differences in this respect. The objective of the applied original research methods was to verify the assumption that sustainable municipal development is closely related with spatial attributes. Some of the applied methods were based on the approaches that are widely described in the literature [13,24,27,28,46–55], and they involved analyses of the current status of historical urban layout. The remaining methods were developed by the authors, and they rely on subjective evaluations of the towns' spatial structure which is defined by the historical urban fabric and architectural features. This approach is largely universal, and it can be applied in towns situated in various regions of the world. The spatial structure of the analyzed towns was assessed with the use of three criteria:

1. Multi-criteria evaluation of the spatial structure of small towns;
2. Evaluation of the influence of spatial attributes on the townscape;
3. Analysis and evaluation of the spatial structure of the studied towns.

3.2.1. Multi-Criteria Evaluation of the Spatial Structure of Small Towns

The aim of this evaluation was to determine the extent to which a slower pace of life in small cities contributes to sustainable development. Due to the large number of analyzed towns (22 towns in the Polish region of Warmia and Mazury, 15 towns in Tuscany, and 11 towns in Umbria, Italy), a quantitative research method was applied to identify and systematize the key elements that determine the quality of urban spaces. The following elements were taken into consideration:

1. Founding date—iconographic sources and an analysis of the literature relating to the history of Italian towns [56,57]. Polish towns were evaluated mainly by reviewing German-language literature and iconographic sources dating back to the 19th century [58,59]. The aim of the analysis was to identify differences in the spatial layout of the examined towns and to determine possible correlations between a town's founding date and its urban layout (Table 1).
2. Topographic features—iconographic sources (maps, satellite images, etc.). The location of the studied towns was analyzed relative to the predominant landform types, including hills, river bends, banks of a river or another water body. Topographic features exerted a considerable influence on the layout of urban structures (Table 2).
3. Urban layout—iconographic sources (maps, satellite images, etc.). The urban planning solutions in the studied towns were analyzed in view of typical Medieval town patterns:

    *   towns that evolved from the existing settlements;
    *   towns built around Roman military camps;
    *   towns built in cruda radice (from a "raw root"), usually along or at the crossroads of major transportation routes.

4. Size and shape of market squares—iconographic sources (maps, satellite images, etc.). The size and shape of market squares in Medieval towns were adapted to the towns' area and economic status in the region. Market squares were assessed based on their size, organization and role in the contemporary urban structure (Table 3).
5. Town area enclosed by fortified walls, present-day area, increase in developed area (IDA) since establishment—iconographic sources (historical maps, satellite images). At the time of their foundation, the area of the studied towns was defined by fortified walls or topographic features (hill, bank of a river or another water body). The original and present area of the evaluated towns was calculated from satellite images with the use of the AutoCad software package. The territorial reach of Medieval towns was determined based on an analysis of satellite maps and the authors' knowledge of the history of urban development in Europe. The analyzed towns had undergone numerous transformations throughout history, and the calculated built-up area within fortified walls was only a rough estimate, but it was sufficient for the needs of this study. An accurate determination of the examined towns' original area would require archeological surveys which fall outside the scope of this study.

Over the centuries, continued economic growth led to the territorial expansion of the studied towns. The towns' original and present area was compared, and the attractiveness of historical urban fabric relative to modern forms of urban development was evaluated with the use of the following indicator:

$$IDA = PA/TAF \tag{1}$$

where: *IDA*—increase in developed area; *PA*—present area; *TAF*—town area at the time of foundation.

Table 1. The influence of founding date on the urban layout of the analyzed towns.

No.	Period of Origin	Urban Layout
1.	Roman era (3rd century BCE—5th century CE)	• towns were built around Roman military camps; • towns developed around the existing settlements (irregular urban layout).
2.	Early Medieval period (8th to 10th century)	• towns were built around Roman military camps; • towns developed around the existing settlements (unplanned urban layout); • towns were built in cruda radice (from a "raw root") with irregular urban layout or a grid street pattern.
3.	Medieval period (11th to 14th century)	• towns developed around the existing settlements (irregular urban layout); • towns were built in cruda radice (from a "raw root") with irregular urban layout or a grid street pattern.

Source: own elaboration based on Tołwiński [48].

Table 2. Correlations between landform and urban planning solutions.

No.	Topographic Feature	Urban Planning Solutions
1.	Hill with steep slopes	• grid street plan or irregular urban layout; • dense urban network; • dense development resulting from space constraints.
2.	River bend	• grid street plan with various degrees of regularity; • uncontrolled urban development in the direct vicinity of the river; • urban expansion in a direction opposite to the river.
3.	Bank of a river or another water body	• grid street plan; • rib pattern or a regular grid patter; • urban expansion in a direction opposite to the river or another water body.

Source: own elaboration based on Czubiel and Domagała [47].

Table 3. Shape and size of market squares in the studied towns.

No.	Shape of Market Square	Size of Market Square
1.	■	The shape and relative surface area of the market square (ratio of market area to town area) influence perceptions of urban structures.
2.	▬	Based on their shape and size, market squares can be perceived as neutral, harmonious, elongated, dynamic, aggressive or cohesive.
3.	◣	The relative surface area of market squares in the analyzed towns
4.	◤	was classified as:
		0.4–2.5%—small:
5.	◢	2.6–10.0%—average:
6.	◣	10.1–21%—large

Source: own elaboration based on Czubiel and Domagała [47].

3.2.2. Evaluation of the Influence of Spatial Attributes on the Townscape

This stage of the research involved cartographic methods as well as field surveys of Polish towns and selected Italian towns. Urban structures were evaluated by directly experiencing the distinctive atmosphere of historical towns. Direct observations were supplemented by analyses of photographs and video footage. Iconographic sources such as maps and satellite images were also used to examine the key elements of the urban structure. These research methods supported assessments of individual spatial attributes which stimulate the senses and enable spectators to experience the unique/distinctive atmosphere of a place (*genius loci*). The following factors were taken into consideration:

1. Urban layout—cartographic sources, observations and field surveys. The extent to which Medieval urban forms have been preserved was evaluated to determine whether and to what degree the development of the studied towns affected their original design. According to Camillo Sitte, an urban theorist whose work exerted a considerable influence on contemporary urban planning, the designers of Medieval towns in Italy appreciated the value of irregularity and esthetics in the urban form [54]. The evolution of Italian towns was influenced by topographic and environmental features as well as their territorial expansion over the centuries. Polish towns were established based on a different set of principles. Their streets followed a strict geometric pattern that was adapted to local conditions, including topographic and environmental features. Towns are living organisms that evolve over time. For this reason, the functions of Medieval towns should be adapted to modern requirements without compromising their original urban layout that has been shaped by historical, topographic and cultural factors. Street patterns were analyzed in:

   - Italian towns with a free-form urban pattern and an irregular street network, where the junction of the main roads acts as a focal point in the town center;
   - Polish towns with a grid street pattern, rigid compositional axes and a clearly defined central area.

   The urban forms in both Polish and Italian towns are aesthetically pleasing, but they evoke completely different perceptions of urban space (Table 4).

Table 4. The influence of urban design on perceptions of space.

No.	Criterion	Perceptions of Space
1.	Free-form urban composition	• authentic historical experience; • enhanced/reduced esthetic experience; • unpredictable urban planning solutions.
2.	Grid street plan	• visual stimuli support the formulation of comprehensive perceptions of urban space; • spatiotemporal series; • predictable urban planning solutions.

Source: own elaboration based on Hall [49], Wejchert [51], Gzell [54].

2. Urban open spaces—urban open spaces such as municipal squares, pocket parks and small public parks were analyzed in Italian and Polish cities. The evaluated open spaces differed in form and design, they were user-friendly and attracted local residents as well as tourists (Table 5).

3. Architecture—local surveys of the quality of urban space. Buildings and structures are the key elements of the townscape, and they build a sense of local identity. Urban architecture is responsible for the distinctive atmosphere of a place and subjective perceptions of the built environment. The unique ambience of Medieval towns can be largely attributed to historical architecture. The form, layout, scale and detail of architectural design, the applied building materials, their texture and color directly influence the human senses. Non-material factors, such as the town's founding date and important historical events, are also powerful stimuli that generate positive perceptions of space [60]. The historical architecture of the studied towns was analyzed to determine its influence on esthetic and sensory perceptions of space. Architecture is the main element of the urban fabric which influences the quality and attractiveness of space. A town's architecture stimulates the senses and creates an esthetic experience (Table 6).

4. Conservation status of historical architecture—iconographic sources, field surveys. The conservation status of historical architecture in the studied towns was evaluated based on source documents and knowledge of the history of architecture. The results

were used to estimate the date of construction, identify the architectural style and assess the authenticity of the analyzed buildings (Table 7).

5. Scale of urban development—the area of the studied towns was analyzed in view of historical changes in their territorial reach relative to their original area. The territorial expansion of Medieval towns was an inevitable consequence of their economic growth. The historical architecture in the center of small towns has to be preserved against the background of modern development to maintain the esthetic composition of the urban structure. This is a necessary prerequisite for preserving the unique identity of historical towns, and it demonstrates that towns are living organisms that evolve over time (Table 8).

Table 5. The influence of urban open spaces on perceptions of space.

No.	Criterion	Perceptions of Space
1.	Squares	• user-friendly spaces; • architecture and green infrastructure contribute to social integration and a sense of local identity.
2.	Municipal parks	• recreational areas; • architecture and green infrastructure contribute to a sense of identity and responsibility for one's place of residence.

Source: own elaboration based on Wejchert [51], Montgomery [53].

Table 6. The influence of architecture on perceptions of space.

No.	Criterion	Perceptions of Space
1.	Individual buildings and structures	• form, scale, building materials, facade color; • architectural style; • positive/negative esthetic experience; • no esthetic experience.
2.	Frontage buildings, urban block	• diverse architectural forms and styles; • form, division, rhythm, roof pattern, building materials, facade color; • enhanced/reduced esthetic experience; • *genius loci*; • architecture contributes to a sense of local identity.

Source: own elaboration based on Burgess [28], Norberg-Schulz [52], Radwan [55].

Table 7. The influence of urban components on perceptions of the townscape.

No.	Conservation Criteria	Assessment
1.	Historical urban design	• analysis of cartographic sources; • field surveys; • knowledge of the history of urban planning.
2.	Historical architecture	• analysis of iconographic sources; • field surveys; • knowledge of the history of architecture.
3.	Identification of the market square	• preservation of the original shape, structure and size; • conservation status of historical buildings flanking the market square; • quality of urban space (urban and architectural design, street furniture).

Table 7. *Cont.*

No.	Conservation Criteria	Assessment
4.	Dominant buildings	• conservation status of dominant buildings in the historical townscape (church, town hall, tower, fortified structures); • composition of townscape elements based on the preserved dominant buildings.
5.	*Genius loci*	• distinctive atmosphere; • subjective emotional perceptions associated with historical architecture.

Source: own elaboration based on Norberg-Schulz [52], Zagroba and Gawryluk [24].

Table 8. Factors responsible for the territorial expansion of the studied towns.

No.	Criterion	Assessment
1.	Historical center	• the urban layout of the historical center is consistent with the original town plan dating back to the period of establishment; • dense, intensive development; • the scale of urban development accentuates the historical center against the backdrop of contemporary urban areas; • historical buildings have a consistent architectural style; • historical centers have a unique atmosphere (*genius loci*).
2.	Developed areas outside the historical center	• unlike historical towns, contemporary urban structures are planned along regional transportation routes; • dispersed development; • the scale of urban development is largely influenced by the historical center; • modern architecture is radically different from historical urban forms; • modern urban development is devoid of the unique atmosphere of historical centers.

Source: own elaboration based on Alexander [27], Benevolo [46], Rossi et al. [13].

The factors listed in Table 8 were used to evaluate the quality of urban space in historical towns. The quality of the historical urban fabric is determined by perceptions of structural archetypes, distinctive features that contrast with contemporary urban structures outside the historical center, as well as compositional factors such as the scale of urban development and urban forms.

3.2.3. Analysis and Evaluation of the Spatial Structure of the Studied Towns

An intuitive (holistic) research method was applied to determine the extent to which the distinctive ambience (*genius loci*) of historical towns shapes the urban landscape in the evaluated regions and contributes to an improvement in the local standard of living. The influence of urban design elements, including structures, buildings, town blocks, streets and squares, on the quality of urban space was evaluated in Polish and Italian towns. Historical architecture and urban design are manifestations of a region's cultural heritage, and they evoke strong emotional responses. The analysis revealed strong links between the quality of local life and the protection of heritage sites. These relationships have evolved differently in Italian and Polish towns, which can be attributed to differences in culture, climate and history (Table 9).

Table 9. Preservation of historical spatial components in the studied towns.

No.	Subjective Evaluation—*Genius loci*	Preservation of Historical Spatial Components of Urban Space
1.	Preserved	• town blocks preserved in original form; • preserved historical architecture; • distinctive atmosphere of a historical town.
2.	Partly preserved	• minor modifications of historical components exert a negligible influence on the overall legibility of urban design; • foreign and modern architectural forms exert a minor effect on historical urban design; • the existing urban structure has a predominantly historical character.
3.	Neutral	• considerable modifications of historical components significantly compromise the overall legibility of urban design; • a predominance of foreign architectural forms significantly detracts from historical urban design; • the structural layout of historical towns is difficult to identify.
4.	Not preserved	• the legibility of the original urban design has been lost; • architectural design does not make a reference to historical forms; • urban spaces do not reflect the town's history.

Source: own elaboration based on Rossi [50], Norberg-Schulz [52].

4. Results

The data collected with the use of the described research method were interpreted based on three criteria for analyzing the spatial structure of the studied towns. The results of the study are presented separately for each stage of the conducted research for greater readability.

4.1. Multi-Criteria Evaluation of the Spatial Structure of Small Towns–Results

Italian towns differ considerably in founding dates because they were established from Roman times to the Medieval period (Figures 2 and 3).

Most of the analyzed Polish towns were established in same century (Figure 4).

The spatial structure of Italian and Polish towns was determined by topographic features and environmental conditions. All towns were founded in areas with favorable terrain characteristics (hill with steep slopes, river bend, bank of a river or another water body) which acted as natural barriers and protected the inhabitants against enemies (Figure 5).

A centrally located market square has always occupied an important place in the spatial structure of the studied towns. The size and shape of historical market squares differ considerably in the analyzed urban structures, but most towns have rectangular market squares (Figure 6).

The size of urban structures in Medieval towns was determined by the towns' economic status and technical capabilities (Tables 10–12). The average area of Medieval towns within fortified walls was determined at 3.89 ha in Tuscany, 6.04 ha in Umbria, and 5.37 ha in Warmia and Mazury. Over the centuries, economic growth fueled the expansion of the analyzed urban structures outside defensive walls. Tables 10–12 present the developed area of Italian and Polish towns, not the area enclosed by their administrative boundaries (which includes undeveloped areas such as forests, water bodies and agricultural land).

Only built-up areas with the accompanying green infrastructure were measured based on orthophotomaps.

The area enclosed by fortified walls was used to calculate the relative area of the market square, which was expressed as the ratio of market square area to developed urban area (Figures 7 and 8). The average relative market square area was determined at 2.15% in Tuscany, 1.91% in Umbria, and 10.92% in Warmia and Mazury.

As demonstrated by Figures 9 and 10, the studied towns continued to expand dynamically throughout history. The greatest territorial expansion was noted in the 19th and 20th centuries. The character and density of urban development also evolved, and this process was accompanied by changes in urban design and architectural style. The average increase in developed area was determined at 19.89% in Tuscan towns, 16.50% in Umbrian towns, and 25.35% in Polish towns in the Region of Warmia and Mazury.

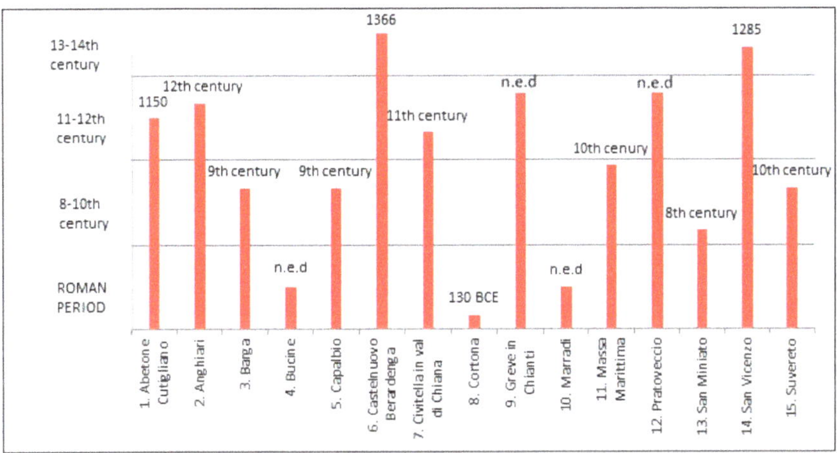

Figure 2. Founding dates of Tuscan towns. Source: own elaboration based on the data published by the Italian National Institute of Statistics [61] and Atlante Storico Iconografico Delle Città Toscane [62]. Key: n.e.d—no exact data available.

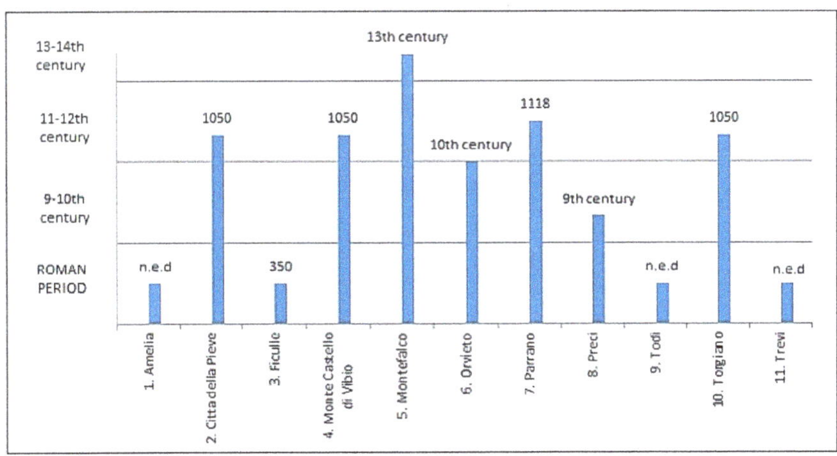

Figure 3. Founding dates of Umbrian towns. Source: own elaboration based on the data published by the Italian National Institute of Statistics [61] and Ross [63]. Key: n.e.d—no exact data available.

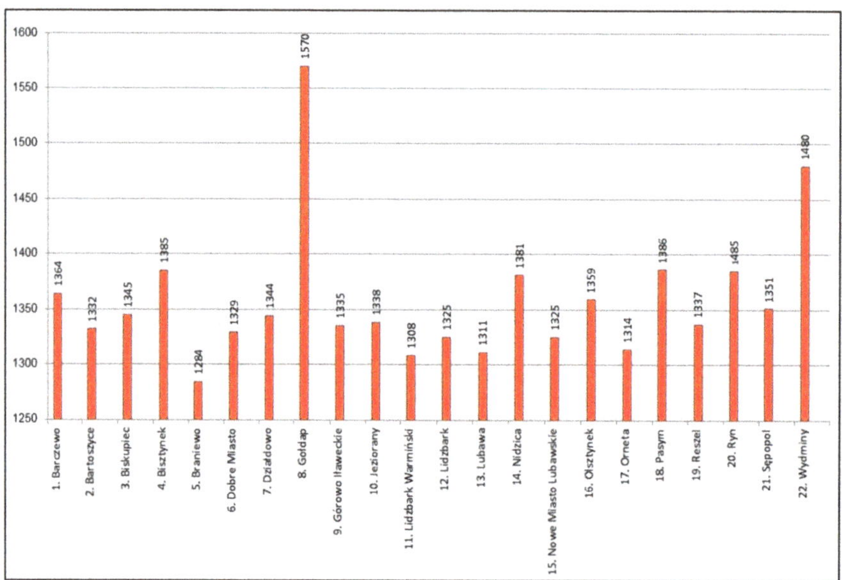

Figure 4. Founding dates of Polish towns. Source: own elaboration based on Czubiel and Domagała [47].

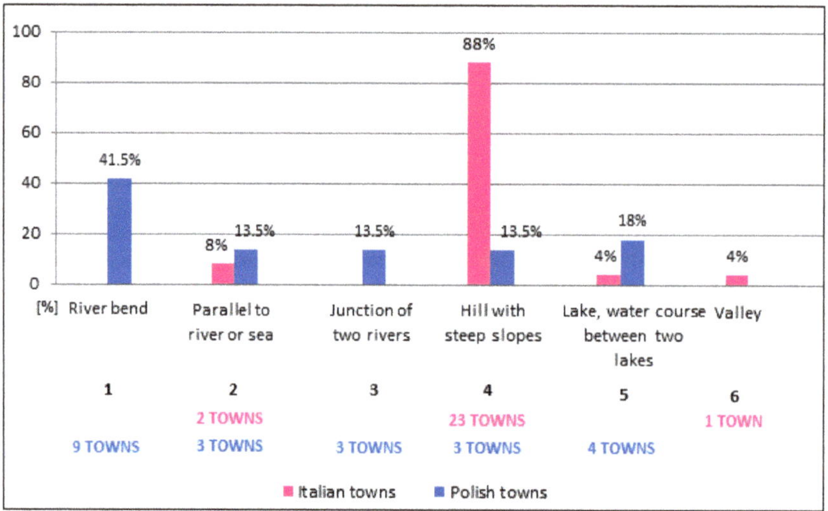

Figure 5. Topographic features in Polish and Italian towns. Source: own elaboration based on cartographic analyses. Key: 1—Polish 9 towns: Bartoszyce, Biskupiec, Braniewo, Lidzbark Warmiński, Nidzica, Nowe Miasto Lubawskie, Olsztynek, Reszel, Sępopol; 2—Italian 2 towns: Maradi (T), San Vicenzo (T); Polish 3 towns: Bisztynek, Działdowo, Gołdap; 3—Polish 3 towns: Barczewo, Dobre Miasto, Lubawa; 4—Italian 23 towns: Abetone Cutigliano (T), Anghiari (T), Barga (T), Bucine (T), Capalbio (T), Castelnuovo Berardenga (T), Civitella in val di Chiana (T), Cortona (T), Massa Marittima (T), Pratovecchio (T), San Miniato (T), Suvereto (T), Amelia (U), Citta della Pieve (U), Ficulle (U), Monte Castello di Vibio (U), Montefalco (U), Orvieto (U), Parrano (U), Preci (U), Todi (U), Torgiano (U), Trevi (U); Polish 3 towns: Jeziorany, Lidzbark, Orneta; 5—Polish 4 towns: Górowo Iławeckie, Pasym, Ryn, Wydminy; 6—Italian 1 town: Greve in Chianti (T); (T)—Tuscany, (U)—Umbria.

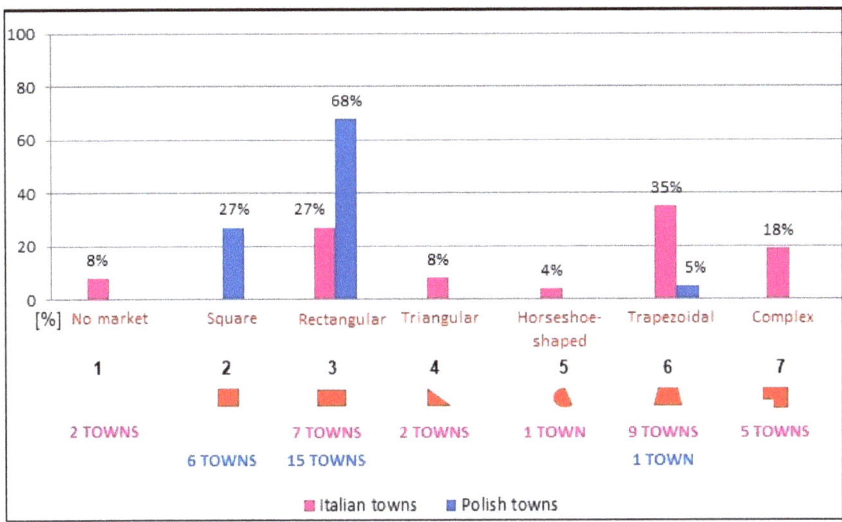

Figure 6. Shape of market squares in Italian and Polish towns. Source: own elaboration. Key: 1—Italian 2 towns: Bucine (T), San Vicenzo (T); 2—Polish 6 towns: Działdowo, Gołdap, Lidzbark Warmiński, Lubawa, Nowe Miasto Lubawskie, Reszel; 3—Italian 7 towns: Capalbio (T), Marradi (T), Suvereto (T), Monte Castello di Vibio (U), Orvieto (U), Todi (U), Torgiano (U); Polish 15 towns: Barczewo, Bartoszyce, Biskupiec, Bisztynek, Braniewo, Dobre Miasto, Górowo Iławeckie, Jeziorany, Lidzbark, Olsztynek, Orneta, Nidzica, Pasym, Ryn, Wydminy; 4—Italian 2 towns: Castelnuovo Berardenga (T), Greve in Chianti (T); 5—Italian 1 town: Montefalco (U); 6—Italian 9 towns: Barga (T), Civitella in val di Chiana (T), Cortona (T), San Miniato (T), Citta della Pieve (U), Preci (U), Ficulle (U), Parrano (U), Trevi (U); Polish 1 town: Sępopol; 7—Italian 5 towns: Abetone Cutigliano (T), Anghiari (T), Massa Marittima (T), Pratovecchio (T), Amelia (U).

Table 10. Tuscan towns—developed area.

No.	Town	Developed Area within City Walls [ha]	Contemporary Built-Up Area [ha]
1.	Abetone Cutigliano	2.2	26
2.	Anghiari	2.3	116
3.	Barga	6.5	140
4.	Bucine	0	71
5.	Capalbio	1.5	19
6.	Castelnuovo Berardenga	2.8	56
7.	Civitella in val di Chiana	2.2	43
8.	Cortona	13.7	55
9.	Greve in Chianti	5.1	72
10.	Marradi	3.3	83
11.	Massa Marittima	6.2	125
12.	Pratovecchio	7.0	75
13.	San Miniato	2.1	150
14.	San Vicenzo	0	274
15.	Suvereto	3.4	57

Key: zero values denote cities without fortified walls; contemporary built-up area—urbanized area (developed area with the accompanying green infrastructure). Source: own elaboration based on Cittaslow Italia [64] and Google Maps [65].

Table 11. Umbrian towns—developed area.

No.	Town	Developed Area Within City Walls [ha]	Contemporary Built-Up Area [ha]
1.	Amelia	12.6	139
2.	Citta della Pieve	5.4	102
3.	Ficulle	2.5	22
4.	Monte Castello di Vibio	2.4	27
5.	Montefalco	3.3	56
6.	Orvieto	9.3	92
7.	Parrano	1.2	19
8.	Preci	1.9	39
9.	Todi	13.4	198
10.	Torgiano	9.0	128
11.	Trevi	5.4	212

Key: contemporary built-up area—urbanized area (developed area with the accompanying green infrastructure). Source: own elaboration based on Cittaslow Italia [64] and Google Maps [65].

Table 12. Towns in Warmia and Mazury—developed area.

No.	Town	Developed Area within City Walls [ha]	Contemporary Built-Up Area [ha]
1.	Barczewo	6.0	458
2.	Bartoszyce	9.0	1179
3.	Biskupiec	6.0	500
4.	Bisztynek	6.0	216
5.	Braniewo	9.0	466
6.	Dobre Miasto	6.0	486
7.	Działdowo	8.4	1147
8.	Gołdap	0	38
9.	Górowo Iławeckie	5.6	332
10.	Jeziorany	4.5	341
11.	Lidzbark Warmiński	6.0	1435
12.	Lidzbark	4.8	568
13.	Lubawa	4.8	1684
14.	Nidzica	6.0	686
15.	Nowe Miasto Lubawskie	8.4	1137
16.	Olsztynek	4.8	769
17.	Orneta	8.0	963
18.	Pasym	4.5	1518
19.	Reszel	6.0	382
20.	Ryn	0	414
21.	Sępopol	4.5	463
22.	Wydminy	0	83

Key: zero values denote cities without fortified walls; contemporary built-up area—urbanized area (developed area with the accompanying green infrastructure). Source: own elaboration based on Cittaslow Polska [66] and Google Maps [65].

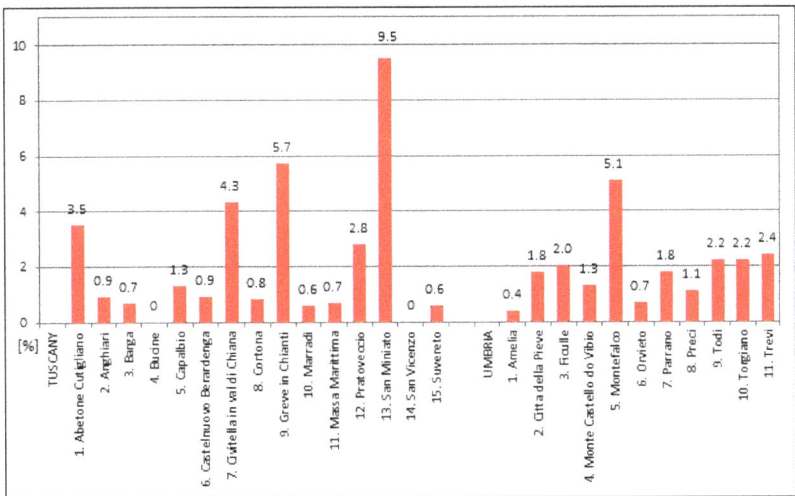

Figure 7. Relative market square area in Italian towns. Source: own elaboration. Key: zero values denote cities without fortified walls.

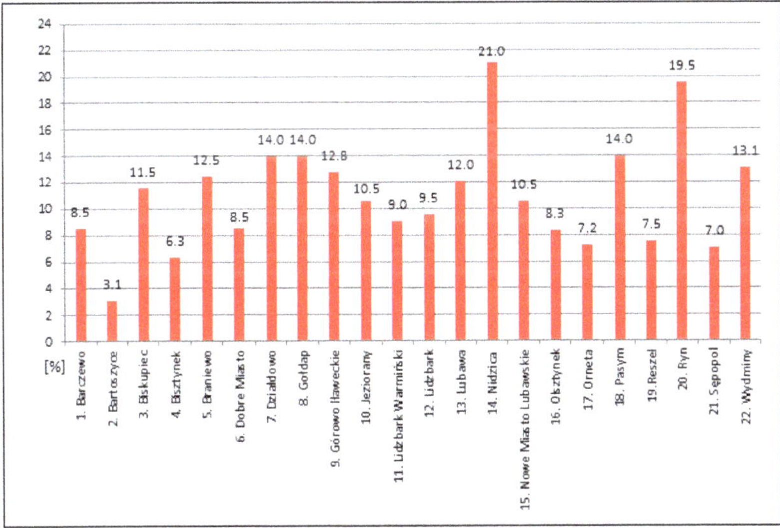

Figure 8. Relative market square area in Polish towns. Source: own elaboration.

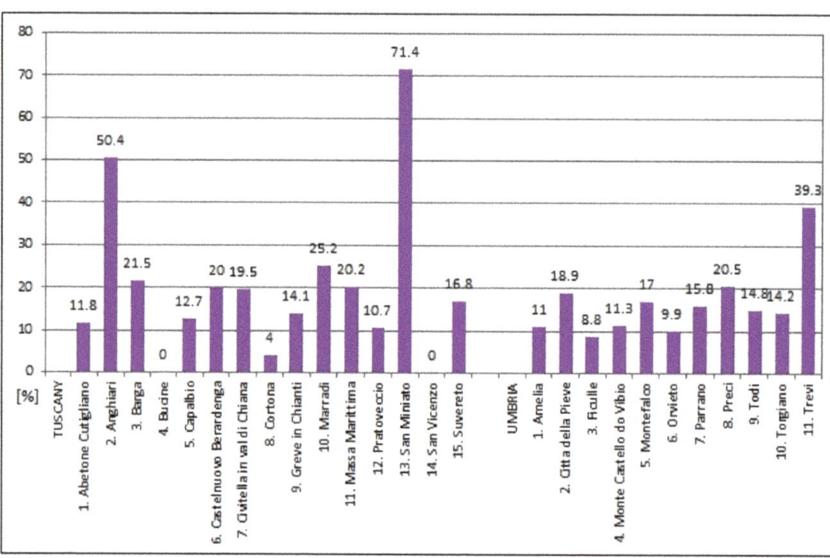

Figure 9. Increase in the developed area of Italian towns. Source: own elaboration. Key: zero values denote towns where the historical urban fabric was not preserved.

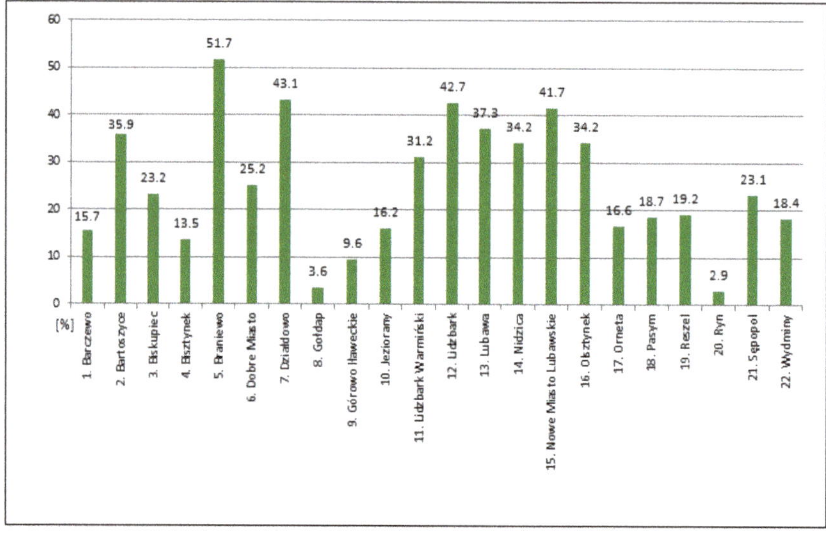

Figure 10. Increase in the developed area of Polish towns. Source: own elaboration.

4.2. Evaluation of the Influence of Spatial Attributes on the Townscape—Results

An analysis based on the adopted criteria (Tables 4–8—urban layout, public urban spaces such as municipal squares, pocket parks and municipal parks, architecture, scale of urban development) revealed that historical solutions considerably influenced urban planning in successive centuries. Historical town centers are characterized by different scale, density and type of development than other built-up areas, and they constitute the focal points in the studied towns (Tables 13–15).

Table 13. Preservation of historical spatial components in Tuscan towns.

TUSCANY

Key: Preserved / Not preserved / Partly preserved (cells are colour-coded in the original).

Criteria for Evaluating the Preservation of Historical Spatial Layout	Abetone Cutigliano	Anghiari	Barga	Bucine	Capalbio	Castelnuovo Berardenga	Civitella in val di Chiana	Cortona
Historical urban layout	Preserved	Preserved	Preserved	Not preserved	Preserved	Preserved	Not preserved	Preserved
Old market square	Preserved	Preserved	Preserved	Not preserved	Preserved	Preserved	Not preserved	Preserved
Historical architecture	Not preserved	Preserved	Partly preserved	Not preserved	Preserved	Preserved	Not preserved	Preserved
Dominant buildings	Not preserved	Preserved	Preserved	Not preserved	Preserved	Preserved	Not preserved	Preserved

Criteria for Evaluating the Preservation of Historical Spatial Layout	Greve in Chianti	Marradi	Massa Marittima	Pratovecchio	San Miniato	San Vicenzo	Suvereto
Historical urban layout	Preserved	Preserved	Preserved	Not preserved	Preserved	Not preserved	Preserved
Old market square	Preserved	Preserved	Preserved	Not preserved	Preserved	Not preserved	Preserved
Historical architecture	Partly preserved	Partly preserved	Partly preserved	Not preserved	Preserved	Not preserved	Preserved
Dominant buildings	Preserved	Preserved	Preserved	Not preserved	Preserved	Not preserved	Preserved

Source: own elaboration. Key: Preserved ▪ Not preserved ▪ Partly preserved.

Table 14. Preservation of historical spatial components in Umbrian towns.

UMBRIA

Criteria for Evaluating the Preservation of Historical Spatial Layout	Amelia	Citta della Pieve	Ficulle	Monte Castello di Vibio	Montefalco	Orvieto	Parrano	Preci	Todi	Torgiano	Trevi
Historical urban layout	Preserved	Preserved	Preserved	Preserved	Preserved	Preserved	Preserved	Preserved	Preserved	Not preserved	Preserved
Old market square	Preserved	Preserved	Preserved	Preserved	Preserved	Preserved	Preserved	Partly preserved	Preserved	Not preserved	Preserved
Historical architecture	Preserved	Preserved	Preserved	Preserved	Preserved	Preserved	Preserved	Partly preserved	Preserved	Not preserved	Preserved
Dominant buildings	Preserved	Preserved	Preserved	Preserved	Preserved	Preserved	Preserved	Preserved	Preserved	Not preserved	Preserved

Source: own elaboration. Key: Preserved ▪ Not preserved ▪ Partly preserved.

Table 15. Preservation of historical spatial components in Polish towns in the Region of Warmia and Mazury.

WARMIA AND MAZURY

Criteria for Evaluating the Preservation of Historical Spatial Layout	Barczewo	Bartoszyce	Biskupiec	Bisztynek	Braniewo	Dobre Miasto	Działdowo	Gołdap	Górowo Iławeckie	Jeziorany	Lidzbark Warmiński
Historical urban layout	Preserved	Preserved	Preserved	Preserved	Preserved	Partly preserved	Preserved	Preserved	Preserved	Preserved	Preserved
Old market square	Preserved	Preserved	Not preserved	Not preserved	Partly preserved	Not preserved	Preserved	Preserved	Preserved	Preserved	Preserved
Historical architecture	Partly preserved	Partly preserved	Partly preserved	Partly preserved	Not preserved	Not preserved	Preserved	Not preserved	Preserved	Preserved	Partly preserved
Dominant buildings	Partly preserved	Partly preserved	Partly preserved	Partly preserved	Partly preserved	Partly preserved	Partly preserved	Preserved	Preserved	Preserved	Partly preserved

Criteria for Evaluating the Preservation of Historical Spatial Layout	Lidzbark	Lubawa	Nidzica	Nowe Miasto Lubawskie	Olsztynek	Orneta	Pasym	Reszel	Ryn	Sępopol	Wydminy
Historical urban layout	Preserved	Preserved	Partly preserved	Preserved	Preserved	Preserved	Preserved	Preserved	Not preserved	Partly preserved	Preserved
Old market square	Preserved	Preserved	Not preserved	Preserved	Not preserved	Preserved	Partly preserved	Preserved	Not preserved	Partly preserved	Partly preserved
Historical architecture	Partly preserved	Partly preserved	Preserved	Not preserved	Partly preserved	Preserved	Preserved	Preserved	Not preserved	Not preserved	Preserved
Dominant buildings	Partly preserved	Preserved	Partly preserved	Preserved	Partly preserved	Partly preserved	Preserved	Preserved	Partly preserved	Preserved	Not preserved

Source: own elaboration. Key: Preserved ▪ Not preserved ▪ Partly preserved.

4.3. Analysis and Evaluation of the Spatial Structure of the Studied Towns—Results

The distinctive atmosphere (*genius loci*) of a town is determined by various components of the urban fabric, including material factors as well as non-material factors which cannot be quantified, but which influence the human senses and are responsible for esthetic perceptions of space. The studied towns differed considerably in subjective perceptions of esthetic appeal (Tables 16–18).

Table 16. *Genius loci* in Tuscan towns.

TUSCANY														
Criteria for Evaluating the Preservation of Historical Spatial Layout	Abetone Cutigliano	Anghiari	Barga	Bucine	Capalbio	Castelnuovo Berardenga	Civitella in val di Chiana	Cortona	Greve in Chianti	Marradi	Massa Marittima	Pratovecchio	San Miniato	San Vicenzo	Suvereto
Genius loci of the urban fabric														

Source: own elaboration. Key: Preserved ■ Not preserved ■ Partly preserved.

Table 17. *Genius loci* in Umbrian towns.

UMBRIA										
Criteria for Evaluating the Preservation of Historical Spatial Layout	Amelia	Citta della Pieve	Ficulle	Monte Castello di Vibio	Montefalco	Orvieto	Parrano	Preci	Todi	Torgiano	Trevi
Genius loci of the urban fabric										

Source: own elaboration. Key: Preserved ■ Not preserved ■ Partly preserved.

Table 18. *Genius loci* in Polish towns in the Region of Warmia and Mazury.

WARMIA AND MAZURY																				
Criteria for Evaluating the Preservation of Historical Spatial Layout	Barczewo	Bartoszyce	Biskupiec	Bisztynek	Braniewo	Dobre Miasto	Działdowo	Gołdap	Górowo Iławeckie	Jeziorany	Lidzbark Warmiński	Lidzbark	Lubawa	Nidzica	Nowe Miasto Lubawskie	Olsztynek	Orneta	Pasym	Reszel	Ryn	Sępopol	Wydminy
Genius loci of the urban fabric																						

Source: own elaboration. Key: Preserved ■ Not preserved ■ Partly preserved.

5. Discussion

Cittaslow towns are conjoined with a rich and long history as well as the charm of places ostensibly frozen in time. While analyzing the localization conditions of the Cittaslow network, it is hard to find any regularity. They represent different geographical, historical and most of all spatial types. They are conjoined mainly by the fact that they have historical centers and/or protected areas. Cultural heritage, valuable landscape, local culture and language are appreciated by Cittaslow towns' authorities, inhabitants and tourists. "Slow" cities lay emphasis on all of the values and owing to that they get unique character [67,68].

Historical planning solutions elicit the strongest emotional responses in perceptions of urban space; they contribute to a sense of local identity and create a sense of responsibility for one's place of residence. The protection and promotion of these values constitute the fundamental tenets of the Cittaslow philosophy [69,70]. It should be mentioned that towns that join Cittaslow network commit themselves to keep the cultural heritage connected with the way of living and living conditions [71,72]. Therefore, they achieve the goals of the movement connected with taking care of the historic urban fabric, renovating monuments and towns' aesthetics [73]. It is connected with maintaining and cultivating their historical identity which is not always connected with the functional demands of modern times. Technical and technological development, fire regulations or the civilization progress in general are a challenge in adjusting historical structures to the modern demands. Unlike big cities, small towns have no possibilities for developing industry, building large shopping centers or thoroughfare [74]. The analyzed Polish and Italian towns have many similarities, such as the size of the original urban settlement. However, they differ in other respects, which can be attributed to cultural and climatic differences. Historical urban planning solutions play an important role in this comparison because they testify to the towns' rich cultural heritage which followed European trends in urban design and architecture. The analyzed urban layout is also indicative of differences in local building traditions [75–77]. Qualitative improvement of the built environment should acknowledge local divergence and models as the manifestation of local identity. The greater awareness of key social factors in urban planning and creating architecture influence the higher potential of social sustainability of residential environment [78]. The extent to which urban design components determine the attractiveness of historical towns was evaluated based on the results of field surveys and analyses of iconographic materials. This is a very important consideration in small towns where the physical attributes of space are closely related with the spiritual needs of public space users. It is particularly visible in Slow Cities that apart from rich cultural heritage also have other important endogenous assets: widely underestimated bond and social awareness, and cooperation skills that stem from a deep sense of belonging to the inhabited place [67,79]. Consequently, the inhabitants feel like taking responsibility for their town which is definitely easier to be done in small local societies, where unlike in cities, the bonds are usually really strong [26,74,80]. The applied research methods supported the identification of links between the quality of historical urban space and the local standards of living. These methods were also successfully used to determine why small historical towns uphold the Cittaslow philosophy and whether different urban layouts in Polish and Italian towns create equal opportunities for implementing the principles of the slow living movement.

The analyzed Italian towns do not differ considerably in urban layout despite the fact that they were founded in different periods of time. The free compositional arrangements are dominant in Italian towns, which could have been connected with transforming former settlements, topography, climate and historical conditioning [81]. The factors, regardless of the widely-acknowledged in the Middle Ages rules of town layout, based on the chessboard layout, made the structure of Umbria and Tuscany to exceed the rules of urban composition created centuries ago [82,83]. Regardless of their founding date, the compared urban structures have many similarities, which suggests that climate and local architectural traditions played a key role in the development of urban forms. The layout of Polish towns is completely different than Italian towns' layouts. It corresponds to the general rules for founding a town that were current in the Medieval Times. Polish towns are characterized by more regular layouts which are based on the chessboard layout with a regular market in the middle and a perpendicular grid of streets. [45,48]. Nearly all of the evaluated towns in the Region of Warmia and Mazury have a regular, geometric street pattern with a centrally located market square, a fortified castle and a church.

Towns founded in different historical periods are also characterized by varied urban layouts. A grid street pattern is rarely encountered in Italian towns. In these towns, narrow streets are chaotically distributed, and dense development offers shade and refuge in the hot summer months. In the Middle Ages, dense settlement also increased security because smaller areas were easier to defend against enemies. That feature characterizes towns in the whole Europe [84]. Most of the evaluated towns abound in historical architecture which makes a reference to local building traditions and creates urban forms that are characteristic of the cultural landscape of Tuscany and Umbria. A completely different urban layout was observed in Polish towns. Warmian and Masurian towns were founded on a rectangular street plan, and they were divided into blocks of dense development with a market square in the center [85]. The original urban layout is still legible in most towns. Despite the above, the historical architecture of Polish towns was largely destroyed at the end of World War II. Consequently, the most valuable and the oldest parts of buildings, sometimes the vivid symbols, buildings acknowledged as the elements that shaped the individual landscape of the towns were degraded. Unfortunately, the post-war reconstruction effort, as well as their development, were very often unsuccessful and they often worked the opposite to the historical functional layout, arrangement and the scale of building that had been functioning in historical Polish towns for centuries [33,86]. At that time Poland belonged to the Eastern Block which negatively influenced the maintenance of historical identity of the towns. The post-war reconstruction was done without respect for historical assumptions which consequently led to losing towns' historical character [87].

Polish and Italian towns were established in areas with different landform and natural features. Most Italian towns were founded on steep hills that are ubiquitous in Tuscany and Umbria [88–90]. Hills are local topographic features which acted as natural barriers and protected towns and their inhabitants. In contrast, north-eastern Poland abounds in rivers and lakes which were the main focal points during the establishment of Medieval towns. Many Polish towns were nested in the bend of a river. It was done mainly because a river was indispensable for life. It provided water for people and animals, allowed for gathering food, watered crops, and fertilized soils. Close vicinity to rivers had also served military purposes [91]. Security was of the utmost concern in Medieval times, and towns were founded in areas where local terrain features offered effective protection. Moreover, a river facilitated trade and communication enabled professions that required the presence of water; it allowed draining the rain and urban pollution or using energy resources of river flow. Its close vicinity to a large extent influenced the economic development and it also had an impact on political, demographical and social relations of towns located close to the river [91].

The size of Medieval towns was determined by their ability to satisfy the residents' basic needs, and the complex process of building fortified walls favored dense development. Throughout the centuries, defensive walls limited the territorial expansion of cities whose size was fairly similar and representative of the level of technological advancement in the Middle Ages [92]. The average area of towns enclosed by fortified walls was similar in all studied regions, although it was somewhat larger in Italian towns (Cortona, Amelia and Todi). Polish towns were characterized by similar size, with the exception of Gołdap and Ryn.

Traders sold their goods in centrally located market squares which fueled the economic growth of Medieval towns. The market square was the heart of the town, and its location influenced the urban development pattern in the remaining parts of the town. The size and shape of the market square were determined by the town's area and street pattern [92]. The vast majority of towns in the Region of Warmia and Mazury have rectangular and square markets. Only three market shapes were identified in this region, which could be attributed to the fact that Polish towns were founded in the same historical period. The shape of market squares in Italian towns is more varied. A trapezoidal shape was most frequently encountered, followed by rectangular markets and markets with a complex shape. A total of seven market shapes were identified in the study, but none of the analyzed Italian towns had square-shaped markets. The observed variations are associated with considerable differences in the founding dates of Italian towns.

The functions of market squares have been expanded in the modern times. Market squares are not only the hubs of commercial and economic activity, but they also serve representative and recreational purposes [93,94]. They are historical landmarks that contribute to a sense of local identity and attract tourists and potential investors [95]. The relative area of the market square differs considerably between Italian and Polish towns. The studied towns in the Region of Warmia and Mazury are characterized by significantly higher relative market area which was determined at 10.74% on average and reached up to 20% in Nidzica and Ryn. In Italian towns, relative market square area approximated 10% only in San Miniato (Tuscany). These variations can be attributed to differences in founding dates. Polish towns were established later than their Italian counterparts, and most of them were not built around the existing settlements, but from a "raw root". From the beginning of their existence, Polish towns were adapted to local needs, and the relative proportions of the market square to the remaining forms of urban development could be freely shaped.

Towns are living organisms that continue to evolve, and their territorial reach changes throughout time. The growth and expansion of towns are determined by their regional and supra-regional status, economic performance, quality of space, and standards of living [96–98]. These factors influence the rate of urban expansion in the course of decades and centuries. In Polish towns, built-up area increased most rapidly in the 19th century when small factories (mills, breweries, lumber mills, brick factories) and railway lines were built in the Region of Warmia and Mazury [47]. The expansion of Italian towns was less dynamic due to topographic barriers (closed-form cities established on hills).

Preservation of historical spatial components significantly influences the shaping and perceptions of urban space. The elements, such as urban layouts, i.e., streets, squares, urban interiors and architecture, create the picture of historical changes. The elements to a large extent influence the spatial identity of towns [99]. In the towns of the analyzed areas, the differences are significant. All preservation criteria have been met in Umbrian towns, excluding Torgiano and, partly, Preci. The historical environment of Tuscan towns has also been successfully preserved (excluding Bucine, Civitella in val di Chiana, Pratovecchio, San Vicenzo and, partly, Abetone Cutigliano). The historical urban layout of Polish towns has been partly maintained or lost. Only the first two criteria in Table 15 have been preserved to a satisfactory degree.

The genius loci notion is indispensably connected with the perception of town space. It is not material, the places are special. They are the source of spiritual experience connected with history, the meaning, symbols and aesthetics which are individually perceived. Urban space, individual in every case, is prestigious, "the tokens" of a town with which the inhabitants identify. This relation equals the sense of belonging to a given cultural circle [100]. The significance of the places is unquestionable as it is consolidated in social awareness by particular events, representational buildings or functions [101]. In urban space, it is significant as town planning and architecture enrich the spiritual sphere, they are born in social space which remains as the space full of meanings and symbols, especially in historical towns [102]. For this reason the genius loci is the most subjective and elusive factor in evaluations of historical urban forms. Historical downtown areas differ considerably from the remaining parts of the urban fabric, and they have distinctive features that have been adapted to contemporary needs without a loss of local identity. The results of the study revealed that the preservation of historical spatial components is directly responsible for the unique ambience of historical towns. This observation is validated by the fact that Torgiano and Preci are the only Umbrian towns without a genius loci. There are no discernible elements that would signify the rich history of the town. The urban layout and architecture do not have hallmarks of historical solutions as space is organized rather in a modern way. In Tuscany, the pervading spirit of a place is also most strongly manifested in towns whose historical environment has been successfully preserved. Polish Medieval towns have been most severely deprived of their historical ambience, and the genius loci has been retained in only six towns (Działdowo, Górowo Iławeckie, Nowe Miasto Lubawskie, Orneta, Pasym, Reszel). In the remaining cases, the spatial solutions do not have any historical features. After World War II their urban layout was changed and the architecture of socialism vividly stood out from the stylistics that had been created for centuries.

To summarize, one can state that towns that ate members of Cittaslow network develop according to the same set of criteria which is at the same time the benchmark of developmental standards. However, it is so profound that it allows for shaping the local specialization in an individual way. Every town has its own unique endogenous potential which under some conditions can contribute to real development and raising its inhabitants' quality of life [67,103], and the same time achieving the goals of functioning of Cittaslow movement.

6. Conclusions

Small historical towns offer a glimpse into Europe's wealthy cultural and architectural heritage. The original spatial structure of these towns has been preserved to a varied extent. Cittaslow International is a network of small towns which prioritizes the quality of life over the hustle and bustle of modern cities. The article analyzes and evaluates historical spatial forms in Cittaslow towns situated in selected European regions with a different culture and history. The relationships between the urban layout and the local quality of life based on sustainable development and protection of cultural heritage were examined.

The study revealed that the Medieval layout of Polish and Italian towns can be successfully incorporated in the modern urban fabric to promote the Cittaslow philosophy of sustainable development and slow living. Cittaslow towns are characterized by small size and population, small-scale urban development and user-friendly public spaces that promote a sense of local identity and foster responsibility for one's place of residence. Historical spatial layouts of the towns (downtown built-up areas) are not adjusted to the modern functional specificity of a big city and its dynamics. There is no possibility for industrial development, building large-areas shopping centers, organizing public transport or building thoroughfare. The small scale of the area in the town center makes social integration an inevitable element of its functioning.

Moreover, historical towns have distinctive spatial features, including small squares, narrow streets that are closed off to vehicular traffic, and historical architecture that triggers emotional responses. These components facilitate the implementation of the Cittaslow concept that advocates harmonious relations between the past, i.e., the preservation and promotion of local traditions and cultural heritage, and the future, i.e., economic growth and an improvement in living standards. Even though Polish and Italian towns have differences in their layouts and character, they still have a common feature—the cozy space organization which being combined with old but well-functioning architecture corresponds to Cittaslow assumptions.

The analyses demonstrate that the components can be applied for spaces developed in a various way, both Polish and Italian towns. Historical urban fabric favors pedestrian movement, which contributes to the formation of stronger social bonds, unlike in large metropolitan areas where social interactions are scarce. Dense development in small towns eliminates physical barriers to human contact, such as extensive open spaces, wide roads and large plots of land zoned for development. These findings indicate that in order to preserve the historical urban fabric, the structure should not be interfered with. Effective planning policies are required to preserve the scale, unique character and identity of historical towns, and they facilitate the achievement of Cittaslow goals. Historical spatial layout and urban structures promote local arts, crafts and services. Preservation of local traditions and cultural heritage contributes to the development of tourism, and modern technology (such as remote work arrangements) improves the quality of local life.

Historical urban fabric is an element of cultural heritage and one of the pillars of the tourism industry which creates ample opportunities for promoting the values of towns and regions. Therefore, the results of this study can be used by local authorities to establish cooperation networks, promote their towns and regions, and exchange experiences relating to the implementation of local development policies. The proposed research methods are largely universal and can be applied to analyze the spatial structure of small towns in other regions of the world. The study adds to the existing body of knowledge relating to the development of historical urban forms and spatial layout. The possibility for comparing Polish and Italian towns, which differ in many above-mentioned aspects, proves the point. The methodology is therefore universal and can contribute to the comprehension and further development of the Slow City idea in various regions in the world. Moreover, it can be used as a key element in building local urban revitalization programs for historical towns. Such practice is commonly applied in urban planning in numerous European regions. However, during conducting the research one should be aware of the limitations connected with the lack of a unified database for given countries and the limitations of the access to iconographic and cartographic resources.

Author Contributions: Conceptualization, M.Z., K.P. and A.S.; methodology, M.Z., K.P. and A.S.; formal analysis, M.Z.; investigation, M.Z. and K.P.; resources, M.Z.; data curation, M.Z. and K.P.; writing—original draft preparation, M.Z., K.P. and A.S.; writing—review and editing, K.P. and A.S.; visualization, M.Z.; supervision, A.S.; project administration, M.Z. All authors have read and agreed to the published version of the manuscript.

Funding: This research received no external funding.

Institutional Review Board Statement: Not applicable.

Informed Consent Statement: Not applicable.

Data Availability Statement: Publicly available datasets were analyzed in this study. This data can be found here: http://asict.arte.unipi.it/index.html/ (accessed on 18 August 2020), http://www.cittaslow. org/ (accessed on 30 August 2020), www.istat.it (accessed on 30 August 2020), www.cittaslow.it (accessed on 30 August 2020), www.google.com/maps (accessed on 30 August 2020), https://cittaslowpolska.pl/ (accessed on 30 August 2020).

Conflicts of Interest: The authors declare no conflict of interest.

References

1. Broadway, M. Implementing the Slow Life in Southwest Ireland: A Case Study of Clonakilty and Local Food. *Geogr. Rev.* **2015**, *105*, 216–234. [CrossRef]
2. Miele, M. CittàSlow: Producing Slowness against the Fast Life. *Space Polity* **2008**, *12*, 135–156. [CrossRef]
3. Servon, L.J.; Pink, S. Cittaslow: Going Glocal in Spain. *J. Urban. Aff.* **2015**, *37*, 327–340. [CrossRef]
4. Cittaslow International. Available online: http://www.cittaslow.org/ (accessed on 30 August 2020).
5. Mayer, H.; Knox, P.L. Slow Cities: Sustainable Places in a Fast World. *J. Urban. Aff.* **2006**, *28*, 321–334. [CrossRef]
6. Orhan, M. Different Approach to Forming Sustainable Cities: Cittaslow. *J. Environ. Prot. Ecol.* **2017**, *18*, 1017–1026.
7. Baldemir, E.; Kaya, F.; Şahin, T.K. A Management Strategy within Sustainable City Context: Cittaslow. *Procedia Soc. Behav. Sci.* **2013**, *99*, 75–84. [CrossRef]
8. Mallet, S. The Cittaslow Label and its Spreading in France: The Slow to Produce Sustainable Spaces? *Territoire en Mouvement. Revue de Géographie et d'Aménagement* **2018**. [CrossRef]
9. Hoeschele, W. Measuring Abundance: The Case of Cittaslow's Attempts to Support Better Quality of Life. *Int. J. Green Econ.* **2010**, *4*, 63–81. [CrossRef]
10. Perano, M.; Abbate, T.; La Rocca, E.T.; Casali, G.L. Cittaslow & Fast-Growing SMEs: Evidence from Europe. *Land Use Policy* **2019**, *82*, 195–203. [CrossRef]

11. Zambon, I.; Serra, P.; Bencardino, M.; Carlucci, M.; Salvati, L. Prefiguring a Future City: Urban Growth, Spatial Planning and the Economic Local Context in Catalonia. *Eur. Plan. Stud.* **2017**, *25*, 1797–1817. [CrossRef]

12. Feliú, E.G.; Vera, J.R.; Villalón, E.P.; de Castro, J.C.P. Urban Modernization and Heritage in the Historic Centre of Santiago de Chile (1818–1939). *Plan. Perspect.* **2020**, *35*, 91–113. [CrossRef]

13. Rossi, D.; Paciotti, D.; Calvano, M. Visionaria. An Open Design Approach for the Regeneration of Historical Urban Heritage. In Proceedings of the Advances in Additive Manufacturing, Modeling Systems and 3D Prototyping, Washington, DC, USA, 24–28 July 2019; Di Nicolantonio, M., Rossi, E., Alexander, T., Eds.; Springer International Publishing: Cham, Switzerland, 2020; Volume 975, pp. 190–201.

14. Abastante, F.; Lami, I.M.; Mecca, B. How to Revitalise a Historic District: A Stakeholders-Oriented Assessment Framework of Adaptive Reuse. In *Values and Functions for Future Cities*; Mondini, G., Oppio, A., Stanghellini, S., Bottero, M., Abastante, F., Eds.; Green Energy and Technology; Springer International Publishing: Cham, Switzerland, 2020; pp. 3–20, ISBN 978-3-030-23786-8.

15. Atkinson, R. The Small Towns Conundrum: What Do We Do about Them? *Reg. Stat.* **2019**, *9*, 1–17. [CrossRef]

16. Senetra, A.; Szarek-Iwaniuk, P. Socio-Economic Development of Small Towns in the Polish Cittaslow Network—A Case Study. *Cities* **2020**, *103*, 102758. [CrossRef]

17. Jaszczak, A.; Kristianova, K.; Pochodyła, E.; Kazak, J.K.; Młynarczyk, K. Revitalization of Public Spaces in Cittaslow Towns: Recent Urban Redevelopment in Central Europe. *Sustainability* **2021**, *13*, 2564. [CrossRef]

18. Farelnik, E.; Stanowicka, A.; Wierzbicka, W. The Effects of Membership in the Polish National Cittaslow Network. *Equilibrium. Q. J. Econ. Econ. Policy* **2021**, *16*, 139–167. [CrossRef]

19. Hatipoglu, B. "Cittaslow": Quality of Life and Visitor Experiences. *Tour. Plan. Dev.* **2015**, *12*, 20–36. [CrossRef]

20. Brodziński, Z.; Kurowska, K. Cittaslow Idea as a New Proposition to Stimulate Sustainable Local Development. *Sustainability* **2021**, *13*, 5039. [CrossRef]

21. Zadęcka, E. Zrównoważony Rozwój Małych Miast w Świetle Warunków i Wymogów Stowarzyszenia Cittaslow. *Zesz. Nauk. Politech. Częstochowskiej Zarządzanie* **2017**, *25*, 35–48. [CrossRef]

22. İlhan, Ö.A.; Karakaş, E.; Özkaraman, B. 'Cittaslow': An Alternative Model for Local Sustainable Development or Just a Myth? Empirical Evidence in the Case of Tarakli (Turkey). *Quaest. Geogr.* **2020**, *39*, 23–37. [CrossRef]

23. Pink, S. Sense and Sustainability: The Case of the Slow City Movement. *Local Environ.* **2008**, *13*, 95–106. [CrossRef]

24. Zagroba, M.; Gawryluk, D. *Revitalisation as a Method of Planning Sustainable Development of Old Town Complexes in Historic Towns*; IOP Publishing: Bristol, UK, 2017; Volume 95, pp. 1–10.

25. Mumford, L. *The Culture of Cities*; Secker & Warburg: London, UK, 1944; ISBN 978-0-15-623301-9.

26. Pawłowska, K. *Idea Swojskości Miasta*; Politechnika Krakowska: Kraków, Poland, 2001; ISBN 978-83-7242-201-9.

27. Alexander, C. A City Is Not Tree. *Archit. Forum* **1965**, *122*, 58–66.

28. Burgess, E.W. *The Urban Community: Selected Papers from the Proceedings of the American Sociological Society, 1925*; University of Chicago Press: Chicago, IL, USA, 1926; ISBN 0-598-60431-6.

29. Geddes, S.P. *Cities in Evolution: An Introduction to the Town Planning Movement and to the Study of Civics*; Ernest Benn Limited: London, UK, 1968; ISBN 0-510-43121-6.

30. Blazy, R. Zagadnienia przestrzenne a postulaty i idee ruchu miast Cittaslow. In *Tendencje w Rozwoju Gospodarczym i Przestrzennym Małych Miast w Polsce*; Bartosiewicz, B., Ed.; Wydawnictwo Uniwersytetu Łódzkiego: Łódź, Poland, 2016; pp. 7–28, ISBN 1733-3180.

31. Mayer, H.; Knox, P.L. Pace of Life and Quality of Life: The Slow City Charter. In *Community Quality-of-Life Indicators: Best Cases III*; Sirgy, M.J., Phillips, R., Rahtz, D., Eds.; Springer: Dordrecht, The Netherlands, 2009; pp. 21–40, ISBN 978-90-481-2257-8.

32. Castells, M. *The Rise of the Network Society*; Wiley-Blackwell: Hoboken, NJ, USA, 2010; ISBN 978-1-4443-1951-4.

33. Dimelli, D. Modern Conservation Principles and Their Application in Mediterranean Historic Centers—The Case of Valletta. *Heritage* **2019**, *2*, 787–796. [CrossRef]

34. Województwa Warmińsko-Mazurskiego. *Ponadlokalny Program Rewitalizacji Sieci Miast Cittaslow*; Województwa Warmińsko-Mazurskiego: Olsztyn, Poland, 2020.

35. Strzelecka, E. Małe miasta a nowoczesne modele rozwoju miast. In *Alternatywne Modele Rozwoju Miast. Sieć Miast Cittaslow*; Strzelecka, E., Ed.; Wydawnictwo Politechniki Łódzkiej: Łódź, Poland, 2017; pp. 13–40, ISBN 978-83-7283-826-1.

36. Sobczak, A. Ośrodki Badawcze Zajmujące Się Inteligentnymi Miastami. Available online: http://inteligentnemiasta.pl/osrodki-badawcze-zajmujace-sie-inteligentnymi-miastami/5210/ (accessed on 14 July 2021).

37. Hermant-de Callataÿ, C.; Svanfeld, C. *Miasta Przyszłości. Wyzwania, Wizje, Perspektywy*; Europejska, K., Generalna, D., Eds.; Polityki Regionalnej: Luxembourg, 2011.

38. Alexandrakis, G.; Manasakis, C.; Kampanis, N.A. Economic and Societal Impacts on Cultural Heritage Sites, Resulting from Natural Effects and Climate Change. *Heritage* **2019**, *2*, 279–305. [CrossRef]

39. Jaszczak, A.; Denekas, J. Znaczenie Cech Krajobrazu w Kreowaniu Przestrzeni Kulturowej w Układach Regionalnych. *Pr. Kom. Kraj. Kult. PTG* **2014**, *25*, 7–18.

40. Kowalczyk, A. Krajobraz kulturowy. In *Uwarunkowania i Plany Rozwoju Turystyki*; Młynarczyk, Z., Zajadacz, A., Eds.; Turystyka i Rekreacja—Studia i Prace; Wydawnictwo Naukowe UAM: Poznań, Poland, 2008; Volume II, pp. 121–137.

41. Wolfram, K. (Ed.) *Zielone Płuca Polski—Walory Przyrodnicze i Kulturowe*; Fundacja Zielone Płuca Polski: Białystok, Poland, 2009; ISBN 978-83-62090-04-4.

42. Wróbel, T. *Zarys Historii Budowy Miast*; Ossolineum: Wrocław, Poland, 1971.
43. Kostof, S. *The City Shaped: Urban. Patterns and Meanings Through History*; Thames And Hudson: London, UK, 1991; ISBN 978-0-8212-1867-9.
44. Matthew, D. *Atlas of Medieval Europe*, 1st ed.; Checkmark Books: New York, NY, USA, 1983; ISBN 978-0-87196-133-4.
45. Słodczyk, J. *Historia Planowania i Budowy Miast*; Wydawnictwo Uniwersytetu Opolskiego: Opole, Poland, 2012; ISBN 978-83-7395-484-7.
46. Benevolo, L. *The History of the City*, 1st ed.; The MIT Press: Cambridge, MA, USA, 1980; ISBN 978-0-262-02146-3.
47. Czubiel, L.; Domagała, T. *Zabytkowe Ośrodki Miejskie Warmii i Mazur*; Pojezierze: Olsztyn, Poland, 1969.
48. Tołwiński, T. *Urbanistyka T.1: Budowa Miasta w Przeszłości*, 2nd ed.; Wydawnictwo Zakładu Urbanistyki Politechniki Warszawskiej: Warszawa, Poland, 1939.
49. Hall, E.T. *Ukryty Wymiar*; Państwowy Instytut Wydawniczy: Warszawa, Poland, 1978.
50. Rossi, A. *The Architecture of the City*; MIT Press: Cambridge, MA, USA, 1984; ISBN 978-0-262-68043-1.
51. Wejchert, K. *Elementy Kompozycji Urbanistycznej*; Arkady: Warszawa, Poland, 1984; ISBN 83-213-3151-3.
52. Norberg-Schulz, C. *Genius Loci: Paesaggio Ambiente Architettura*; Edizioni Electa: Milan, Italy, 1986; ISBN 978-88-435-1961-3.
53. Montgomery, C. *Happy City: Transforming Our Lives through Urban Design*; Farrar Straus Giroux: New York, NY, USA, 2013; ISBN 978-1-84614-320-5.
54. Gzell, S. *O Architekturze. Szkice Pisane i Rysowane*; Blue Bird: Warszawa, Poland, 2014; ISBN 978-83-64870-02-6.
55. Radwan, A. Color in Architecture Is It Just an Aesthetic Value or a True Human Need? *Int. J. Eng. Res. Technol. (IJERT)* **2015**, *4*, 523–533. [CrossRef]
56. Brogiolo, G.P. *Le Origini Della Città Medievale*; SAP: Walldorf, Germany, 2011; ISBN 978-88-87115-68-0.
57. Ward-Perkins, B. The towns of northern Italy: Rebirth or renewal? In *The Rebirth of Towns in the West AD 700–1050*; Hodges, R., Hobley, B., Eds.; Council for British Archaeology: London, UK, 1988; pp. 16–27.
58. Boetticher, A. *Die Bau- Und Kunstdenkmäler Der Provinz Ostpreußen. Bd. 4: Die Bau- Und Kunstdenkmäler in Ermland*; Die Bau- und Kunstdenkmäler der Provinz Ostpreußen; Kommissionsverlag von Bërnh. Teichert: Königsberg, Preußen, 1894; Volume 4.
59. Bonk, H. *Die Städte und Burgen in Altpreussen (Ordensgründungen) in Ihrer Beziehung zur Bodengestaltung*; Ferd. Beyer's Buchhandlung (Thomas & Oppermann): Königsberg, Preußen, 1895.
60. Maciejczak-Kwiatkowska, Z.; Retko-Bernatowicz, M.; Wiśniewski, R. *W Dialogu z Otoczeniem? Społeczne Postrzeganie Przestrzeni Publicznej i Architektury w Polsce*; Narodowe Centrum Kultury: Warszawa, Poland, 2018; p. 71.
61. Italian National Institute of Statistics. Available online: www.istat.it (accessed on 30 August 2020).
62. Atlante Storico Iconografico Delle Città Toscane. Available online: http://asict.arte.unipi.it/index.html/ (accessed on 18 August 2020).
63. Ross, I.C. *Umbria: A Cultural History (Cultural Guide)*; Signal Books Ltd.: Oxford, UK, 2013; ISBN 978-1-908493-85-9.
64. Cittaslow Italia. Available online: www.cittaslow.it (accessed on 30 August 2020).
65. Google Maps. Available online: www.google.com/maps (accessed on 30 August 2020).
66. Cittaslow Polska. Available online: https://cittaslowpolska.pl/ (accessed on 30 August 2020).
67. Mazur-Belzyt, K. Współczesne podstawy rozwoju małych miast na przykładzie sieci miast Cittaslow. *Probl. Rozw. Miast* **2014**, *XI*, 39–45.
68. Mehanna, W.A.E.-H. Urban Renewal for Traditional Commercial Streets at the Historical Centers of Cities. *Alex. Eng. J.* **2019**, *58*, 1127–1143. [CrossRef]
69. Knox, P.L. Creating Ordinary Places: Slow Cities in a Fast World. *J. Urban Des.* **2005**, *10*, 1–11. [CrossRef]
70. Presenza, A.; Abbate, T.; Micera, R. The Cittaslow Movement: Opportunities and Challenges for the Governance of Tourism Destinations. *Tour. Plan. Dev.* **2015**, *12*, 479–488. [CrossRef]
71. Karatosun, M.; Çakar, D. Effects of Cittaslow Movement on Conservation of Cultural Heritage: Case of Seferihisar & Halfeti, Turkey. *Civ. Eng. Archit.* **2017**, *5*, 71–82. [CrossRef]
72. Jaszczak, A.; Kristianova, K. Social and Cultural Role of Greenery in Development of Cittaslow Towns. *IOP Conf. Ser. Mater. Sci. Eng.* **2019**. [CrossRef]
73. Farelnik, E.; Stanowicka, A.; Wierzbicka, W. *Cittaslow—Model Rozwoju i Współpracy Małych Miast*; UWM w Olsztynie: Olsztyn, Poland, 2020.
74. Pulawska, S.; Starowicz, W. Ecological Urban Logistics in the Historical Centers of Cities. *Procedia Soc. Behav. Sci.* **2014**, *151*, 282–294. [CrossRef]
75. Krassowski, W. *Dzieje Budownictwa i Architektury na Ziemiach Polskich*; Arkady: Warszawa, Poland, 1990; Volume 2.
76. Pevsner, N. *An Outline of European Architecture*; Thames & Hudson: London, UK, 2009; ISBN 0-500-34241-5.
77. Della Spina, L. Multidimensional Assessment for "Culture-Led" and "Community-Driven" Urban Regeneration as Driver for Trigger Economic Vitality in Urban Historic Centers. *Sustainability* **2019**, *11*, 7237. [CrossRef]
78. Hargreaves, A. Building Communities of Place: Habitual Movement around Significant Places. *J. Hous. Built Environ.* **2004**, *19*, 49–65. [CrossRef]
79. Park, E.; Kim, S. The Potential of Cittaslow for Sustainable Tourism Development: Enhancing Local Community's Empowerment. *Tour. Plan. Dev.* **2015**, 351–369. [CrossRef]

80. Sztabiński, P.B.R.; Sztabiński, F. Przywiązanie do miejsca zamieszkania jako wymiar polskiego tradycjonalizmu. In *Elementy Nowego Ładu*; Domański, H., Rychard, A., Eds.; Wydaw. IFiS PAN: Warszawa, Poland, 1997; pp. 256–269, ISBN 978-83-86166-87-9.
81. Piccinato, L. *Urbanistica Medievale*; Dedalo: Bari, Italy, 1993; ISBN 978-88-220-3306-2.
82. Valenti, M. Architecture and Infrastructure in the Early Medieval Village: The Case of Tuscany. *Technol. Transit. A.D. 300–650* **2006**, 451–489. [CrossRef]
83. Randelli, F.; Romei, P.; Tortora, M.; Mossello, M. Rural Tourism Driving Regional Development in Tuscany. The Renaissance of the Countryside. *Dip. Di Sci. Econ. Univ. Degli Studi Di Firenze Work. Pap. Ser.* **2012**, *11*, 1–21.
84. Bogdanowski, J. *Architektura Obronna w Krajobrazie Polski. Od Biskupina do Westerplatte*; Wydawnictwo Naukowe PWN: Warszawa, Poland, 2002; ISBN 83-01-12223-4.
85. Paszkowski, Z. *Historia Idei Miasta: Od Antyku Do Renesansu*; Wydawnictwo hogben: Szczecin, Poland, 2015; ISBN 83-63868-51-5.
86. Salm, J. Idea Cittaslow a tożsamość miast województwa warmińsko-mazurskiego. Między Umbrią a Warmią i Mazurami. In *Alternatywne Modele Rozwoju Miast. Sieć Miast Cittaslow*; Strzelecka, E., Ed.; Wydawnictwo Politechniki Łódzkiej: Łódź, Poland, 2017; pp. 74–81, ISBN 978-83-7283-826-1.
87. Basista, A. *Betonowe Dziedzictwo: Architektura w Polsce Czasów Komunizmu*; Wydawn. Nauk. PWN: Warszawa, Poland, 2001; ISBN 978-83-01-13224-8.
88. Cencetti, C.; Conversini, P.; Tacconi, P. The Rock of Orvieto (Umbria, Central Italy). *Giornale di Geologia Applicata* **2005**, *1*, 103–112. [CrossRef]
89. Falco, E. Transferable Development Rights in Regeneration Schemes for Historic City Centres. Legislation in the Umbria Region. *Ital. J. Plan. Pract.* **2012**, *2*, 4–14.
90. Paba, G.; Perrone, C.; Lucchesi, F.; Zetti, I. Territory matters: A regional portrait of Florence and Tuscany. In *Post-Metropolitan Territories*; Balducci, A., Fedeli, V., Curci, F., Eds.; Routledge: London, UK; New York, NY, USA, 2017; pp. 95–116, ISBN 978-1-315-62530-0.
91. Pancewicz, A. Rola Rzek w Rozwoju Przestrzennym Historycznych Miast Nadrzecznych. *Woda W Przestrz. Przyodniczej I Kult.* **2003**, *11*, 275–286.
92. Kostrzewska, M. *Miasto Europejskie na Przestrzeni Dziejów: Wybrane Przykłady*; Akapit-DTP: Gdańsk, Poland, 2013; ISBN 978-83-64333-02-6.
93. Czerner, R. *Zabudowy Rynków: Średniowieczne Bloki Śródrynkowe Wybranych Dużych Miast Śląska*; Oficyna Wydawnicza Politechniki Wrocławskiej: Wrocław, Poland, 2002; ISBN 978-83-7085-671-7.
94. Przesmycka, E. "Serce" współczesnego miasta. *Czas. Techniczne. Archit.* **2008**, *105*, 77–87.
95. Sarzyński, P. Polacy wobec krajobrazu miejskiego. Co nas porusza? In *Krajobraz Kulturowo-Przyrodniczy z Perspektywy Społecznej*; Ratajski, S., Ziółkowski, M., Eds.; Polski Komitet do spraw UNESCO Narodowe Centrum Kultury: Warszawa, Poland, 2015; pp. 213–228, ISBN 978-83-902939-9-8.
96. Bisaga, A. Małe i średnie miasta jako lokalne centra rozwoju regionalnego. In *Rozwój Miast i Zarządzanie Gospodarką Miejską*; Słodczyk, J., Ed.; Wydawnictwo Uniwersytetu Opolskiego: Opole, Poland, 2004; pp. 225–234, ISBN 83-7395-069-9.
97. Orchowska, A. Man, as a User and Observer of the Urban Space. *Środowisko Mieszk.* **2014**, *13*, 180–185.
98. Besson, R. Living cities in two cycles and three movements. In *Living Cities. Contributions to the Theme Europan E16 Europan France*; Europan France: Montreuil, France, 2019; pp. 15–20.
99. Ostrowski, W. *Wprowadzenie do Historii Budowy Miast. Ludzie i Środowisko*, 2nd ed.; Oficyna Wydawnicza Politechniki Warszawskiej: Warszawa, Poland, 2001; ISBN 83-86569-28-X.
100. Durrell, L. *Spirit of Place. Letters and Essays on Travel*; Faber & Faber: London, UK, 1969; ISBN 0-571-08994-1.
101. Canter, D.V. *The Psychology of Place*, 1st ed.; Palgrave Macmillan: New York, NY, USA, 1977; ISBN 978-0-312-65322-4.
102. Szymski, A.M. Genius loci—Czyli o odkrywaniu i na nowo definiowaniu znaczeń w istniejącej przestrzeni miejskiej (trzy przykłady). *Czas. Techniczne. Archit.* **2008**, *105*, 162–166.
103. Oppido, S.; Ragozino, S.; Esposito De Vita, G. Exploring Territorial Imbalances: A Systematic Literature Review of Meanings and Terms. In *New Metropolitan Perspectives*; Springer: Berlin/Heidelberg, Germany, 2020; pp. 90–100.

land

Case Report

Toward a Dualistic Growth? Population Increase and Land-Use Change in Rome, Italy

Leonardo Bianchini [1], Gianluca Egidi [1], Ahmed Alhuseen [2], Adele Sateriano [3], Sirio Cividino [4], Matteo Clemente [5] and Vito Imbrenda [6,*]

1   Department of Agricultural and Forestry Sciences (DAFNE), Tuscia University, Via San Camillo de Lellis, I-01100 Viterbo, Italy; l.bianchini@unitus.it (L.B.); egidi.gianluca@unitus.it (G.E.)
2   Global Change Research Institute CAS, Lipova 9, 370 05 Ceske Budejovice, Czech Republic; alhussen.a@czechglobe.cz
3   Independent Researcher, I-00184 Rome, Italy; adele.sateriano.pul@gmail.com
4   Department of Agriculture, University of Udine, Via del Cotonificio 114, I-33100 Udine, Italy; sirio.cividino@crslaghi.net
5   Department of Planning, Design and Architectural Technologies, University of Rome 'La Sapienza', Via E. Gianturco 2, I-00186 Rome, Italy; matteo.clemente@uniroma1.it
6   IMAA—CNR (Institute of Methodologies for Environmental Analysis—Italian National Research Council), c.da Santa Loja snc, I-85050 Tito Scalo, Italy
*   Correspondence: vito.imbrenda@imaa.cnr.it

Abstract: The spatial mismatch between population growth and settlement expansion is at the base of current models of urban growth. Empirical evidence is increasingly required to inform planning measures promoting urban containment in the context of a stable (or declining) population. In these regards, per-capita indicators of land-use change can be adopted with the aim at evaluating long-term sustainability of urbanization processes. The present study assesses spatial variations in per-capita indicators of land-use change in Rome, Central Italy, at five years (1949, 1974, 1999, 2008, and 2016) with the final objective of quantifying the mismatch between urban expansion and population growth. Originally specialized in agricultural productions, Rome's metropolitan area is a paradigmatic example of dispersed urban expansion in the Mediterranean basin. By considering multiple land-use dynamics, per-capita indicators of landscape change delineated three distinctive waves of growth corresponding with urbanization, suburbanization, and a more mixed stage with counter-urbanization and re-urbanization impulses. By reflecting different socioeconomic contexts on a local scale, urban fabric and forests were identified as the 'winner' classes, expanding homogeneously over time at the expense of cropland. Agricultural landscapes experienced a more heterogeneous trend with arable land and pastures declining systematically and more fragmented land classes (e.g., vineyards and olive groves) displaying stable (or slightly increasing) trends. The continuous reduction of per-capita surface area of cropland that's supports a reduced production base, which is now insufficient to satisfy the rising demand for fresh food at the metropolitan scale, indicates the unsustainability of the current development in Rome and more generally in the whole Mediterranean basin, a region specialized traditionally in (proximity) agricultural productions.

Keywords: metropolitan expansion; per-capita urban area; per-capita cropland; land mismatch; population; Italy

Citation: Bianchini, L.; Egidi, G.; Alhuseen, A.; Sateriano, A.; Cividino, S.; Clemente, M.; Imbrenda, V. Toward a Dualistic Growth? Population Increase and Land-Use Change in Rome, Italy. *Land* **2021**, *10*, 749. https://doi.org/10.3390/land10070749

Academic Editors: Luca Salvati, Iwona Cieśla and Andrzej Biłozor

Received: 30 June 2021
Accepted: 13 July 2021
Published: 17 July 2021

Publisher's Note: MDPI stays neutral with regard to jurisdictional claims in published maps and institutional affiliations.

1. Introduction

Land use/land cover changes (LULCCs) are due to a large number of factors encompassing climate change, occurrence of natural risk, regulation/conservation policies, and human disturbance mainly consisting in various forms of urbanization [1–9]. In this perspective, in urban areas, settlement expansion and landscape transformations were intimately associated, involving multiple research dimensions encompassing ecology, planning, economic issues, and social aspects [10]. For a comprehensive study of urbanization-driven landscape changes, defining urban growth in terms of density, spatial morphology, socioeconomic implications, and environmental impacts is a difficult task [11–13]. Focusing on agglomeration, contemporary urbanization has been progressively shifted beyond the typical ideal city models in which the density of population, settlements, and economic activities display a significant decline moving away from downtown [14]. Urban sprawl was proposed as a comprehensive definition of this evolution: while sprawl comes across as a matter of degree not easily quantifiable, it is argued that in this case the density gradient of any urban function (population, buildings, and businesses) became less steep over time [15–17]. However, low-density development differs from place to place, thus defining of such a process is a mostly place-specific exercise [18]. Impacts of sprawl on fringe landscapes are also mixed and hardly predictable as land-use change in metropolitan regions is often the result of a complex interplay of socioeconomic factors and land constraints (e.g., accessibility, infrastructural development, land/house prices, and access to credit) [19–22].

The dynamic linkage between landscape transformations and population dynamics in advanced economies became increasingly multifaceted because of the mutual interplay of environmental, social, and planning dimensions that influence the level of sustainable development at different spatial scales from regions to local communities [23,24]. It was demonstrated how population dynamics have influenced sprawl less intensively than what occurred under more compact and dense stages of urban expansion in the past [25]. In fact, the current model of urban development (i.e., low-density settlements scattered on a natural or agricultural matrix) results in spatially articulated settlement morphologies leading to high rates of land consumption [26–28].

Following Couch et al. [2], urban sprawl has been increasingly treated as a process of urban change rather than a pattern of urbanization. In this sense, the definition given by Glaster and other scholars nearly twenty years ago [1] is the one that best allows sprawl to be intended as a process and not merely a spatial pattern. Within this approach, the simplest indicator for measuring the extent of sprawl in a metropolitan area is a ratio of two growth rates: the rate at which land development near the outer suburbs has increased divided by the rate at which the population of the metropolitan area has grown [29]. Measuring urban and suburban expansion in this context, a recent study defined exurban development as a process in which "the spread of development across the landscape far outpaces population growth" [30–32]. Similarly, another study notes that "if land is being consumed at a faster rate than population growth, then a metropolitan area can be characterized as 'sprawling'. If population is growing more rapidly than land is being consumed for urbanization, then a metropolitan area can be characterized as 'densifying'" [3]. Taking this approach, Wiewel and Persky's work on Chicago's suburban sprawl [13] demonstrated that between 1970–1990, urbanized land grew by 46%, while the population increased by only 4% (a ratio of 11.5 to 1). This is only an example of the possible use of per-capita indicators of land-use [33–38].

A large imbalance between a place's spatial expansion and its population change (where the former increases much more rapidly than the latter) is not unusual in advanced economies [39]. Growth of this sort produces a lower density outcome with people and their residential and commercial buildings using more space often at the expense of forests and farmland [40]. Examining landscapes from the other side of the fringe, exurban development has produced fragmented and chaotic natural landscapes that are progressively losing their characteristic traits (e.g., as far as the composition, configuration, and structure are concerned) [41]. While land-use change in metropolitan regions was the subject of earlier studies delineating knock-on effects on the environmental matrix [42–44], the relationship between land-use change and urban cycles has not yet been adequately studied, especially in regions where human pressure has increased more rapidly in recent years [45].

Based on earlier studies [46], a complete urban cycle is usually represented as a sequence of four development stages with different demographic characteristics from urbanization to suburbanization and from counter-urbanization to re-urbanization [47].

The downtown population increased in both the first and the fourth stage; the suburban population increased in the second stage; and the rural population in metropolitan regions increased in the third stage [48]. Per-capita land-use indicators seem to be an appropriate solution to better delineate the temporal outline of different stages of a complete urban cycle, in turn quantifying the incipient mismatch between population growth and urban expansion [49]. This evidence indicates, likely better than other indicators, the shift from compact models to a more dispersed settlements' structure [50]. By linking (urban) form and (ecological) functions in peri-urban areas, a long-term analysis of per-capita land-use indicators provides the necessary informative base to any policy aimed at containing urban expansion [51–54]. In this direction, spatial planning is urgently required to finely tune urban expansion with population increase and growth in economic activities, limiting settlement diffusion (e.g., in shrinking cities) [55].

The drastic spread of official statistics, land-use maps, and geo-spatial databases support a refined landscape analysis investigating per-capita land-use indicators over a sufficiently long-time interval [56]. Per-capita indicators of land-use change represent a suitable tool to test assumptions on the relationship between urban cycles and landscape transformations over different waves of metropolitan expansion [57]. In this direction, exploratory multivariate data analysis contributes to reveal the complex linkage between compositional changes in relict, fringe landscapes, and the increase of resident population, considered an anticipatory indicator of the specific stage of a given city life cycle [58,59].

Grounded on this theory, our study provides a long-term analysis of per-capita land cover changes at the fringe of a metropolitan region in southern Europe. In light of different models of urban expansion from compact and radio-centric urbanization to dispersed expansion of low-density settlements, the present study documents urban and peri-urban changes over nearly 70 years (1949–2016). As a novel contribution to land-use science, the present work delineates the prevalent mode of urban expansion in Rome (Italy) considering both socioeconomic functions (whose changes are reflected in the sequential stages of the city life cycle) and morphological shifts (e.g., compact vs. dispersed). The Rome metropolitan area is a typical semi-compact and dense city in southern Europe experiencing different waves of settlement expansion and represents an example of metropolitan transformations in the Mediterranean [60–62]. The empirical results of this study allow for the discussion of the (supposed) unsustainability of current urban expansion compared with past settlement structures as far as land fragmentation and loss of relict habitats and traditional crops at the fringe are concerned.

2. Materials and Methods

2.1. Study Area

The investigated area covers a large part of Rome's province (Latium region, Central Italy) encompassing the municipalities of Rome and Fiumicino (1500 km^2, see Figure 1). The study area is mostly flat and corresponds with the so called 'Agro Romano', a cultivated district with traditional rural landmarks, biodiversity, and cultural heritage surrounding the historical city of Rome along the alluvial plain of the Tiber river [60]. Climate regime in Rome is typically Mediterranean with rainfalls concentrated in the fall and spring and mild temperatures in the winter. Average annual rainfalls and mean daily temperatures were nearly 700 mm and 16 °C in the period between 1960–1990, although in the last decades the study area is becoming drier and warmer as a consequence of reduced precipitation and increased temperatures.

During the investigated period (1949–2016), this area has been subjected to considerable transformations due to a mix of drivers. Among them, Common Agricultural Policy, local market conditions, demographic trends, and conservation measures (e.g., Natura 2000 network) have stimulated on the one hand a strong intensification and simplification of the agricultural productions; on the other, an expansion of land abandonment phenomena, all contributing to an increased environmental fragility of these areas [63,64]. A pivotal role has been played by urban growth impacting agricultural lands and compromising their distinctive socioecological features. The low-density discontinuous settlement that developed on the edges of Rome (Figure 2) penetrates into the Agro Romano, fragmenting its agro-forestry matrix. The negative effects of this expansion can be cumulative and can increasingly affect large areas as shown by the regional distribution of per-capita land consumption computed at the municipal scale.

In particular, despite the fact that Rome exhibits intermediate values for the extent of urban areas on the total municipal area and for per-capita built-up areas partially due to a huge municipal area, the neighboring municipalities exhibit a sprawling urbanization with variable settlement densities that are frequently higher than those observed in the central municipality. Settlements traditionally organized on radial axes around the consolidated city are characterized by variable sealing rates with very heterogeneous percentages in the regional area, particularly high around the county seats (not only in Rome but also in the other Latium provinces), representing sub-centers in expansion. Simultaneously, the original settlement structure in the rural districts of the Agro Romano has been transformed into a medium and low-rise fabric spatially decomposed as a result of urban de-concentration processes only partially devised with consequent spatial colonization of areas located at 20–40 km from the central city.

Urban settlements in Rome can be divided into 2 main districts: (1) the consolidated urban fabric hosting 63.4% of the population (as the percentage share in the total population) with a population density of 4873 inhabitants/km^2 and (2) the suburbs hosting the 36.6% of the population (as the percentage share in the total population) with a population density of 1037 inhabitants/km^2 (2011 census).

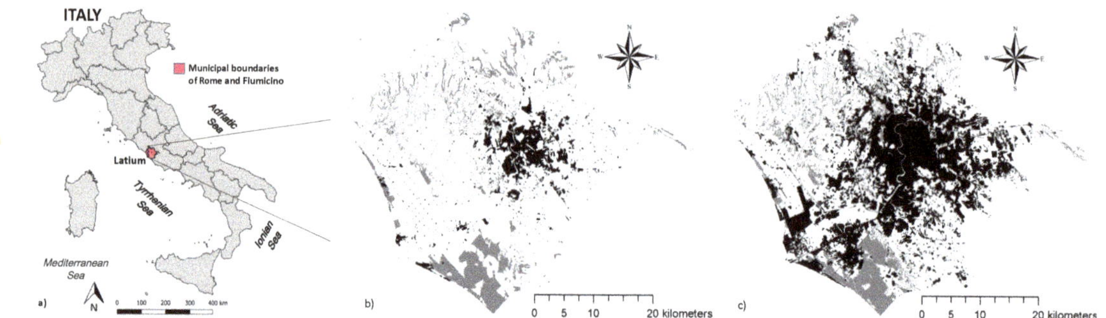

Figure 1. A map indicating the position of the study area (Rome and Fiumicino municipality) in Central Italy (**a**) and two maps illustrating the spatial distribution of basic land-use classes (black: urban settlements; grey: woodlands; and white: agricultural areas) in Rome 1949 (**b**); 2016 (**c**). Urban settlements include impervious land and urban parks; agricultural areas include arable land, crop mosaic, vineyards, and olive groves; and woodlands include forests, pastures, and water bodies.

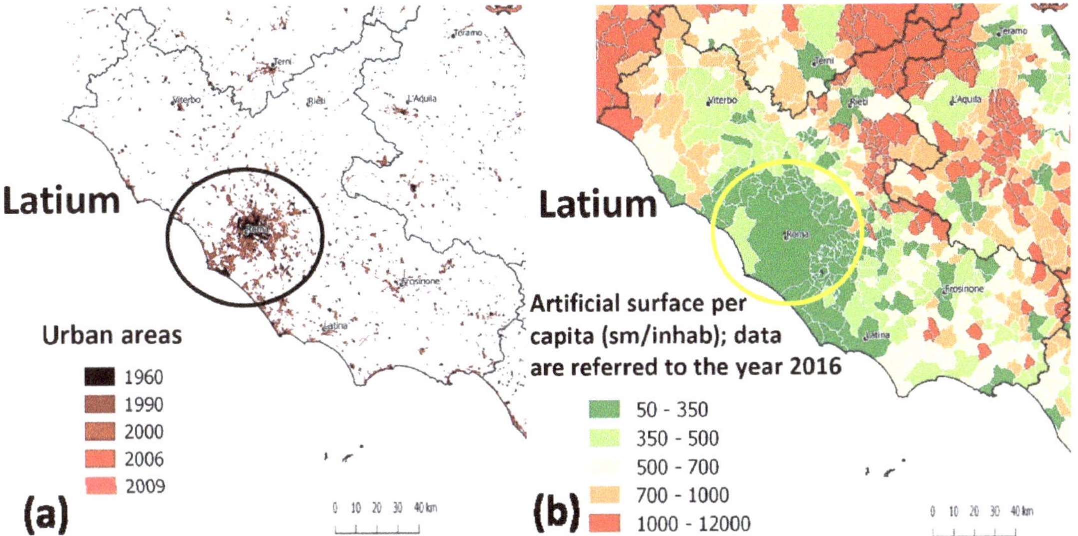

Figure 2. Boundaries of the Latium region located at the heart of Italy. Within the circle the municipalities of Rome and Fiumicino are shown with their corresponding urban areas (1960–2009, sub-figure (**a**)) and 2016 artificial surface per-capita (sub-figure (**b**)). These figures are taken from the annual report (2017) that ISPRA (Italian National Institute for Environmental Protection and Research) draws up on soil consumption, territorial dynamics and ecosystem services [65].

2.2. Land-Use Maps

Following earlier studies [26], a database reporting the surface area of several land-use classes was derived from elaboration of 3 compatible maps with a land nomenclature based on a simplified Corine Land Cover classification system [57–59,61,62]: (i) the Italian Istituto Geografico Militare topographic map (1:25,000 scale) produced in 1949, (ii) a land-use map (1:25,000) produced by the Cartographical Service of the Rome's province authority derived from field surveys and in-house photo-interpretation of digital ortho-photographs taken for cadastral purposes, and (iii) land-use maps (1:25,000) realized by the Cartographical Service of Latium regional authority from photo-interpretation of digital ortho-images released from the Italian National Geoportal. In addition, a first map was originally produced for 1999 and subsequently updated for 2008 and 2016 with new images referring to the same years. Eight homogeneous classes with a minimum mapping unit of 1 hectare have been considered: (i) arable land, (ii) crop mosaic, (iii) vineyards, (iv) olive groves, (v) woodlands, (vi) pastures, (vii) water bodies and wetlands, (viii) impervious land, and (ix) urban parks.

2.3. Land-Use Indicators

In this work, we used per-capita indicators as suggested by the United Nations [66] for the achievement of the Sustainable Development Goals (SDGs). In particular, the 2030 Agenda strongly encourages the adoption of the indicator "relationship between land consumption and population growth". In addition, Italy has adopted this indicator in the annual report of soil consumption drawn up by ISPRA (Italian National Institute for Environmental Protection and Research, see Reference [65]). For each investigated year (1949, 1974, 1999, 2008, and 2016), a per-capita indicator of surface area by land-use class was constructed considering (i) the total extension (ha) of land devoted to a given class and (ii) the total population residing in the study area (municipalities of Rome and Fiumicino). Total population was derived from integration of population census data (conducted every 10 years in Italy on behalf of the National Statistical Institute, Istat) and results from the national population registry available for each country's municipality.

Annual rates of change over time were also calculated separately for each time interval (1949–1974, 1974–1999, 1999–2008, and 2008–2016) and land-use class.

To summarize long-term land-use patterns and trends in the study area, a Principal Component Analysis (PCA) was run on a matrix composed of two indicators' sets: (i) the relative proportion of the 9 classes (see above) by year (i.e., class area in total landscape, labelled with 'c') and (ii) per-capita surface area (ha) of each land-use class by year (labelled with 'p'). PCA is a multivariate exploratory technique aimed at analyzing two-way data matrices, inspecting the latent relationship among variables organized by rows and columns together. In our case, PCA provides delineates a dynamic representation of land-use change in the study area considering year-by-year class transformations by evaluating both landscape composition (class area) and land-use intensity (per-capita class area). The input matrix was constituted of 9 land-use classes (columns) and 2 indicators (percent class share in the total landscape and per-capita surface area by class) quantified at five years each (rows) and occupying ten rows. Components were extracted on the base of the absolute eigenvalue. Components with eigenvalue > 1 were regarded as significant and analyzed further for loadings and scores' structure. A biplot was realized with the aim of illustrating the statistical distribution of component loadings (land-use classes) and scores (indicators and years), evidencing the apparent linkages between land-use composition and landscape structure. This plot provided a refined and summary outlook of similarities and differences in landscape transformations across the study area.

3. Results

The study area has experienced land-use changes over time (Figure 3). Urban growth with sequential acceleration and deceleration waves has been associated with a moderate increase in the forest area. Landscape transformations have negatively impacted the traditional agricultural matrix that has contracted significantly. Urban fabric, initially compact and concentrated around the central city, has gradually expanded into rural areas occupying more than 30% of the surface area administered by Rome's municipality in 2016 and developing a spatially heterogeneous shape. Based on the landscape structure in 1949, the agricultural matrix formed a natural interface between urban areas and forests. In 2016, it was extremely fragmented, no longer acting as a buffer zone between settlements and woods, and now placed in close contact in most cases. These changes have transformed Rome's morphology, initially dense and organized mono-centrically around the historical town. During the study period, urban development progressively lost its radio-centric and additive character. In 2016, Rome's morphology was found more dispersed and spatially discontinuous. Taken together, these preliminary results highlight a transition from a compact form to a more dispersed and fragmented morphology common with other Mediterranean cities.

The long-term transformation of Rome's landscape (1949–2016) coincided with intense population growth in the study area (Figure 3). However, combining urban expansion and population growth rates, different rates of change were observed in the four sub-periods of investigation. The first period (1949–1974) was characterized by a sustained rate of urban expansion (about 4% per year) associated with an equally high rate of population growth (about 3% per year) and their ratio is about 1.5. This period reflects compact, dense, and radio-centric urbanization leveraged by demographic growth. The second time interval (1974–1999) delineated a massive urban expansion (about 3% per year) that did not correspond to a similar demographic trend, though, as the population grew very weakly with a ratio reaching the tremendous value of about 188. This period coincides with a phase of suburbanization, the result of which was the suburban expansion of low-density and spatially dispersed settlements in the metropolitan area. The largest mismatch between urban expansion and population growth was observed at this stage. The two subsequent periods (1999–2008 and 2008–2016) were rather similar, characterized by moderate urban growth in the face of modest demographic dynamics. In relative terms, the mismatch between the two growth rates remained high with ratio values of about 6 and 3.5, respectively.

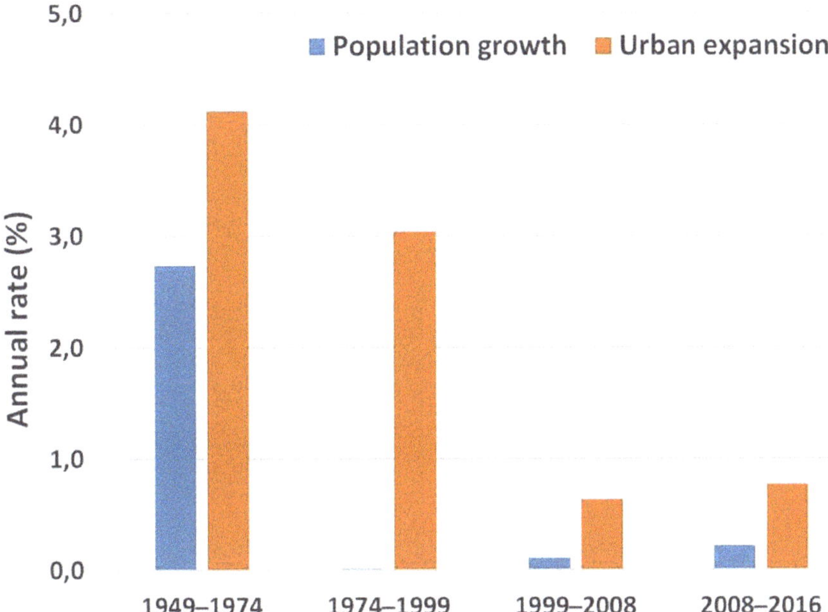

Figure 3. Percent annual rate of growth (population: blue; settlements: orange) by time interval in Rome. Urban expansion includes impervious land and parks/gardens within the city.

Considering the availability of different land-uses per-capita as an indicator of long-term transformations in urban, agricultural, and forest landscapes (Table 1), a substantial reduction in per-capita arable land was observed over time. Impervious lands (urban settlements) exhibit opposite dynamics similar to what has been observed for urban parks and gardens, a class physically associated with settlements and therefore dependent on their dynamics over time.

Table 1. Per-capita surface area (hectares) by land-use class in Rome, 1949–2016.

Class	1949	1974	1999	2008	2016
Arable Land	6.277	2.777	2.505	2.342	2.199
Crop Mosaic	0.161	0.178	0.070	0.111	0.112
Vineyards	0.221	0.126	0.068	0.069	0.070
Olive Groves	0.039	0.056	0.056	0.062	0.063
Woodlands	1.018	0.551	0.634	0.631	0.631
Pastures	0.549	0.770	0.433	0.425	0.422
Water Bodies	0.062	0.036	0.035	0.036	0.035
Impervious Land	0.594	0.800	1.388	1.460	1.530
Urban Parks	0.144	0.090	0.172	0.174	0.174

Following a significant decrease in the first time interval (1949–1974), forests have exhibited slight increases in more recent periods. Despite the significant contraction of agricultural areas, the cropland mosaic has undergone an important spatial reorganization. Some agricultural systems exhibited a decrease (vineyards and pastures), while other classes exhibited a slight growth (olive groves). In every case, trends over time have been non-linear, highlighting a substantial heterogeneity that corresponds to very different dynamics over space according to the different phases of urban expansion as better demonstrated in the following analysis. The empirical analysis of per-capita land-use change in the four phases of metropolitan development outlined above highlights different dynamics in the first two time periods, corresponding with two sequential stages of the urban cycle in Europe (urbanization–suburbanization). By contrast, land-use change rates are fairly homogeneous in the two most recent periods (Table 2). These periods represent a rather heterogeneous phase of the urban cycle that combines elements of counter-urbanization (slow but constant land consumption at progressively greater distances from urban centers) and re-urbanization (slow recovery of the population even in central areas). More specifically, urbanization in Rome (1949–1974) coincided with a significant contraction of the per-capita surfaces of almost all agricultural and forest land-use (except olive groves, pastures, and crop mosaics) in favor of urban settlements. In the context of suburbanization (1974–1999), a slower (per-capita) contraction of all agricultural land-use was observed in favor of both urban settlements and forests. In the face of a continuous growth of urban settlements, the natural landscape responded in a very different way in the most recent phases. Arable lands continued to decline significantly. This extensive use of agricultural land characteristic of the study area has constituted a stock of buildable area over the entire study period considering the modest value added.

Table 2. Percent change over time (annual rate) in the per-capita surface area by land-use class in Rome, 1949–2016.

Class	1949–1974	1974–1999	1999–2008	2008–2016
Arable Land	−2.23	−0.39	−0.72	−0.77
Crop Mosaic	0.41	−2.42	6.42	0.05
Vineyards	−1.72	−1.86	0.24	0.09
Olive Groves	1.79	−0.06	1.22	0.21
Woodlands	−1.84	0.61	−0.05	−0.01
Pastures	1.61	−1.75	−0.21	−0.10
Water Bodies	−1.70	−0.05	0.09	−0.23
Impervious Land	1.39	2.94	0.57	0.60
Urban Parks	−1.51	3.67	0.11	0.03

The agricultural mosaic together with vineyards and olive groves demonstrated evident signs of recovery. This result is in line with a generalized fragmentation of rural spaces at the fringe. Expansion of tree crops coincided with the consolidation of agricultural mosaics, a highly fragmented landscape that retains important elements of naturalness. The future challenge for fragmented landscapes will be to preserve this type of land-use from further isolation, fragmentation, and degradation resulting from urban growth. On the contrary, pastures have undergone a slow but steady decrease in line with what has been observed for arable lands. In the Mediterranean, fringe pastures have traditionally represented a reservoir of land potentially buildable or convertible to other (economically more profitable) uses. From this perspective, the expansion of pastures in 1949–1974 confirms the massive urban pressure typical of that period when pastures, often representing a human-driven evolution of arable lands, were frequently regarded as a 'temporary' land use pending 'speculative' conversion to more profitable uses. In the subsequent periods, pastures mostly acted as the largest stock of land to be converted, sometimes along with arable land.

Based on the previous results, a PCA was developed to summarize the main transformations in fringe landscapes and compare the latent trends in terms of the total area and area per-capita for each land-use class (Figure 4). The PCA extracted two axes cumulatively explaining more than 90% of the total variance. Component 1 (71.9%) clearly discriminates the total dynamics (negative scores) from the per-capita land-use indicators (positive scores), differentially depicting trends in the agricultural landscape. In fact, arable lands are considered in opposition to other agricultural land uses. This result highlights how the greatest transformations of the agricultural landscape in the study area concerned the arable lands, which have undergone a strong downsizing in per-capita terms. In contrast, all other uses of agricultural land have experienced much more modest changes. Component 2 (18.7%) demonstrates a marked similarity between urban settlements and forest areas, the two land uses that exhibited increases (more or less strong) during the survey period and to the detriment of agricultural areas. Component 2 also discriminates between earlier periods (p49 and p74) and more recent times (p99, p08, and p16), suggesting a substantial diversification in the three phases of urban development previously outlined (urbanization, suburbanization, and a mixed stage between counter-urbanization and latent re-urbanization).

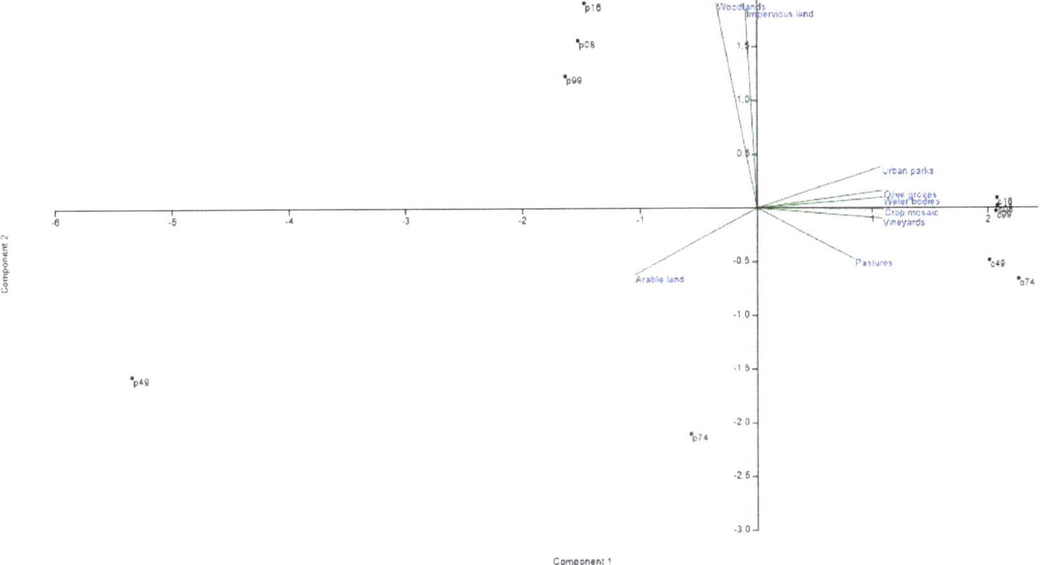

Figure 4. Results of a Principal Component Analysis run on the matrix including percent rates of change in land-use and per-capita land-use in Rome, 1949–2016.

4. Discussion

The analysis of landscape transformations conducted through the use of per-capita surface indicators by land use class provided an integrated view of environmental and socioeconomic nature. As far as built settlements are concerned, per-capita land-use indicators quantified the mismatch between urban expansion and population growth, outlining distinctive phases of the metropolitan cycle [67]. In the study area, this indicator has assumed profoundly different values in correspondence with (i) compact urbanization and (ii) low-density suburbanization, assuming intermediate values in the most recent period characterized by the coexistence of re-urbanization and counter-urbanization impulses. With regard to non-urban land-use, per-capita indicators provide an even more significant view; in the case of agricultural landscapes, the negative evolution of per-capita cropland indicated the impossibility to satisfy the growing (local) demand for fresh food. These contexts, however, were known for centuries for their specialization in primary production, which experienced a significant decline in the most recent decades [68]. Therefore, a recent analysis [69] argues that there is a need to focus more on the form and quality of urbanization processes rather than simply on the volume and speed of urbanization. Thus, the measure of the amounts of land in urban use and their spatial configuration remains crucial to provide useful information on anticipating future needs and to ensure an adequate supply of public goods such as infrastructure, open spaces, and common facilities for urban expansion [70].

The multivariate analysis clearly indicates the long-term competition between land-use, highlighting the relationship between 'predators' (urban settlements and more recently forests) and 'prey' (arable land and more recently pastures). This interpretation, in line with earlier studies developed in the Mediterranean basin, highlights the substantial dichotomy between 'predators' with a competitive advantage (urban settlements) and more marginal actors (forests) that opportunistically occupy landscape niches left free from abandonment of agricultural land as a result of real estate speculation [70]. This dynamic is visible in marginal areas, while being occasionally observed at the fringe, for example in contexts where physical (e.g., steepness and accessibility) and/or regulatory (planning/zoning) constraints to building are more pronounced [71–73]. Simultaneously, our analysis establishes how agricultural landscapes do not experience homogeneous dynamics [74–76]. In the face of the originally dominant arable lands that systematically acted as a 'prey' between 1949–2016, other land-uses have consolidated their spatial distribution with a significant increase in per-capita surface area. Among these, per-capita increase in olive groves, a 'synanthropic' crop in the Mediterranean basin, appears very significant, having occupied peri-urban voids generated by a fragmented growth of low-density settlements [77,78]. The rapid suburbanization process during the 1960s, 1970s, and the following period became a dominant form of urban growth and mainly contributed to the spatial dispersion of urban centers towards the urban fringes. However, remarkably, the dispersal of the employment was slower than that of the city during this period, therefore the commuters' distances have increased. During 1980s and 1990s, settlement accesses have grown along road infrastructures, encouraging the emergence of new commercial, industrial, and leisure areas. The expansion of urban towards peripheries preying on more arable lands was fueled by the increase of car ownership, the expansion of transport infrastructure, and the emergence of new commercial and industrial areas that have served as new suburban job hubs [79]. The local context was at the base of landscape transformations, having land speculation, second homes, tourism development, and migration as characteristic roles in landscape changes at the fringe [80]. In these regards, socioeconomic factors have demonstrated to influence metropolitan growth in Rome. Assuming the relevance of a comparative analysis of land-use change over a sufficiently long term, a joint monitoring of landscape transformations and population supports sustainable development policies in metropolitan regions [79]. In comparison with northern, central, and western counterparts, southern European metropolises have experienced landscape fragmentation in a context of increasing ecological fragility. Urban sprawl has demonstrated to influence fringe landscapes with changes in the intimate structure of both agricultural and forest mosaics [81]. The intrinsic relationship between landscape transformations and different waves of urban expansion was also illustrated, distinguishing at least three developmental waves, namely urbanization, suburbanization, and a more mixed phase with impulses of counter-urbanization and re-urbanization. Per-capita indicators of land-use change clearly documented the complete unsustainability of recent urban expansion. The intrinsic mismatch between population growth and urban expansion enlarged in recent times, particularly evident during both suburbanization and the most recent developmental wave.

5. Conclusions

Comparative analysis delineating latent patterns and trends in land-use has recently benefited from indicators that assess the intrinsic mismatch between landscape transformations and population growth. Rising availability of high-resolution datasets supports such investigation, allowing for the construction of new indicators and refining empirical approaches to land-use science. An extensive use of per-capita land-use indicators will contribute to monitor the effectiveness of adopted policy measures and forecast possible future land-use changes in a spatial planning perspective. When evaluating long-term transformations of fringe landscapes, the empirical results of our study indicate the increasing mismatch between settlement expansion and population growth that can be easily captured through the ratio between the two growth rates. This ratio goes from the value of just 1.5 for the period of 1949–1974 to the enormous peak of 188 in the time frame of 1974–1999, and then reaches the values of about 6 and 3.5 in the subsequent periods (1999–2008 and 2008–2016, respectively). In combination, this developmental mode reveals all its traits of unsustainability, leading to landscape fragmentation and abnormal consumption of soil resources. Spatial planning (and primarily town master plans) should envisage practical action to contain the unavoidable effects of recent urban growth on the capacity of ecosystems to maintain a complete provision of services and goods in metropolitan region.

Author Contributions: Supervision, M.C.; project administration and funding acquisition, M.C.; conceptualization and writing—original draft preparation, A.S. and A.A.; writing—review and editing, S.C. and L.B.; software and data curation, S.C.; conceptualization, methodology, visualization, and supervision, G.E. and L.B.; writing—review and editing, V.I. All authors have read and agreed to the published version of the manuscript.

Funding: This research received no external funding.

Conflicts of Interest: The authors declare no conflict of interest.

References

1. Bruegmann, R. *Sprawl: A Compact History*; University of Chicago Press: Chicago, IL, USA, 2005.
2. Couch, C.; Petschel-Held, G.; Leontidou, L. (Eds.) *Urban Sprawl in Europe: Landscape, Land-Use Change and Policy*, 1st ed.; Wiley-Blackwell: Oxford, UK; Malden, MA, USA, 2007; ISBN 978-1-4051-3917-5.
3. Brouwer, F.M.; Thomas, A.J.; Chadwick, M.J. *Land Use Changes in Europe-Processes of Change, Environmental Transformations and Future Patterns*; luwer AcademicPublishers: Dordrecht, The Netherlands, 1991.
4. Lambin, E.F.; Turner, B.L.; Geist, H.J.; Agbola, S.B.; Angelsen, A.; Bruce, J.W.; Oliver, T.C.; Rodolf, D.; Gunther, F.; Carl, F.; et al. The causes of land-use and land-cover change: Moving beyond the myths. *Glob. Environ. Chang.* **2001**, *11*, 261–269. [CrossRef]
5. Pelorosso, R.; Leone, A.; Boccia, L. Land cover and land use change in the Italian central Apennines: A comparison of assessment methods. *Appl. Geogr.* **2009**, *29*, 35–48. [CrossRef]
6. Simoniello, T.; Coluzzi, R.; Imbrenda, V.; Lanfredi, M. Land cover changes and forest landscape evolution (1985–2009) in a typical Mediterranean agroforestry system (high Agri Valley). *Nat. Hazards Earth Syst. Sci.* **2015**, *15*, 1201–1214. [CrossRef]
7. Imbrenda, V.; Coluzzi, R.; Lanfredi, M.; Loperte, A.; Satriani, A.; Simoniello, T. Analysis of landscape evolution in a vulnerable coastal area under natural and human pressure. *Geomat. Nat. Hazards Risk* **2018**, *9*, 1249–1279. [CrossRef]
8. Coluzzi, R.; D'Emilio, M.; Imbrenda, V.; Giorgio, G.A.; Lanfredi, M.; Macchiato, M.; Simoniello, T.; Telesca, V. Investigating climate variability and long-term vegetation activity across heterogeneous Basilicata agroecosystems. *Geomat. Nat. Hazards Risk* **2019**, *10*, 168–180. [CrossRef]
9. Quaranta, G.; Salvia, R.; Salvati, L.; Paola, V.D.; Coluzzi, R.; Imbrenda, V.; Simoniello, T. Long-Term Impacts of Grazing Management on Land Degradation in a Rural Community of Southern Italy: Depopulation Matters. *Land Degrad. Dev.* **2020**, *31*, 1–16. [CrossRef]
10. Salvati, L.; Zambon, I.; Chelli, F.M.; Serra, P. Do Spatial Patterns of Urbanization and Land Consumption Reflect Different Socioeconomic Contexts in Europe? *Sci. Total Environ.* **2018**, *625*, 722–730. [CrossRef]
11. O'Sullivan, A. *Urban Economics*; McGraw-Hill/Irwin: New York, NY, USA, 2002.
12. Oueslati, W.; Alvanides, S.; Garrod, G. Determinants of Urban Sprawl in European Cities. *Urban Stud.* **2015**, *52*, 1594–1614. [CrossRef]
13. Capello, R.; Nijkamp, P. *Urban Dynamics and Growth: Advances in Urban. Economics: 266*; North Holland: Amsterdam, Holland, 2004; ISBN 978-0-444-51481-3.
14. Coulson, N.E. Really Useful Tests of the Monocentric Model. *Land Econ.* **1991**, *67*, 299–307. [CrossRef]
15. Ahlfeldt, G. If Alonso Was Right: Modeling Accessibility and Explaining the Residential Land Gradient. *J. Reg. Sci.* **2011**, *51*, 318–338. [CrossRef]
16. Kazemzadeh-Zow, A.; Shahraki, S.Z.; Salvati, L.; Samani, N.N. A Spatial Zoning Approach to Calibrate and Validate Urban Growth Models. *Int. J. Geogr. Inf. Sci.* **2017**, *31*, 763–782. [CrossRef]
17. Salvati, L.; Ciommi, M.T.; Serra, P.; Chelli, F.M. Exploring the Spatial Structure of Housing Prices Under Economic Expansion and Stagnation: The Role of Socio-Demographic Factors in Metropolitan Rome, Italy. *Land Use Policy* **2019**, *81*, 143–152. [CrossRef]
18. Schwarz, N. Urban Form Revisited—Selecting Indicators for Characterising European Cities. *Landsc. Urban Plan.* **2010**, *96*, 29–47. [CrossRef]
19. Kasanko, M.; Barredo, J.I.; Lavalle, C.; McCormick, N.; Demicheli, L.; Sagris, V.; Brezger, A. Are European Cities Becoming Dispersed? A Comparative Analysis of 15 European Urban Areas. *Landsc. Urban Plan.* **2006**, *77*, 111–130. [CrossRef]

20. Chelleri, L.; Schuetze, T.; Salvati, L. Integrating Resilience with Urban Sustainability in Neglected Neighborhoods: Challenges and Opportunities of Transitioning to Decentralized Water Management in Mexico City. *Habitat Int.* **2015**, *48*, 122–130. [CrossRef]

21. Salvati, L.; Ferrara, A.; Chelli, F. Long-Term Growth and Metropolitan Spatial Structures: An Analysis of Factors Influencing Urban Patch Size under Different Economic Cycles. *Geogr. Tidsskr./Dan. J. Geogr.* **2018**, *118*, 56–71. [CrossRef]

22. Imbrenda, V.; Quaranta, G.; Salvia, R.; Egidi, G.; Salvati, L.; ProkopovÁ, M.; Coluzzi, R.; Lanfredi, M. Land degradation and metropolitan expansion in a peri-urban environment. *Geomat. Nat. Hazards Risk.* (in press). [CrossRef]

23. Di Feliciantonio, C.; Salvati, L. 'Southern' Alternatives of Urban Diffusion: Investigating Settlement Characteristics and Socio-economic Patterns in Three Mediterranean Regions. *Tijdschr. voor Econ. en Soc. Geogr.* **2015**, *106*, 453–470. [CrossRef]

24. De Rosa, S.; Salvati, L. Beyond a 'Side Street Story'? Naples from Spontaneous Centrality to Entropic Polycentrism, towards a 'Crisis City'. *Cities* **2016**, *51*, 74–83. [CrossRef]

25. Cuadrado-Ciuraneta, S.; Durà-Guimerà, A.; Salvati, L. Not Only Tourism: Unravelling Suburbanization, Second-Home Expansion and "Rural" Sprawl in Catalonia, Spain. *Urban Geogr.* **2017**, *38*, 66–89. [CrossRef]

26. Cecchini, M.; Zambon, I.; Pontrandolfi, A.; Turco, R.; Colantoni, A.; Mavrakis, A.; Salvati, L. Urban Sprawl and the 'Olive' Landscape: Sustainable Land Management for 'Crisis' Cities. *GeoJournal* **2019**, *84*, 237–255. [CrossRef]

27. Schneider, A.; Woodcock, C.E. Compact, Dispersed, Fragmented, Extensive? A Comparison of Urban Growth in Twenty-Five Global Cities Using Remotely Sensed Data, Pattern Metrics and Census Information. *Urban Stud.* **2008**, *45*, 659–692. [CrossRef]

28. Burger, M.; Meijers, E. Form Follows Function? Linking Morphological and Functional Polycentricity. *Urban Stud.* **2012**, *49*, 1127–1149. [CrossRef]

29. Salvati, L. Agro-Forest Landscape and the 'Fringe' City: A Multivariate Assessment of Land-Use Changes in a Sprawling Region and Implications for Planning. *Sci. Total Environ.* **2014**, *490*, 715–723. [CrossRef]

30. Veneri, P. The Identification of Sub-Centres in Two Italian Metropolitan Areas: A Functional Approach. *Cities* **2013**, *31*, 177–185. [CrossRef]

31. Turok, I.; Mykhnenko, V. The Trajectories of European cities, 1960–2005. *Cities* **2007**, *24*, 165–182. [CrossRef]

32. Pili, S.; Grigoriadis, E.; Carlucci, M.; Clemente, M.; Salvati, L. Towards Sustainable Growth? A Multi-Criteria Assessment of (Changing) Urban Forms. *Ecol. Indic.* **2017**, *76*, 71–80. [CrossRef]

33. Di Feliciantonio, C.; Salvati, L.; Sarantakou, E.; Rontos, K. Class Diversification, Economic Growth and Urban Sprawl: Evidences from a Pre-Crisis European city. *Qual. Quant.* **2018**, *52*, 1501–1522. [CrossRef]

34. Salvati, L. The Dark Side of the Crisis: Disparities in Per Capita Income (2000–12) and the Urban–Rural Gradient in Greece. *Tijdschr. Econ. Soc. Geogr.* **2016**, *107*, 628–641. [CrossRef]

35. Samat, N.; Hasni, R.; Elhadary, Y. Modelling Land Use Changes at the Peri-Urban Areas Using Geographic Information Systems and Cellular Automata Model. *J. Sustain. Dev.* **2011**, *4*, 72. [CrossRef]

36. Chorianopoulos, I.; Pagonis, T.; Koukoulas, S.; Drymoniti, S. Planning, Competitiveness and Sprawl in the Mediterranean City: The Case of Athens. *Cities* **2010**, *27*, 249–259. [CrossRef]

37. Terzi, F.; Bolen, F. Urban Sprawl Measurement of Istanbul. *Eur. Plan. Stud.* **2009**, *17*, 1559–1570. [CrossRef]

38. Alphan, H. Land-Use Change and Urbanization of Adana, Turkey. *Land Degrad. Dev.* **2003**, *14*, 575–586. [CrossRef]

39. Matteucci, S.D.; Morello, J. Environmental Consequences of Exurban Expansion in an Agricultural Area: The Case of the Argentinian Pampas Ecoregion. *Urban Ecosyst.* **2009**, *12*, 287–310. [CrossRef]

40. Kane, K.; Connors, J.P.; Galletti, C.S. Beyond Fragmentation at the Fringe: A Path-Dependent, High-Resolution Analysis of Urban Land Cover in Phoenix, Arizona. *Appl. Geogr.* **2014**, *52*, 123–134. [CrossRef]

41. Anselm, N.; Brokamp, G.; Schütt, B. Assessment of Land Cover Change in Peri-Urban High Andean Environments South of Bogotá, Colombia. *Land* **2018**, *7*, 75. [CrossRef]

42. York, A.M.; Shrestha, M.; Boone, C.G.; Zhang, S.; Harrington, J.A.; Prebyl, T.J.; Swann, A.; Agar, M.; Antolin, M.F.; Nolen, B.; et al. Land Fragmentation under Rapid Urbanization: A cross-site analysis of southwestern cities. *Urban Ecosyst.* **2011**, *14*, 429–455. [CrossRef]

43. Çakir, G.; Ün, C.; Baskent, E.Z.; Köse, S.; Sivrikaya, F.; Keleş, S. Evaluating Urbanization, Fragmentation and Land Use/Land Cover Change Pattern in Istanbul City, Turkey from 1971 TO 2002. *Land Degrad. Dev.* **2008**, *19*, 663–675. [CrossRef]

44. Catalán, B.; Saurí, D.; Serra, P. Urban Sprawl in the Mediterranean? Patterns of Growth and Change in the Barcelona Metropolitan Region 1993–2000. *Landsc. Urban. Plan.* **2008**, *85*, 174–184. [CrossRef]

45. Salvati, L.; Serra, P. Estimating Rapidity of Change in Complex Urban Systems: A Multidimensional, Local-Scale Approach. *Geogr. Anal.* **2016**, *48*, 132–156. [CrossRef]

46. Zambon, I.; Serra, P.; Sauri, D.; Carlucci, M.; Salvati, L. Beyond the 'Mediterranean City': Socioeconomic Disparities and Urban Sprawl in Three Southern European Cities. *Geogr. Ann. Ser. B Hum. Geogr.* **2017**, *99*, 319–337. [CrossRef]

47. Jarah, S.H.A.; Zhou, B.; Abdullah, R.J.; Lu, Y.; Yu, W. Urbanization and Urban Sprawl Issues in City Structure: A Case of the Sulaymaniah Iraqi Kurdistan Region. *Sustainability* **2019**, *11*, 485. [CrossRef]

48. Zambon, I.; Benedetti, A.; Ferrara, C.; Salvati, L. Soil Matters? A Multivariate Analysis of Socioeconomic Constraints to Urban Expansion in Mediterranean Europe. *Ecol. Econ.* **2018**, *146*, 173–183. [CrossRef]

49. Hassan, M.M. Monitoring Land Use/Land Cover Change, Urban Growth Dynamics and Landscape Pattern Analysis in Five Fastest Urbanized Cities in Bangladesh. *Remote Sens. Appl. Soc. Environ.* **2017**, *7*, 69–83. [CrossRef]

50. Serra, P.; Vera, A.; Tulla, A.F.; Salvati, L. Beyond Urban–Rural Dichotomy: Exploring Socioeconomic and Land-Use Processes of Change in Spain (1991–2011). *Appl. Geogr.* **2014**, *55*, 71–81. [CrossRef]

51. Attorre, F.; Bruno, M.; Francesconi, F.; Valenti, R.; Bruno, F. Landscape Changes of Rome through Tree-Lined Roads. *Landsc. Urban Plan.* **2000**, *49*, 115–128. [CrossRef]

52. Tsai, Y.-H. Quantifying Urban Form: Compactness versus "Sprawl". *Urban Stud.* **2005**, *42*, 141–161. [CrossRef]

53. Carlucci, M.; Grigoriadis, E.; Rontos, K.; Salvati, L. Revisiting a Hegemonic Concept: Long-Term 'Mediterranean Urbanization' in between City Re-Polarization and Metropolitan Decline. *Appl. Spat. Anal.* **2017**, *10*, 347–362. [CrossRef]

54. Gavalas, V.S.; Rontos, K.; Salvati, L. Who Becomes an Unwed Mother in Greece? Sociodemographic and Geographical Aspects of an Emerging Phenomenon. *Popul. Space Place* **2014**, *20*, 250–263. [CrossRef]

55. Weber, C.; Petropoulou, C.; Hirsch, J. Urban Development in the Athens Metropolitan Area Using Remote Sensing Data with Supervised Analysis and GIS. *Int. J. Remote Sens.* **2005**, *26*, 785–796. [CrossRef]

56. Yu, Z.; Zhang, J.; Yang, G.; Schlaberg, J. Reverse Thinking: A New Method from the Graph Perspective for Evaluating and Mitigating Regional Surface Heat Islands. *Remote Sens.* **2021**, *13*, 1127. [CrossRef]

57. Frenkel, A.; Ashkenazi, M. Measuring Urban Sprawl: How Can We Deal With It? *Environ. Plann. B Plann. Des.* **2008**, *35*, 56–79. [CrossRef]

58. Ioannidis, C.; Psaltis, C.; Potsiou, C. Towards a Strategy for Control of Suburban Informal Buildings Through Automatic Change Detection. *Comput. Environ. Urban Syst.* **2009**, *33*, 64–74. [CrossRef]

59. Lanfredi, M.; Coppola, R.; Simoniello, T.; Coluzzi, R.; D'Emilio, M.; Imbrenda, V.; Macchiato, M. Early Identification of Land Degradation Hotspots in Complex Bio-Geographic Regions. *Remote. Sens.* **2015**, *7*, 8154–8179. [CrossRef]

60. Longhi, C.; Musolesi, A. European Cities in the Process of Economic Integration: Towards Structural Convergence. *Ann. Reg. Sci.* **2007**, *41*, 333–351. [CrossRef]

61. Romano, B.; Zullo, F.; Fiorini, L.; Marucci, A.; Ciabò, S. Land Transformation of Italy Due to Half a Century of Urbanization. *Land Use Policy* **2017**, *67*, 387–400. [CrossRef]

62. Fiorini, L.; Zullo, F.; Marucci, A.; Romano, B. Land Take and Landscape Loss: Effect of Uncontrolled Urbanization in Southern italy. *J. Urban Manag.* **2019**, *8*, 42–56. [CrossRef]

63. Salvati, L. Agricultural Land-Use Changes and Soil Quality: Evaluating Long-Term Trends in a Rural Mediterranean Region. *Int. Sch. Res. Not.* **2013**, *2013*, 18240. [CrossRef]

64. Salvati, L.; De Zuliani, E.; Sabbi, A.; Cancellieri, L.; Tufano, M.; Caneva, G.; Savo, V. Land-cover changes and sustainable development in a rural cultural landscape of central Italy: Classical trends and counter-intuitive results. *Int. J. Sustain. Dev. World Ecol.* **2017**, *24*, 27–36. [CrossRef]

65. Marchetti, M.; Marino, D.; De Toni, A.; Giaccio, V.; Giannelli, A.; Mastronardi, L. Soil Consumption, Territorial Dynamics and Ecosystem Services (in Italian); ISPRA (Italian National Institute for Environmental Protection and Research, see Reference), Annual Report 2017. Available online: https://www.isprambiente.gov.it/en/publications/reports/soil-consumption-in-italy-1 (accessed on 16 July 2021).

66. Union Nations. Transforming Our World: The 2030 Agenda for Sustainable Development. A/RES/70/1.2015. Available online: http://www.un.org/en/ga/search/view_doc.asp?symbol=A/RES/70/1 (accessed on 16 July 2021).

67. Antrop, M. Landscape Change and the Urbanization Process in Europe. *Landsc. Urban Plan.* **2004**, *67*, 9–26. [CrossRef]

68. Anarfi, K.; Hill, R.A.; Shiel, C. Highlighting the Sustainability Implications of Urbanisation: A Comparative Analysis of Two Urban Areas in Ghana. *Land* **2020**, *9*, 300. [CrossRef]

69. OECD. *Trends in Urbanisation and Urban Policies in OECD Countries: What Lessons for China?* Organisation for Economic Co-Operation and Development and China Development Research Foundation. 2010. Available online: http://www.oecd.org/dataoecd/2/18/45159707.pdf (accessed on 16 July 2021).

70. Robert, S.; Fox, D.; Boulay, G.; Grandclément, A.; Garrido, M.; Pasqualini, V.; Prévost, A.; Schleyer-Lindenmann, A.; Trémélo, M.-L. A Framework to Analyse Urban Sprawl in the French Mediterranean Coastal Zone. *Reg. Environ. Chang.* **2019**, *19*, 559–572. [CrossRef]

71. Pozoukidou, G.; Ntriankos, I. Measuring and Assessing Urban Sprawl: A Proposed Indicator System for the City of Thessaloniki, Greece. *Remote Sens. Appl. Soc. Environ.* **2017**, *8*, 30–40. [CrossRef]

72. Herrschel, T. City Regions, Polycentricity and the Construction of Peripheralities Through Governance. *Urban Res. Pract.* **2009**, *2*, 240–250. [CrossRef]

73. Brezzi, M.; Veneri, P. Assessing Polycentric Urban Systems in the OECD: Country, Regional and Metropolitan Perspectives. *Eur. Plan. Stud.* **2015**, *23*, 1128–1145. [CrossRef]

74. Paül, V.; Tonts, M. Containing Urban Sprawl: Trends in Land Use and Spatial Planning in the Metropolitan Region of Barcelona. *J. Environ. Plan. Manag.* **2005**, *48*, 7–35. [CrossRef]

75. Volgmann, K.; Münter, A. Understanding Metropolitan Growth in German Polycentric Urban Regions. *Reg. Stud.* **2020**, 1–14. [CrossRef]

76. Lagarias, A.; Sayas, J. Urban Sprawl in the Mediterranean: Evidence from coastal medium-sized cities. *Reg. Sci. Inq.* **2018**, *10*, 15–32.

77. Guastella, G.; Oueslati, W.; Pareglio, S. Patterns of Urban Spatial Expansion in European Cities. *Sustainability* **2019**, *11*, 2247. [CrossRef]

78. Fertner, C.; Jørgensen, G.; Nielsen, T.A.S.; Nilsson, K.S.B. Urban Sprawl and Growth Management-Drivers, Impacts and Responses in Selected European and us Cities. *Future Cities Environ.* **2016**, *2*, 1–13. [CrossRef]
79. Meijers, E. Measuring polycentricity and its promises. *Eur. Plan. Stud.* **2008**, *16*, 1313–1323. [CrossRef]
80. *European Environment Agency Urban Sprawl in Europe–The Ignored Challenge*. Copenhagen, Denmark. 2006. Available online: https://www.eea.europa.eu/publications/eea_report_2006_10/eea_report_10_2006.pdf/view (accessed on 16 July 2021).
81. Patacchini, E.; Zenou, Y.; Henderson, J.V.; Epple, D. Urban Sprawl in Europe. *Brook. Whart. Pap. Urban Aff.* **2009**, 125–149.

 *land*

Technical Note

Forest Transition and Metropolitan Transformations in Developed Countries: Interpreting Apparent and Latent Dynamics with Local Regression Models

Leonardo Bianchini [1], Rosanna Salvia [2], Giovanni Quaranta [2], Gianluca Egidi [1], Luca Salvati [3,*] and Alvaro Marucci [1]

[1] Department of Agricultural and Forestry Sciences (DAFNE), University of Tuscia, Via S. Camillo De Lellis snc, I-01100 Viterbo, Italy; l.bianchini@unitus.it (L.B.); egidi.gianluca@unitus.it (G.E.); marucci@unitus.it (A.M.)
[2] Department of Mathematics, Computer Science and Economics, University of Basilicata, Viale dell'Ateneo Lucano, I-85100 Potenza, Italy; rosanna.salvia@unibas.it (R.S.); giovanni.quaranta@unibas.it (G.Q.)
[3] Department of Economics and Law, University of Macerata, Via Armaroli 43, I-62100 Macerata, Italy
* Correspondence: luca.salvati@crea.gov.it; Tel.: +39-06-61-57-10; Fax: +39-06-61-57-10-36

Abstract: Metropolitan fringes in Southern Europe preserve, under different territorial contexts, natural habitats, relict woodlands, and mixed agro-forest systems acting as a sink of biodiversity and ecosystem services in ecologically vulnerable landscapes. Clarifying territorial and socioeconomic processes that underlie land-use change in metropolitan regions is relevant for forest conservation policies. At the same time, long-term dynamics of fringe forests in the northern Mediterranean basin have been demonstrated to be rather mixed, with deforestation up to the 1950s and a subsequent recovery more evident in recent decades. The present study makes use of Forest Transition Theory (FTT) to examine spatial processes of forest loss and expansion in metropolitan Rome, Central Italy, through local regressions elaborating two diachronic land-use maps that span more than 80 years (1936–2018) representative of different socioeconomic and ecological conditions. Our study evaluates the turnaround from net forest area loss to net forest area gain, considering together the predictions of the FTT and those of the City Life Cycle (CLC) theory that provides a classical description of the functioning of metropolitan cycles. The empirical findings of our study document a moderate increase in forest cover depending on the forestation of previously abandoned cropland as a consequence of tighter levels of land protection. Natural and human-driven expansion of small and isolated forest nuclei along fringe land was demonstrated to fuel a polycentric expansion of woodlands. The results of a Geographically Weighted Regression (GWR) reveal the importance of metropolitan growth in long-term forest expansion. Forest–urban dynamics reflect together settlement sprawl and increased forest disturbance. The contemporary expansion of fringe residential settlements and peri-urban forests into relict agricultural landscapes claims for a renewed land management that may reconnect town planning, reducing the intrinsic risks associated with fringe woodlands (e.g., wildfires) with environmental policies preserving the ecological functionality of diversified agro-forest systems.

Keywords: land-use change; metropolitan gradient; spatial econometrics; agricultural mechanization; Mediterranean

check for **updates**

Citation: Bianchini, L.; Salvia, R.; Quaranta, G.; Egidi, G.; Salvati, L.; Marucci, A. Forest Transition and Metropolitan Transformations in Developed Countries: Interpreting Apparent and Latent Dynamics with Local Regression Models. *Land* **2022**, *11*, 12. https://doi.org/10.3390/land11010012

Academic Editor: Alexandru-Ionuţ Petrişor

Received: 30 November 2021
Accepted: 17 December 2021
Published: 22 December 2021

Publisher's Note: MDPI stays neutral with regard to jurisdictional claims in published maps and institutional affiliations.

1. Introduction

Providing a global interpretation of socio-environmental changes across a development gradient, Forest Transition Theory (FTT) predicts the inherent shift from net forest area loss, typical of emerging economies, to net forest area expansion [1]. This shift is characteristic of advanced economies and is coupled with a generalized recovery of mixed agricultural–wildland ecosystems [2–4]. This process has occurred in recent times via natural regeneration [5], active planting [6–8], or a combination of the two [9]. From this perspective, the notion of 'forest transition' is associated intimately with underlying socioeconomic forces [10], such as urbanization, late industrialization, tourism growth, and infrastructural development [11]. The different processes leading to forest transitions clearly depend on the local (territorial) context [12]. Although some generic processes can be identified for affluent countries (Figure 1), regions and districts do not necessarily experience a regular pattern of forest cover change over time, and the causes and effects of forest transitions may vary largely over space [12–14]. For instance, urbanization can determine a forest decline from direct clearcutting or wildfires [15], basically answering to increased housing demand (e.g., because of population growth) or real estate speculation, especially in contexts with less rigid planning rules or adopting new liberal urbanism schemes [16].

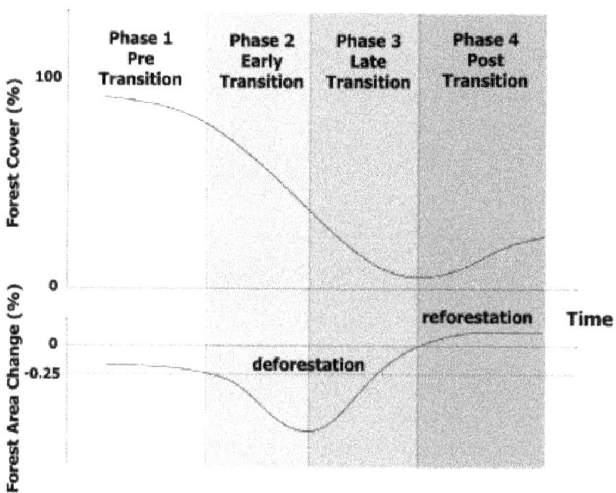

Figure 1. A framework illustrating the Forest Transition Theory (FTT) in advanced economies (Source: authors' elaboration from Rudel et al., 2005).

The predictions of the FTT have been tested empirically at various spatial scales [17–21]. Although the main forces of deforestation act locally, forest transitions in developed countries are far less investigated at that scale, especially in metropolitan regions [22–24]. As a result of urbanization, these regions represent a challenging opportunity to investigate the response of agro-forest landscapes to ecological disturbance and socioeconomic shocks [15,25–27]. Continuous urbanization has reflected the long-term interactions between nature and humans characteristic of the Mediterranean region [7,15,20,25–29]. Following urbanization, the abandonment of cropland, clearcutting, but also wood recolonization in some cases, may result in sequential stages of forest decline and recovery, depending on the local context [30–32]. Wildland–Urban Interfaces (WUIs) in Southern Europe frequently mix low-density settlements with natural habitats [17]. As a result, fringe landscapes feature irregular boundaries with discontinuous, low-density settlements that reflect suburbanization (i.e., population relocation to suburbs) and the delocalization of economic activities in peripheral areas [33]. These fragmented and mixed landscapes derive from a latent (and quite recent) process of peri-urban forest expansion into cropland, shrubland, and pastures [34], which has followed a significant decline in fringe forest cover following compact urban expansion [35]. As a matter of fact, economic development was demonstrated to fuel, in Italy as in Spain, and in Portugal as in Greece, compact urban growth in the 1950s and the 1960s and the uneven expansion of residential, low-density settlements in the 1970s and the 1980s [36–39]. From this perspective, FTT may appropriately predict cyclical trends in forest decline and expansion in complex peri-urban landscapes including relict agro-forest systems [40]. At the same time, predictions of the FTT were compared with those delineated by the City Life Cycle (CLC) theory. This theory provided a classical description of metropolitan cycles in advanced economies [41], explaining—at least indirectly—the socioeconomic dynamics at the base of peri-urban forest decline and expansion [38]. CLC indicates different, sequential stages of urban growth forming a cycle of urbanization, suburbanization, counter-urbanization, and re-urbanization [33]. The first two stages, urbanization and suburbanization, have been particularly well studied in Mediterranean Europe [16], and they respectively reflect a compact-dense/mono-centric growth and a low-density, spatially heterogeneous expansion of residential settlements [42].

For the first time, to the best of our knowledge, the present study mixes these two dynamic theories (CLC and FTT) with the aims of (i) improving forest assessment in Mediterranean fringe districts and (ii) informing dedicated planning strategies for the sustainable management of WUIs [2,18,43,44]. Mediterranean forests are widely recognized as an invaluable sink of biodiversity [45], forming green infrastructures that may (indirectly) contain urbanization and the negative externalities of economic activity [34]. However, earlier studies have demonstrated that built-up areas around forests expanded in parallel with a decrease in cropland. In other words, residential settlements and forests act as competing land uses with negative impacts on agricultural systems [5].

Going beyond a model of forest expansion from small nuclei of pristine, high-quality woodlands (concentrated in peripheral districts), recent land-use dynamics reflect a more complex path with the growth of scattered and fragmented forest nuclei embedded in a consolidated network of relict peri-urban and rural woodlands [31,46,47]. The shift from a traditional model of forestation irradiating from a few peripheral locations (hereafter defined as 'mono-centric') to a 'polycentric' structure of forest growth (irradiating from multiple locations even close to cities), may be seen at the base of FTT in Mediterranean contexts [48]. This pattern may inform a refined interpretation of forest transitions in metropolitan regions [39]. This 'polycentric' network of forests is also at the base of green infrastructures, assuring the efficient protection of natural habitats in fringe districts [48].

The present study introduces a refined analysis of long-term changes in forest cover as a result of urban expansion in a large metropolitan area (Rome, Central Italy) representative of compact, mono-centric cities in Southern Europe [33]. Reflecting socio-ecological dynamics that impact ecologically fragile Mediterranean environments, land-use complexity warrants further investigation of forest trends (1936–2018) in light of the FTT and the parallel urban cycle. A particularly long time interval was investigated in this study covering a productive cycle of forests and a period that encompassed various stages of metropolitan growth including urbanization and suburbanization [35]. Peri-urban landscapes surrounding Rome (Central Italy) have been identified as the most dynamic across the country because of city size and human pressure [34], resulting in sequential waves of forest decline and recovery [47]. While 55 Natura 2000 sites involving 22% of the study area have been recently established in Rome [48], natural land converted to settlements grew by 0.4% per year during the last half-century, endangering fragile coastal ecosystems more than mountain systems [49].

To check the validity of the polycentric model applied to forest expansion, local econometric estimations based on a Geographically Weighted Regression (GWR) contributed to ascertain the relationship between settlement expansion and changes over time in the composition and morphology of fringe forest landscapes [45]. Being deliberately simplified, our approach tests the shift from a mono-centric model of forest distribution over space toward a more polycentric structure, as hypothesized by Colantoni et al. (2015) [35]. The empirical verification of this hypothesis benefits from a spatially explicit analysis of a dependent variable (percent share of forest cover in total landscape area) and of a predictor, the distance from a central location, which is assumed to represent a metropolitan gradient from central to peripheral locations typical of mono-centric regions [41]. To account for the residual variability that the predictor is unable to explain [16], space was explicitly considered in the econometric specification. The empirical results of this study may inform policy and planning strategies preserving relict agro-forest systems in ecologically sensitive contexts typical of Mediterranean metropolises.

2. Methodology

2.1. Study Area

The investigated area is the metropolitan region of Rome (5355 km^2), which is an administrative partition of the Latium region (Central Italy) encompassing the previously established province of Rome (NUTS-3 level of the European Nomenclature of Territorial Statistics). The area includes 15% flat land (<100 m at the sea level) and 20% mountainous land (>600 m at the sea level). The Simbruini mountains (highest elevation: 1820 m above sea level), which belong to the Apennine district, are the most important relief in the area (Figure 2, left); the alluvial flat area of the Tiber river is made up of lowlands (the so-called "Agro Romano" district; Figure 2, right). The climate is typically Mediterranean, with rainfalls concentrated in autumn and spring, and mild winter temperatures [35]. The average annual rainfall and mean daily temperature in Rome (1971–2000) were nearly 700 mm and 17 °C, respectively [47].

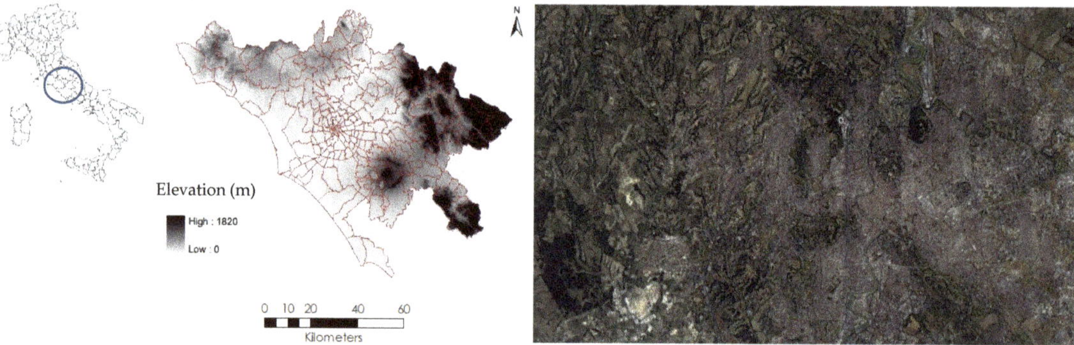

Figure 2. Left: elevation of the study area (meters on the sea level) overlapped with municipalities and urban districts; the insert map represents the position of Rome in Italy; **right**: a zenith photograph of inner Rome with peri-urban forests (green areas) concentrating on the left side of the landscape scene, courtesy of Google Earth.

The total population almost doubled during the study period from 2 million to 4 million inhabitants [5]. The percent share of the population living in inner Rome in the total population of the study area increased from 20% to 40% between 1936 and 2020 [39]. Even though urban settlements are becoming prevalent in the study area, the majority of the province's land is still comprised of forests, pastures, and cultivated fields [45]. Summer wildfires and human pressure, because of Rome expansion, have degraded the pristine forests, but relict woodlands have been sometimes preserved in both flat and hilly areas [48]. Based on official statistics [49], forest composition was changed significantly in the study area. Chestnuts and conifers increased considerably between the mid-1930s and late 2010s (respectively from 13.8% to 18.5% and from 1.7% to 3.0% of total forest stock); beech experienced a moderate decline (from 10.1% to 9.8% of total forest stock). Cropland abandonment around the central city was the base of natural afforestation uplands [47]; urbanization determined, at the same time, a progressive fragmentation of relict forests around the city, especially along the coastal rim [45]. Industrial areas were traditionally located in the eastern part of the 'Agro Romano' district, while traditional rural areas were (and still are) relatively abundant in the Western side of the study area [49].

Land-use maps were reproduced in Figure 3 to illustrate the mono-centric organization of Rome's metropolitan region, with compact settlements located downtown and a moderate spread of low-density settlements around central locations [34]. The shift toward dispersed urbanization in 2018 was not altering the polarization in central places (downtown Rome) and peripheral districts West and East of Rome [16]. Overall, the study area is administered by 122 municipalities, including the municipality of Rome that encompasses the historical city ('within the Aurelian Walls') and the surrounding area (1285 km^2) within and outside the main Ring Road ('Grande Raccordo Anulare', GRA). In this study, the central municipality of Rome was further partitioned in 115 local districts (a sort of formal neighborhood known as 'suddivisioni toponomastiche' in Italian, and it is mainly used for statistical reporting at a sub-municipal scale). Following earlier studies [33], the present study made use of a specific (territorial) partition of the area into 236 elementary spatial units that derive from the union (i.e., spatial merging) of 121 municipalities (excluding Rome) and 115 local districts in Rome.

Figure 3. The spatial distribution of the compact and dispersed urban fabric in the study area in the early 1960s (**left**) and late 2010s (**right**), based on the elaboration on a local land cover map for 1960 and Corine Land Cover database for 2018.

2.2. Forest Maps

Two forest maps spanning the time frame between 1936 and 2018 and presenting a comparable spatial resolution and forest definition were used here. The former source is the Forest Map realized at 1:100,000 scale by 'Milizia Forestale', the Italian Forest Service during the Fascist period, as part of a forest inventory dated 1936 and recently disseminated in digital format (shapefile). The latter source is a Corine Land Cover (CLC) digital map dated 2018 and adopting a legend of 44 land-use classes at a 1:100,000 scale. Polygons classified as CLC 3.1 type ('forests') were extracted and elaborated (Figure 4). The information accuracy of the two maps and the reliability of the forest cover measured for both 1936 and 2018 were internally checked considering additional data that provided estimates of forest area in Rome: (i) the long-term annual forest survey carried out by Italian National Institute of Statistics (Istat), (ii) the two national forest inventories run in 1985 and 2003, (iii) a 25,000 topographic map produced by Italian Military Geographic Institute (Florence) and referring to 1949, (iv) a land map produced by National Research Council (CNR) at 1:200,000 scale and referring to the early 1960s [50], (v) four CLC maps dated 1990, 2000, 2006, and 2012, and (vi) the 1:25,000 land-use maps of Latium produced for 1999 and 2016 by the Cartographic Service of the Regional Authority of Latium region through the interpretation of digital ortho-photographs.

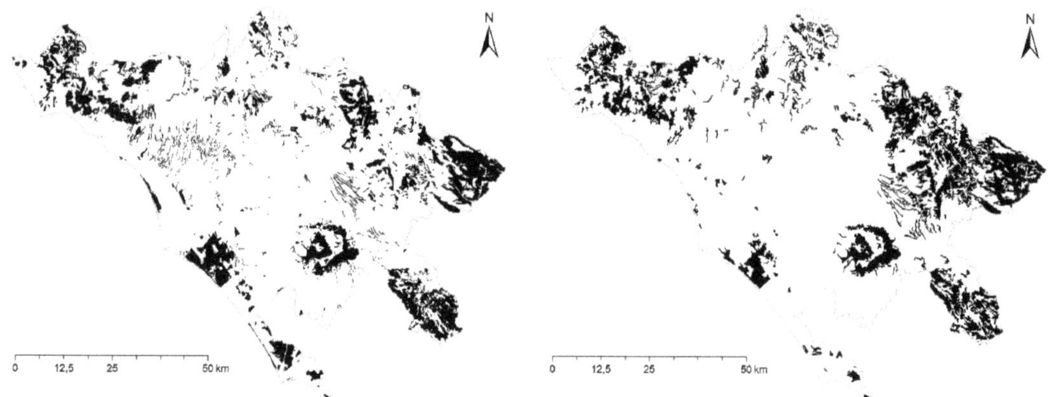

Figure 4. The spatial distribution of forest cover in the study area in 1936 (**left**) and in 2018 (**right**), based on elaboration on the forest maps described above.

2.3. Logical Framework

To test the inherent shift toward a 'polycentric' model of forest expansion (Colantoni et al. 2015), the percent of forest land area in the total landscape (i.e., municipal/district area) was derived, for each elementary analysis' unit, from the spatial overlay of a shapefile map illustrating the boundaries of the selected domains (n = 236 municipalities/local districts, see above) separately with each forest map (1936 and 2018). The average distance of each spatial unit (polygon) from downtown Rome was computed as the linear distance of the polygon centroid from a central place in the historical city—considered the heart of political, social, economic, touristic, and cultural life [16]. Adopting administrative boundaries as the elementary unit of investigation gave room to reliable comparisons with official statistics and constitutes a relevant domain for planning purposes [34]. In this study, we assumed Rome's metropolitan region as a laboratory of forest transitions [47], e.g., hypothesizing a classical ('mono-centric') model of forest decline around settlements and forest expansion in peripheral districts, and testing a possible shift toward a more 'polycentric' model of forest expansion around suburban settlements [48]. To verify such assumptions, changes over time in the spatial distribution of forest land along the urban gradient were analyzed comparing the results of spatially implicit and spatially explicit regressions that test the predictions of the FTT at a local scale [5]. More specifically, global and local regressions assessed the empirical relationship between percent forest area (dependent variable) and the distance from downtown Rome (predictor) at two times (1936 and 2018). We assumed a linear and spatially homogeneous relationship between the two variables in a 'mono-centric' model of forest expansion (i.e., a low percentage of forest area close to the inner city that increases with the distance from the city itself). With 'poly-centric' forest expansion [35], the relationship between the two variables is expected to be less polarized along the metropolitan gradient, with the predictor's impact varying largely across space [33]. The highest impact can be reflective of a local context (directly or indirectly) supporting forest expansion [5].

2.4. Statistical Analysis

Following the logical framework exposed above, both spatially implicit and explicit models, estimated separately for two time points (1936 and 2018) have been adopted to interpret long-term forest expansion/decline and the underlying territorial dynamics along the metropolitan gradient in Rome. A preliminary check of the linear or non-linear relationship between forest cover and distance from downtown Rome was performed using parametric (Pearson) and non-parametric (Spearman) pair-wise correlation coefficients, testing for significance at $p < 0.05$. A detailed description of the regression techniques subsequently adopted to test the polycentric expansion of forests has been provided in the following two paragraphs, distinguishing global from local approaches.

2.4.1. Global Approaches

Spatially implicit global approaches adopted in this study included Ordinary Least Square (OLS) regressions using linear and non-linear specifications, namely first (linear), second-order (squared), and third-order (cubic) forms. More specifically, the mono-centric structure was first tested adopting a linear specification [33] estimated through logarithmic transformations of both dependent and predictor variables as follows:

$$F = a + b(D) + e \qquad (1)$$

where F is the percent share of forest cover in total landscape, D is the distance of each spatial unit from downtown Rome (km, log-transformed), a and b are the regression coefficients, and e is the regression error. In this case, forest cover increases (or decreases) linearly along the metropolitan gradient, satisfying the assumption of the mono-centric model [33]. To verify the possible impact of the varying size (i.e., surface area) of each spatial partition adopted as the elementary analysis' unit in this study (i.e., municipalities/urban districts), we run OLS linear estimations of Equation (1) both un-weighted and weighted with the surface area of each elementary unit (km^2). District area was the result of the application of an ArcGIS 'calculate area' tool on the shapefile depicting municipalities/urban districts provided by the National Institute of Statistics (ISTAT) and statistical office of Rome municipality (Figure 2, left).

A squared specification was also tested with the aim at defining a quadratic relationship of the metropolitan gradient on forest cover:

$$F = a + b(D) + c(D^2) + e \qquad (2)$$

where F is the percent share of forest cover in the total landscape, D is the distance of each spatial unit from downtown Rome (km, log-transformed), a, b, and c are the regression coefficients, and e is the regression error. More complex patterns based on a cubic specification were tested to highlight place-specific responses of forest cover as the distance from downtown Rome increases [48], as follows:

$$F = a + b(D) + c(D^2) + d(D^3) + e \qquad (3)$$

where F is the percent share of forest cover in the total landscape, D is the distance of each spatial unit from downtown Rome (km, log-transformed), a, b, c, and d are the regression coefficients, and e is the regression error. Significant specifications were assessed using standard diagnostics, namely the adjusted R^2.

2.4.2. Local Approaches

We adopted a spatially explicit approach referring to the Geographically Weighted Regression (GWR) introduced by Fotheringham et al. (2002) to interpret complex spatial patterns of environmental and socioeconomic processes along a given geographical gradient. A refined understanding of the spatial variability related to the mono-centric expansion of metropolitan regions contributes to identify significant, local deviations from global patterns [48], informing spatial policies better adapted to specific territorial conditions. The use of administrative districts, such as municipalities, taken as policy-relevant spatial domains, allows a comprehensive investigation of context-based interactions that are frequently demised in 'centralized' (spatially implicit) interpretative models [51]. Based on these premises, GWRs estimate local regression models for each spatial domain in a given area, accounting for spatial dependence and heterogeneity [38], i.e., controlling for spatial structures that characterize the relationship between the dependent variable and the predictor (Ali et al., 2007). We adopted a standard GWR specification for a given location s = 1 to n, as follows:

$$F(s) = B(s)D(s) + \mathbf{e}(s) \tag{4}$$

where F(s) and D(s) are, respectively, the dependent variable (percent forest cover) and the predictor (distance from downtown Rome) illustrated in Equation (1) above, which were both measured at each location s; B(s) means the column vector of the regression coefficients at location s, and $\mathbf{e}$(s) is the random error at location s. Regression parameters were estimated separately for 1936 and 2018 at each location by weighted least squares and are thus spatially explicit, i.e., a function of s. Spatial coordinates of each elementary unit considered in this study were determined based on a shapefile map of administrative boundaries (see Figure 2, left) managed in ArcGIS (ESRI Inc., Redwoods, CA, USA) environment. A Kernel spline function was adopted to calculate weights for the estimation of local regressions [38]. Limitations regarding the use of GWR only arise when drawing conclusions based on a reduced number of sample observations, which is not the case of this study. GWRs were run separately for each time interval. For each regression, the model's results included (i) a global measure of goodness-of-fit (adjusted R^2) that was compared with the same index obtained from the respective OLS regression (both weighted and un-weighted for the surface area of each spatial partition, see above) and (ii) a spatial distribution of local coefficients (R^2, slope coefficient) illustrated through maps (Salvati et al. 2016). The GWR results clarify the role of spatial heterogeneity as a novel contribution to the empirical test of the mono-centric model [33].

The intrinsic shift from a classical 'mono-centric' model of forest expansion toward a more 'poly-centric' structure was reflected in a different spatial distribution of both slope and R^2 regression coefficients [16]. In fact, we expect (i) a more fragmented/heterogeneous distribution of R^2 coefficients (possibly indicating the increase in sparse nuclei of forest expansion) and (ii) an expansion of the spatial domains with high slope coefficients, reflecting a latent, but generalized, increase in forest cover [35]. With this perspective in mind, comparing the intensity and direction of the relationship between percent forest cover and the distance from downtown Rome for different econometric models and specifications (e.g., global vs. local, spatially implicit vs. spatially explicit) provides an intrinsic measure of the role of territorial factors [48]. Such contextual aspects should be more explicitly considered in any environmental policy targeting the mitigation of (and adaptation to) urbanization processes [49].

3. Results

Forest cover in Rome increased from 18.3% in 1936 to 19.9% in 2018. The spatial distribution of forest area (Figure 5) reflects the intense landscape transformations between 1936 and 2018, with forest expansion in relict croplands East of Rome and moderate decline in flat districts surrounding the 'Agro Romano' and along the sea coast. These districts included, up to now, the main remnants of the forest cover extending along the Tyrrhenian Sea coast from Tuscany to Campania in ancient times. Accelerated rates of forest recovery have been observed in rural, upland districts west and north of Rome, which were originally devoted to shrublands and olive groves (e.g., 'Sabina' district). Forest cover was relatively stable in the 'Castelli Romani' district, a peri-urban, upland area with high human pressure (population density around 500 inhabitants/km^2) and mostly devoted to chestnut production.

3.1. Global Approaches

All statistical models testing changes in the spatial relationship between forest cover and the distance from downtown Rome (Table 1) indicated how the correlation between the two variables intensified over time. Non-parametric Spearman rank coefficients delineate a positive and strong correlation between forest cover and distance from a central place in Rome (r_s = 0.76 and 0.86, respectively for 1936 and 2018). However, while the first-order (linear) Ordinary Least Square (OLS) model performed rather low for 1936 and 2018, second-

and third-order (i.e., squared and cubic) OLS regressions performed better for both years (adjusted R^2 = 0.58 for 2018). With global R^2 coefficients increasing from 0.65 in 1936 to 0.75 in 2018, local regressions outperformed OLS (linear and non-linear) models, documenting the importance of the spatial dimension in long-term forest transitions characteristic of the study area. Moran's spatial autocorrelation indexes demonstrate that the percent share of forest cover in the total municipal/district area was not randomly distributed over space (1936: $I(z)$ = 4.29, $p < 0.05$; 2018: $I(z)$ = 3.69, $p < 0.05$). Spatial autocorrelation justifies the use of a local regression approach when assessing the relationship between percent share of forest cover in the total municipal/district area as the dependent variable and the distance from downtown Rome as an intrinsic test of the mono-centric (or poly-centric) forest structure.

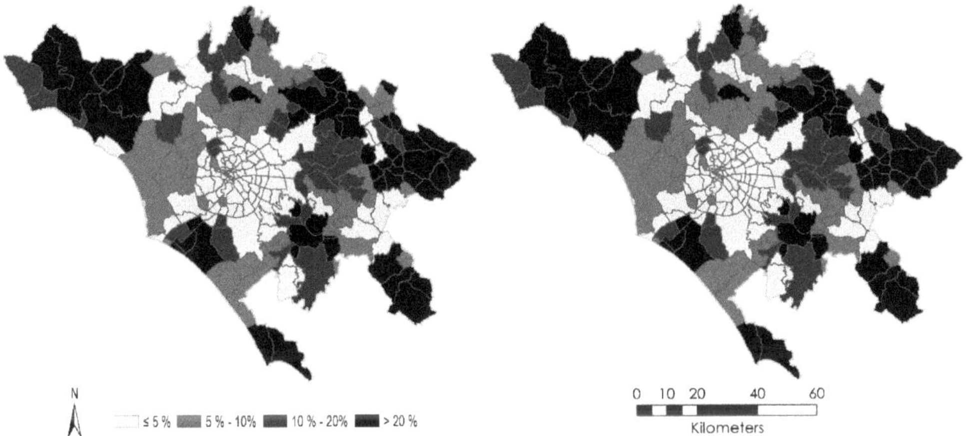

Figure 5. The spatial distribution of percent share of forest cover in the total municipal (or district) area in Rome (**left**: 1936; **right**: 2018).

Table 1. Correlation coefficients and best-fit R^2 of global econometric models (Ordinary Least Squares (OLS) with linear and non-linear specifications; Geographically Weighted Regression (GWR) estimating the percentage of forest area in each spatial domain (dependent variable) from the distance to downtown Rome (predictor) by year.

Model/Specification	1936	2018
Pearson moment–product correlation	0.616	0.750
Non-parametric Spearman r_s rank correlation	0.755	0.852
Un-weighted, linear Ordinary Least Square (OLS)	0.380	0.563
Weighted, linear OLS †	0.382	0.555
Square (second-order polynomial) OLS	0.402	0.573
Cubic (third-order polynomial) OLS	0.439	0.582
Un-weighted Geographically Weighted Regression (GWR)	0.643	0.769
Weighted GWR†	0.703	0.754

All models are significant at $p < 0.05$. † weighted by surface area of each spatial domain (i.e., municipalities/urban districts).

3.2. Local Approaches

The satisfactory result of local regressions suggests how territorial contexts distinctive of municipalities and urban districts in the study area shape forest dynamics at different distances from downtown Rome. Local regressions modeled the spatial dimension of both 1936 and 2018 data by incorporating a location distance matrix that allowed the local estimation of both slope (measuring the predictor's impact) and R^2 (measuring the model's goodness-of-fit) regression coefficients for each spatial domain. Vector maps were used to illustrate the spatial distribution of local regression slope and local R^2 coefficients. The results of local regression models for forest cover as the dependent variable and the distance from downtown Rome as the predictor were illustrated in maps considering separately the spatial distribution of regression slopes and local R^2 goodness-of-fit coefficients (Figure 6). Slope coefficients depicted the (more or less intense) contribution of specific areas to a mono-centric spatial structure.

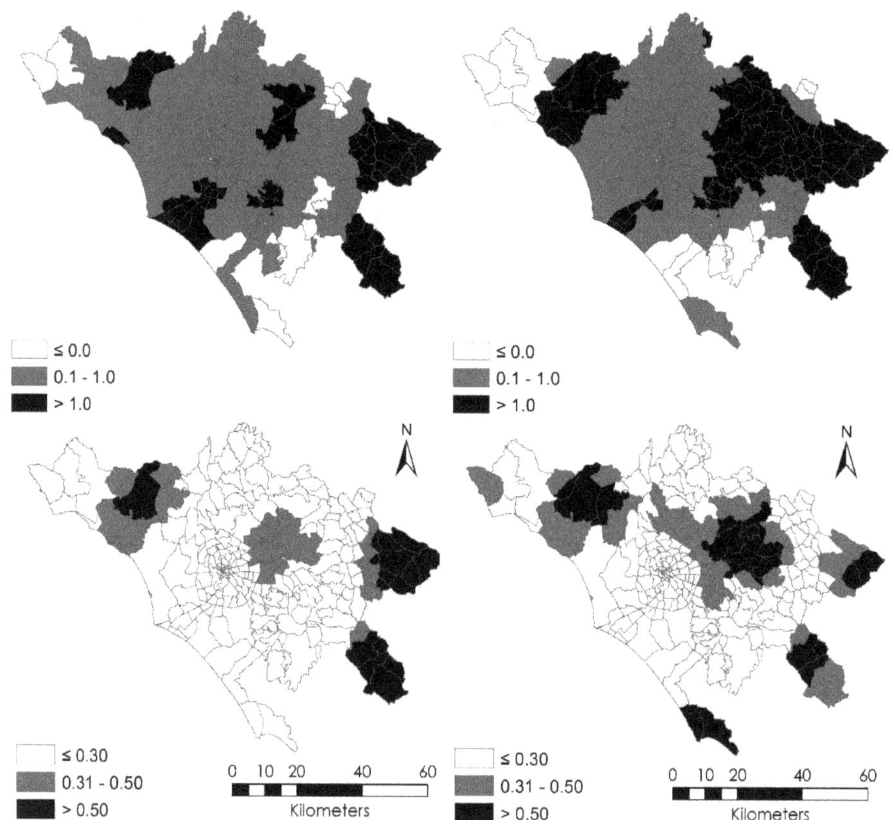

Figure 6. Results of a Geographically Weighted Regression assessing the relationship between the percent share of forest cover in total landscape and the distance from downtown Rome by year (**left**: 1936; **right**: 2018; **upper** panel: local slope (regression) coefficients; **lower** panel: local R^2 goodness-of-fit coefficients).

The empirical results of local regressions indicate a distinctive spatial distribution of forest cover for 1936 and 2018, in turn reflecting a substantially different relationship with the distance from a central place in Rome. Comparing the models' goodness-of-fit, the regression for the last observation year (2018) was performing slightly better than the model for the first observation year (1936). Spatial units (municipalities/districts) with local $R^2 > 0.5$ (regarded as a satisfactory model's fit at that spatial scale) increased during the study period. The spatial distribution of local R^2 coefficients in 1936 delineated particularly high values in economically peripheral municipalities of the Apennine mountain districts ('Lucretili' and 'Lepini' districts), as well as in rural, remote municipalities in the northwest of Rome ('Tolfa-Bracciano' district). In 2018, the highest R^2 coefficients were found either in upland/mountainous municipalities East of Rome ('Lucretili', 'Simbruini' and 'Lepini' districts); the same pattern was observed in sparse coastal/lowland municipalities featuring human disturbance and low ecological quality ('Anzio-Nettuno' coastal district South of Rome; 'Bracciano Lake' district). In both cases, spatial units with high R^2 were found closer to Rome in 2018 than in 1936. As far as the impact of the distance from downtown, the highest regression slopes for 1936 were recorded in remote districts East of Rome and in more sparse flat areas close to the city. High slopes were also recorded in a rural, remote district West of Rome, and they decreased in intensity along the sea coast and in flat districts around the city. Eighty years later, while peripheral municipalities in the Apennine district still had the highest slope coefficients, more heterogeneous, peri-urban districts both west and east of Rome displayed high (or very high) slope coefficients. Taken together, local slope and R^2 coefficients reflect the polarization in districts with forest recovery and decline, which is likely because of wildfires, clearcutting, or other human/ecological disturbances. The gradual disappearance of coastal woodlands because of progressive clearcutting made the spatial distribution of forests more heterogeneous, in turn contributing to a less polarized landscape in urban and rural areas. At the same time, urban sprawl and the rising human pressure driven by sprawled urbanization has consolidated some smaller nuclei of forest expansion, acting as an 'endogenous source' of forest transitions (i.e., fueling the shift from net decline to net recovery) at the local scale.

4. Discussion

During the last century, a significant decrease in cropland and a contemporary increase in residential settlements and forests has been observed in Italy [50] and, more generally, in Southern Europe [23]. Land-use change was responsible for the expansion of 'contact areas' between developed (i.e., residential and/or industrial settlements) and natural/semi-natural landscapes in Mediterranean Europe [24,25,52,53]. By confirming the predictions of the Forest Transition Theory, our study documents the relevance of a comprehensive landscape analysis along urban–forest interfaces to inform policies containing urbanization and mitigating the negative impact of human disturbance on the surrounding natural habitats [7]. Barbati et al. (2013) found that the spatial distribution of natural landscapes in recent times became more polarized along the metropolitan continuum [5]. A thorough understanding of the intrinsic mechanisms of urban growth and the consequent dynamics of natural landscapes—mainly forests—along this gradient provides an integral vision and future perspective to landscape research [48]. From this perspective, our study contributes to regional science suggesting how the notion of mono-centric (or poly-centric) urban expansion can be (more or less directly) extended to dynamics related to other land use, namely forests, underlying socioeconomic mechanisms and territorial processes that are governed (and are in turn influenced by) the intrinsic linkage between humans and nature at the base of regional landscape dynamics [8]. A global analysis failed in assessing these complex dynamics, suggesting how spatially explicit approaches are appropriate tools to investigate the importance of heterogeneous local contexts [33], especially when they assume a stronger role than regional drivers of change and local socioeconomic pressures [9].

The results of local regressions presented in our study provided evidence on the importance of a mono-centric structure of urban growth in Rome, which is in line with the empirical findings of earlier studies [5,16,42]—as plastically illustrated in Figure 2 earlier in this paper with the parallel consolidation of a moderately polycentric model of forest expansion [35]. This complex spatial pattern is assumed to be the joint impact of two processes: (i) intense (natural) forestation in mountainous and economically depressed areas and (ii) moderate (and mostly human-induced) forest recovery in flat and coastal areas, depending on the intensity of conflicts over alternative land-use, i.e., on the basis of the 'economic strength' of the agricultural sector [31,38,45,47]. As a matter of fact, cropland abandonment typical of economically marginal districts was observed also in some peri-urban areas originally devoted to agriculture because of particularly fertile soils [51]. In such areas, forest expansion occurred more likely than in other socioeconomic contexts of the same geographical region [5,39,48].

Being representative of more general dynamics at the regional scale, distinctive transformations of peri-urban landscapes were described in the study area considering together the empirical results of this study and earlier literature: (i) economic decline of remote districts resulting in re-forestation [28,50,54], (ii) crop intensification in more accessible rural districts [5,29,34], and (iii) urban sprawl, agricultural land consumption, and forest fragmentation in flat districts [35,55,56]. As a counter-intuitive result, forest increase was intrinsically associated with reduced proximity between settlements and natural land [51], as a result of forestation of abandoned cropland in accessible districts indirectly fueled by urban expansion [37–39]. From a socioeconomic standpoint, such transformations fueled the heterogeneous expansion of forest land cover [25], with ecosystem dynamics remaining essentially under-investigated [30].

From the spatial planning perspective, the results of our study highlight the need to adopt a multi-target strategy for the conservation of natural and semi-natural areas in peri-urban contexts [35]. This strategy should recognize forest expansion in peri-urban areas together as a positive process of greening [31] and a potentially negative phenomenon, as the increase in forest cover as well as in the length of the margin directly in contact with residential settlements causes an increased risk of wildfires in already vulnerable areas [47]. Future studies should better interpret long-term changes in metropolitan landscapes [37], clearly distinguishing the impact of socioeconomic drivers and planning/biophysical constraints to building, in turn clarifying the possible role of different forest and agro-forest types in that change [5,45,48]. More generally, the analysis of changes in the forest landscape in peri-urban areas [46] cannot be separated from an accurate investigation of the causes of expansion and decline of agriculture in the facing rural contexts.

5. Conclusions

The present study refers to long-term changes in forest area that encompass a sufficiently long time period covering the productive cycle of a high forest. Long-term changes in forest cover include both natural and human-induced transformations and may account for a truly comprehensive picture of landscape changes. In this perspective, we tried to estimate the synergic impact of both drivers. The complex results of this study (i.e., the emergence of a polycentric model of forest expansion in a context of mono-centric expansion of urban settlements) outlines the urgent need for policy strategies managing urban–forest interactions along the fringe. Effective strategies governing peri-urban landscape dynamics should include (i) urban containment policies reducing the environmental impact of sprawl on fringe land and (ii) measures controlling the unwanted expansion of peri-urban woodlands (starting from the multiplication of a sort of 'hotspots' indirectly acting as nuclei of forest spread) at the expense of abandoned cropland.

Based on the empirical findings of this study and a pertinent review of the recent literature, these actions should target the integral preservation of the traditional diversity of agro-forest mosaics in Southern European fringe lands. Our study highlights the importance of monitoring techniques for the quantification of urban–forest ecosystem services, starting from a recent field experience in the Mediterranean context. New logical approaches and information technologies for the 'holistic' study of peri-urban forests are requested to integrate operational tools, methodologies, and analysis' techniques, with increasing, comparative evidence from case studies. Assessing the quality and health of peri-urban forests implies a reduction of monitoring costs, the improvement of sampling efficiency and accuracy, and a more effective integration of the existing measurement networks in a long-term vision that refers to the paradigms of ecological resilience and socioeconomic sustainability.

Author Contributions: Conceptualization, R.S. and G.Q.; methodology, L.S.; software, A.M.; validation, L.B. and G.E.; formal analysis, L.S.; investigation, R.S.; resources, G.Q.; data curation, G.E.; writing—original draft preparation, R.S.; writing—review and editing, L.S. and A.M.; visualization, L.B.; supervision, L.B.; project administration, A.M.; funding acquisition, G.Q. All authors have read and agreed to the published version of the manuscript.

Funding: This research received no external funding.

Data Availability Statement: Official statistics derived from public websites were used in this study.

Conflicts of Interest: The authors declare no conflict of interest.

References

1. Mather, A.S. Forest transition theory and the reforesting of scotland. *Scott. Geogr. J.* **2004**, *120*, 83–98. [CrossRef]
2. Perz, S.G. Grand theory and context-specificity in the study of forest dynamics: Forest transition theory and other directions. *Prof. Geogr.* **2007**, *59*, 105–114. [CrossRef]
3. Barbier, E.B.; Burgess, J.C.; Grainger, A. The forest transition: Towards a more comprehensive theoretical framework. *Land Use Policy* **2010**, *27*, 98–107. [CrossRef]
4. Meyfroidt, P.; Lambin, E.F. Global Forest Transition: Prospects for an End to Deforestation. *Annu. Rev. Environ. Resour.* **2011**, *36*, 343–371. [CrossRef]
5. Barbati, A.; Corona, P.; Salvati, L.; Gasparella, L. Natural forest expansion into suburban countryside: Gained ground for a green infrastructure? *Urban For. Urban Green.* **2013**, *12*, 36–43. [CrossRef]
6. Pagnutti, C.; Bauch, C.T.; Anand, M. Outlook on a Worldwide Forest Transition. *PLoS ONE* **2013**, *8*, e75890. [CrossRef]
7. Baines, O.; Wilkes, P.; Disney, M. Quantifying urban forest structure with open-access remote sensing data sets. *Urban For. Urban Green.* **2020**, *50*, 126653. [CrossRef]
8. Bonilla-Bedoya, S.; Mora, A.; Vaca, A.; Estrella, A.; Herrera, M.Á. Modelling the relationship between urban expansion processes and urban forest characteristics: An application to the Metropolitan District of Quito. *Comput. Environ. Urban Syst.* **2020**, *79*, 101420. [CrossRef]
9. Alvarez, S.; Soto, J.R.; Escobedo, F.J.; Lai, J.; Kibria, A.S.M.G.; Adams, D.C. Heterogeneous preferences and economic values for urban forest structural and functional attributes. *Landsc. Urban Plan.* **2021**, *215*, 104234. [CrossRef]
10. Rudel, T.K.; Coomes, O.T.; Moran, E.; Achard, F.; Angelsen, A.; Xu, J.; Lambin, E. Forest transitions: Towards a global understanding of land use change. *Glob. Environ. Chang.* **2005**, *15*, 23–31. [CrossRef]
11. Fagan, C.B.; Yackulic, M.; Jain, M.; Jina, A.; Lim, Y.; Marlier, M.; Muscarella, R.; Adame, P.; DeFries, R.; Uriarte, M. Biophysical and Socioeconomic Factors Associated with Forest Transitions at Multiple Spatial and Temporal Scale. *Ecol. Soc.* **2011**, *16*, 15. [CrossRef]
12. Mather, A.S.; Needle, C.L. The forest transition: A theoretical basis. *Area* **1998**, *30*, 117–124. [CrossRef]
13. Grainger, A. The Forest Transition: An Alternative Approach. *Area* **1995**, *27*, 242–251.
14. Mather, A.S.; Fairbairn, J. From floods to reforestation: The forest transition in Switzerland. *Environ. Hist. Camb.* **2000**, *6*, 399–421. [CrossRef]
15. van Vliet, J. Direct and indirect loss of natural area from urban expansion. *Nat. Sustain.* **2019**, *2*, 755–763. [CrossRef]
16. Salvati, L.; Ciommi, M.T.; Serra, P.; Chelli, F.M. Exploring the spatial structure of housing prices under economic expansion and stagnation: The role of socio-demographic factors in metropolitan Rome, Italy. *Land Use Policy* **2019**, *81*, 143–152. [CrossRef]
17. Plieninger, T.; Schaich, H.; Kizos, T. Land-use legacies in the forest structure of silvopastoral oak woodlands in the Eastern Mediterranean. *Reg. Environ. Chang.* **2011**, *11*, 603–615. [CrossRef]
18. Nowak, D.J.; Walton, J.T.; Dwyer, J.F.; Kaya, L.G.; Myeong, S.; Myeong, S. The Increasing Influence of Urban Environments on US Forest Management. *J. For.* **2005**, *103*, 377–382.
19. Kauppi, P.E.; Ausubel, J.H.; Fang, J.; Mather, A.S.; Sedjo, R.A.; Waggoner, P.E. Returning forests analyzed with the forest identity. *Proc. Natl. Acad. Sci. USA* **2006**, *103*, 17574–17579. [CrossRef]
20. Redo, D.J.; Grau, H.R.; Aide, T.M.; Clark, M.L. Asymmetric forest transition driven by the interaction of socioeconomic development and environmental heterogeneity in Central America. *Proc. Natl. Acad. Sci. USA* **2012**, *109*, 8839–8844. [CrossRef]
21. Rodríguez, L.G.; Pérez, M.R. Recent changes in Chinese forestry seen through the lens of Forest Transition theory. *Int. For. Rev.* **2013**, *15*, 456–470. [CrossRef]
22. Baptista, S.R.; Rudel, T.K. A re-emerging Atlantic forest? Urbanization, industrialization and the forest transition in Santa Catarina, southern Brazil. *Environ. Conserv.* **2006**, *33*, 195–202. [CrossRef]
23. Blondel, J.; Aronson, J.; Bodiou, J.-Y.; Boeuf, G. *The Mediterranean Region—Biological Diversity in Space and Time*, 2nd ed.; Oxford University Press: Oxford, UK, 2010; ISBN 9780199557998.
24. Sirami, C.; Nespoulous, A.; Cheylan, J.P.; Marty, P.; Hvenegaard, G.T.; Geniez, P.; Schatz, B.; Martin, J.L. Long-term anthropogenic and ecological dynamics of a Mediterranean landscape: Impacts on multiple taxa. *Landsc. Urban Plan.* **2010**, *96*, 214–223. [CrossRef]
25. Vince, S.W.; Duryea, M.; Macie, E.A.; Hermansen, L.A. *Forests at the Wildland-Urban Interface: Conservation and Management*; CRC Press: Boca Raton, FL, USA, 2005; ISBN 9780367578213.
26. Stewart, S.I.; Radeloff, V.C.; Hammer, R.B.; Hawbaker, T.J. Defining the Wildland-Urban Interface. *J. For.* **2007**, *105*, 201–207.
27. Endreny, T.A. Strategically growing the urban forest will improve our world. *Nat. Commun.* **2018**, *9*, 1160. [CrossRef] [PubMed]
28. Antrop, M. Changing patterns in the urbanized countryside of Western Europe. *Landsc. Ecol.* **2000**, *15*, 257–270. [CrossRef]
29. Antrop, M. Landscape change and the urbanization process in Europe. *Landsc. Urban Plan.* **2004**, *67*, 9–26. [CrossRef]

30. Scarascia-Mugnozza, G.; Oswald, H.; Piussi, P.; Radoglou, K. Forests of the Mediterranean region: Gaps in knowledge and research needs. *For. Ecol. Manag.* **2000**, *132*, 97–109. [CrossRef]
31. Tomao, A.; Quatrini, V.; Corona, P.; Ferrara, A.; Lafortezza, R.; Salvati, L. Resilient landscapes in Mediterranean urban areas: Understanding factors influencing forest trends. *Environ. Res.* **2017**, *156*, 1–9. [CrossRef]
32. Bianchini, L.; Marucci, A.; Sateriano, A.; Di Stefano, V.; Alemanno, R.; Colantoni, A. Urbanization and long-term forest dynamics in a metropolitan region of southern europe (1936–2018). *Sustainability* **2021**, *13*, 12164. [CrossRef]
33. Salvati, L.; Carlucci, M. Distance matters: Land consumption and the mono-centric model in two southern European cities. *Landsc. Urban Plan.* **2014**, *127*, 41–51. [CrossRef]
34. Salvati, L.; Zitti, M. Territorial disparities, natural resource distribution, and land degradation: A case study in southern Europe. *GeoJournal* **2007**, *70*, 185–194. [CrossRef]
35. Colantoni, A.; Mavrakis, A.; Sorgi, T.; Salvati, L. Towards a 'polycentric' landscape? Reconnecting fragments into an integrated network of coastal forests in Rome. *Rend. Lincei* **2015**, *3*, 615–624. [CrossRef]
36. Çakir, G.; Ün, C.; Baskent, E.Z.; Köse, S.; Sivrikaya, F.; Keleş, S. Evaluating urbanization, fragmentation and land use/land cover change pattern in Istanbul city, Turkey from 1971 to 2002. *Land Degrad. Dev.* **2008**, *19*, 663–675. [CrossRef]
37. Pili, S.; Grigoriadis, E.; Carlucci, M.; Clemente, M.; Salvati, L. Towards sustainable growth? A multi-criteria assessment of (changing) urban forms. *Ecol. Indic.* **2017**, *76*, 71–80. [CrossRef]
38. Duvernoy, I.; Zambon, I.; Sateriano, A.; Salvati, L. Pictures from the other side of the fringe: Urban growth and peri-urban agriculture in a post-industrial city (Toulouse, France). *J. Rural Stud.* **2018**, *57*, 25–35. [CrossRef]
39. Perrin, C.; Nougarèdes, B.; Sini, L.; Branduini, P.; Salvati, L.; Perrin, C.; Nougarèdes, B.; Sini, L.; Branduini, P.; Salvati, L. Governance changes in peri-urban farmland protection following decentralisation: A comparison between Montpellier (France) and Rome (Italy). *Land Use Policy* **2018**, *70*, 535–546. [CrossRef]
40. Žlender, V. Characterisation of peri-urban landscape based on the views and attitudes of different actors. *Land Use Policy* **2021**, *101*, 105181. [CrossRef]
41. Di Feliciantonio, C.; Salvati, L.; Sarantakou, E.; Rontos, K. Class diversification, economic growth and urban sprawl: Evidences from a pre-crisis European city. *Qual. Quant. Int. J. Methodol.* **2018**, *52*, 1501–1522. [CrossRef]
42. Ciommi, M.; Chelli, F.M.; Carlucci, M.; Salvati, L. Urban Growth and Demographic Dynamics in Southern Europe: Toward a New Statistical Approach to Regional Science. *Sustainability* **2018**, *10*, 2765. [CrossRef]
43. Theobald, D.M.; Romme, W.H. Expansion of the US wildland-urban interface. *Landsc. Urban Plan.* **2007**, *83*, 340–354. [CrossRef]
44. Bianchini, L.; Egidi, G.; Alhuseen, A.; Sateriano, A.; Cividino, S.; Clemente, M.; Imbrenda, V. Toward a dualistic growth? Population increase and land-use change in Rome, Italy. *Land* **2021**, *10*, 749. [CrossRef]
45. Ferrari, B.; Corona, P.; Mancini, L.D.; Salvati, R.; Barbati, A. Taking the pulse of forest plantations success in peri-urban environments through continuous inventory. *New For.* **2017**, *48*, 527–545. [CrossRef]
46. Camarretta, N.; Puletti, N.; Chiavetta, U.; Corona, P. Quantitative changes of forest landscapes over the last century across Italy. *Plant Biosyst.* **2017**, *152*, 1011–1019. [CrossRef]
47. Grotti, M.; Mattioli, W.; Ferrari, B.; Tomao, A.; Merlini, P.; Corona, P. A multi-temporal dataset of forest mensuration of reforestations: A case study in peri-urban Rome, Italy. *Ann. Silvic. Res.* **2019**, *43*, 97–101. [CrossRef]
48. Pili, S.; Serra, P.; Salvati, L. Landscape and the city: Agro-forest systems, land fragmentation and the ecological network in Rome, Italy. *Urban For. Urban Green.* **2019**, *41*, 230–237. [CrossRef]
49. Biasi, R.; Brunori, E.; Smiraglia, D.; Salvati, L. Linking traditional tree-crop landscapes and agro-biodiversity in central Italy using a database of typical and traditional products: A multiple risk assessment through a data mining analysis. *Biodivers. Conserv.* **2015**, *24*, 3009–3031. [CrossRef]
50. Falcucci, A.; Maiorano, L.; Boitani, L. Changes in land-use/land-cover patterns in Italy and their implications for biodiversity conservation. *Landsc. Ecol.* **2007**, *22*, 617–631. [CrossRef]
51. Cecchini, M.; Zambon, I.; Pontrandolfi, A.; Turco, R.; Colantoni, A.; Mavrakis, A.; Salvati, L. Urban sprawl and the 'olive' landscape: Sustainable land management for 'crisis' cities. *GeoJournal* **2019**, *84*, 237–255. [CrossRef]
52. Zipperer, W.C. Species composition and structure of regenerated and remnant forest patches within an urban landscape. *Urban Ecosyst.* **2002**, *6*, 271–290. [CrossRef]
53. Foley, J.A.; DeFries, R.; Asner, G.P.; Barford, C.; Bonan, G.; Carpenter, S.R.; Chapin, F.S.; Coe, M.T.; Daily, G.C.; Gibbs, H.K.; et al. Global consequences of land use. *Science* **2005**, *309*, 570–574. [CrossRef] [PubMed]
54. Bajocco, S.; Dragoz, E.; Gitas, I.; Smiraglia, D.; Salvati, L.; Ricotta, C. Mapping Forest Fuels through Vegetation Phenology: The Role of Coarse-Resolution Satellite Time-Series. *PLoS ONE* **2015**, *10*, e0119811. [CrossRef]
55. Attorre, F.; Bruno, M.; Francesconi, F.; Valenti, R.; Bruno, F. Landscape changes of Rome through tree-lined roads. *Landsc. Urban Plan.* **2000**, *49*, 115–128. [CrossRef]
56. Cavallo, A.; Marino, D. Understanding Changing in Traditional Agricultural Landscapes: Towards a Framework. *J. Agric. Sci. Technol.* **2012**, *2*, 971–978.

 *land* 

Article

Research on Green Innovation of the Great Changsha-Zhuzhou-Xiangtan City Group Based on Network

Lu Wang [1], Wenzhong Ye [1] and Lingming Chen [1,2,*]

[1] School of Business, Hunan University of Science and Technology (HNUST),
Xiangtan 411201, China; 19011501016@mail.hnust.edu.cn (L.W.); 1300030@hnust.edu.cn (W.Y.)

[2] School of Economics and Management, Xinyu University (XYU), Xinyu 338004, China

* Correspondence: lingming1016@mail.hnust.edu.cn; Tel.: +86-158-7009-2285

Abstract: This article aims to promote the high-quality development of the Great Changsha-Zhuzhou-Xiangtan City Group and improve the green innovation efficiency of urban agglomeration. This article takes green innovation in networked urban agglomerations as its research subject. Furthermore, it analyzes the impact of network structure characteristics such as network scale and network structure hole on green innovation in urban agglomerations. Moreover, this study uses the unexpected output SBM model to measure green innovation efficiency of the eight prefecture-level cities in the Great Changsha-Zhuzhou-Xiangtan City Group from 2012 to 2018 and analyzes its influencing factors using the panel Tobit model. The results show that the overall green innovation efficiency of the Great Changsha-Zhuzhou-Xiangtan City Group is stable. The distribution of urban green innovation efficiency in the region is characterized by urban gradient and mid-stream drive. In the process of networked innovation, economic development, which has a positive impact on green innovation, promotes the overall effectiveness of the network structure. The low efficiency of urban educational resources, which has a negative impact on green innovation, leads to the redundancy of a network scale. The unapparent advantage of industrial structures, which have a negative impact on the development of green innovation, leads to the insufficient depth and breadth of network openness. Lastly, government support and the level of infrastructure have no impact on green innovation.

Keywords: Great Changsha-Zhuzhou-Xiangtan City Group; green innovation; network structure; unexpected output SBM model

Citation: Wang, L.; Ye, W.; Chen, L. Research on Green Innovation of the Great Changsha-Zhuzhou-Xiangtan City Group Based on Network. *Land* **2021**, *10*, 1198. https://doi.org/10.3390/land10111198

Academic Editors: Luca Salvati, Iwona Cieślak and Andrzej Biłozor

Received: 5 October 2021
Accepted: 3 November 2021
Published: 5 November 2021

Publisher's Note: MDPI stays neutral with regard to jurisdictional claims in published maps and institutional affiliations.

1. Introduction

The Fifth Plenary Session of the 19th Central Committee of the Communist Party of China put forward a new goal of promoting green development and the harmonious coexistence between humans and nature. To launch the new development policy, the power institution proposed that the regional economic development takes green innovation as the new driving force, protects the environment, and considers the overall economic development situation to achieve sustainable economic growth. Due to the economic growth resulting from the traditional extensive economic development model, China is facing problems related to resource shortage, environmental pollution, and ecological destruction. The realization of sustainable economic and social development has become the focus of scholars. In the traditional economic and social development model, the model of high energy consumption and high pollution in exchange for economic benefits has caused damage to the ecological environment. Ignoring the environmental benefits generated by innovation behavior is not conducive to sustainable development. It is important to overcome the barriers of the traditional development model, take green innovation as the core driving force, and realize the leap of economic development and innovation. Green innovation is also known as environmental innovation, ecological innovation, or sustainable innovation. Compared with traditional innovation, green innovation aims at reducing environmental impact, achieving economic development while reducing resource consumption and environmental costs, which is beneficial to building an efficient and green economic and social development environment.

Urban agglomerations are the main form of the global division of labor, new urbanization, and rural revitalization. In the context of China entering a new era, urban agglomerations face problems such as the continued expansion of population and infrastructure construction scale, prominent contradictions in the economy, resources, environment, and increased pressure on urban bearing. To fully implement the *Outline of Changsha-Zhuzhou-Xiangtan Regional Integrated Development Planning*, we argue for providing full play to the driving role of Changsha-Zhuzhou-Xiangtan in implementing the strategy of "Three High and Four New", improving the speed of urban development, and striving to build a modern city with harmonious coexistence between humans and nature. At the same time, we should improve the linkage mechanism of ecological environment co-protection, jointly solve environmental pollution problems, jointly protect the Changsha-Zhuzhou-Xiangtan ecology, and strive for the Changsha-Zhuzhou-Xiangtan national ecological civilization pilot zone. The construction of the Changsha-Zhuzhou-Xiangtan metropolitan area is a significant achievement contributing to the rise of central China and the development of the Yangtze River Economic Belt. In September 2020, General Secretary Xi Jinping visited Hunan and emphasized that the integrated development of Changsha, Zhuzhou, and Xiangtan should continue achieving exceptional results and insisted on leading innovation and green growth. Thus, we intend to research green innovation and its influencing factors in the Great Changsha-Zhuzhou-Xiangtan City Group.

The article is arranged as follows: Section 2 presents the literature review; Section 3 introduces the theoretical mechanism, including the effect of network structure characteristics on green innovation in the urban agglomeration, and the influencing mechanism of green innovation in the Great Changsha-Zhuzhou-Xiangtan City Group based on a network of green innovation in the urban agglomeration. Section 4 introduces the model methodology and variable selection, including the index system and the selection of the index. Section 5 uses the unexpected output SBM model to measure the efficiency of green innovation. Section 6 uses data from the Great Changsha-Zhuzhou-Xiangtan City Group (from 2012 to 2018) to empirically analyze the impact of network structure on green innovation, and test its theoretical mechanism. Section 7 summarizes the conclusions and makes recommendations, and proposes future research directions.

2. Literature Review

An urban agglomeration is a vital innovation space and carrier. Some scholars have constructed the innovation cooperation network model among urban agglomerations to optimize the innovation strategy of urban agglomerations [1–3]. At present, many scholars are investigating the relationship between urban agglomeration and innovation performance of the network structure. For example, Sheng Yanwen et al. [4] used social network analysis and the DEA model to analyze the impact of central position, intermediary position, and structural holes of urban agglomeration innovation network on innovation efficiency. Ren [5] constructed a regional innovation network system to explore the impact of network size, network openness, structural holes, strong ties, and weak ties on innovation capability. Other scholars, based on stochastic frontier analysis [6], data envelopment analysis [7], or canonical correlation analysis [8], analyzed the impact of network size, network openness, network structure holes, and network density on regional innovation.

Research on green innovation primarily includes the concept definition, index measurement, and driving factors. So far, scholars in various countries have gained a profound understanding of the concept of green innovation [9,10]. For example, Kemp [11] argued that green innovation is an activity directed toward reducing environmental pollution and resource waste. The methods commonly used for green innovation efficiency are generally divided into parametric methods and non-parametric methods, such as stochastic frontier analysis and data envelopment analysis (DEA) [12,13]. Meanwhile, there are two standard methods for green innovation evaluation: One is to construct the evaluation index of green innovation [14] and the other is to use data envelopment analysis (DEA). For example, Tone [15] proposed the non-angle and non-radial DEA-SBM model, which solves the slack variables problem that the traditional model does not consider. Teng [16] verified the spatial correlation of green innovation development among cities in the Yangtze River Delta by combining the Super-SBM model with the GML index. The drivers of green innovation are mainly economic development [17], government support [18], and policy system considerations [19].

Green innovation research of the Changsha-Zhuzhou-Xiangtan urban agglomeration is mainly from green ecology and development, including: to evaluate the safety of the ecosystem of Changsha-Zhuzhou-Xiangtan urban agglomeration [20], to protect and construct green heart ecology [21], and to research the relationship between the construction of a two-oriented society ' comprehensive reform area and promote regional green development [22].

Compared with simply studying the green innovation of urban agglomerations, it is more appropriate to study the green innovation effect of the Changsha-Zhuzhou-Xiangtan City Group. In general, most of the existing literature suggests that green innovation can influence regional innovation development; however, conclusions based on the green innovation effect are not uniform. Moreover, the literature ignores the internal mechanism of the green innovation effect of urban agglomerations from the perspective of the network and the mechanism of the effect of other economic and social factors on the network structure in the process of green innovation. In terms of research samples, most of the literature is limited to provincial or national data, with macro sample size and coverage of the survey area or relatively mature urban agglomerations area, which has certain geographical achievements that ignore the micro aspects. Compared with the previous literature, the main contributions of this article are as follows: First, the Great Changsha-Zhuzhou-Xiangtan City Group is a micro-level region with vast space for innovation and progress regarding the economic development period, innovation period, and green environment construction period. In this regard, this article uses data from the Great Changsha-Zhuzhou-Xiangtan City Group from 2012 to 2018. The research selected eight prefecture-level cities in the Great Changsha-Zhuzhou-Xiangtan City Group specific to a certain degree of representativeness. Second, this article investigates the impact of network structure on green innovation using the unexpected output SBM model to measure the green innovation efficiency and the panel Tobit regression method of the re-centered influence function, which directly examines the marginal impact from the perspective of the network.

3. Theoretical Mechanism Analysis

3.1. Analysis of the Effect of Network Structure Characteristics on Green Innovation in Urban Agglomeration

The urban agglomeration network is the main form of urban spatial structure. For the development of urban agglomeration, the region is necessary to establish a network structure system. The network development of urban agglomeration is not only determined by internal attributes but also dominated by external dynamic factors. The dynamic mechanism of urban agglomeration network development focuses on green innovation. The key to researching networked green innovation rests within the high frequency of learning and communication between regional innovation subjects to promote the free flow of knowledge, technology, talents, and other innovation resources. By placing focus on the synergistic effect among innovation subjects in the urban agglomeration, this effect can improve the service of scientific and technological innovation and the level of achievement transformation to reduce the negative impacts on the environment and achieve green innovation. On the measurement of network characteristics for further characterization, domestic and foreign research mainly selects network scale, network openness, network structure hole, network density (network connection strength) [23]. Hence, the structural characteristics will be analyzed from four aspects: network scale, network openness, network structure hole, and network density. Its functions are as follows.

3.1.1. Network Scale

Network scale in the network is the number of innovation subjects in urban agglomeration. The larger the number of creative subjects, the larger the network scale. The more substantial the heterogeneity between subjects, the richer the innovation resources. Heterogeneous network members have a high degree of complementary knowledge, which is conducive to reducing transaction costs and achieving results pertaining to the transformation of innovation.

3.1.2. Network Openness

Network openness is the capacity of network entities to absorb external information, related technologies, and other innovative resources. The higher the degree of openness, the stronger the ability to acquire external information, knowledge, technology, and other creative resources conducive to innovation.

3.1.3. Network Structure Hole

The more structural holes, the more social relations carry resources, providing opportunities for different innovation subjects to communicate and interact. The junction in the hub position occupies a structural hole and connects the creative subjects that lack direct contact to build a bridge of communication and cooperation, reduce information asymmetry effectively, and stimulate the realization of innovative achievements.

3.1.4. Network Density

Network density reflects the connection degree of innovation nodes. The closer is the connection between nodes in the network, the stronger the intention of innovation subjects to carry out innovation cooperation, and the more conducive to exchange and collaboration. High network density provides an effective platform for building and maintaining trust and cooperation among innovation subjects, conducive to obtaining external innovation resources within the innovation network, realizing the transformation of innovation achievements.

3.2. Influencing Mechanism of Green Innovation in the Great Changsha-Zhuzhou-Xiangtan City Group Based on Network

Regional knowledge production and technology transformation achieve development through innovation networks. Typically, innovation motivation stems from the connection between innovation subjects. Economic development provides a conducive environment for the flow of regional knowledge, technology, and other elements; government funding significantly contributes to the smooth running of innovative activities; improving the information infrastructure is a prerequisite for the flow of innovative talents among regions; the tertiary industry enables innovation subjects to have a competitive advantage, expanding the scale of innovation; urban education to promote regional innovation activities to carry out the study provides a platform for exchange. Hence, based on network structure characteristics, this paper arranges the influencing mechanisms of green innovation in the Great Changzhou-Zhuzhou-Xiangtan area, as shown in Figure 1.

3.2.1. Economic Development Level

Economic development is the strategic basis for building an innovative city and carrying out creative activities. Usually, the size of the regional economy and resources constrain innovation elements and innovation activities, and innovation communities are jointly built by innovation subjects and governments within the region, creating an implementable innovation network structure [24].

3.2.2. Government Support

The results of government support for innovation activities are affected by the intensity of investment. The reasons are as follows: First, the increase in government input stimulates enterprises and other innovation subjects to carry out scientific and technological innovation collaborations; second, the tendency of government investment, which crowds out private investment, leads to the crowding-out effect caused by the excessive accumulation of social capital. The government provides fiscal and tax subsidies for enterprises to build innovation network platforms. Enterprises in the innovation network rely on the support of the government to integrate innovation resources effectively and promote the development of innovative networks to enhance competitive advantages for regional innovation cooperation. Excessive capital tilt, which is not conducive to the transformation of research results, triggers inefficient capital utilization, inhibits the effect of large-scale innovation, imbalances the distribution of innovation elements, and hinders the flow of talents, knowledge, technology, and other elements in the innovation network.

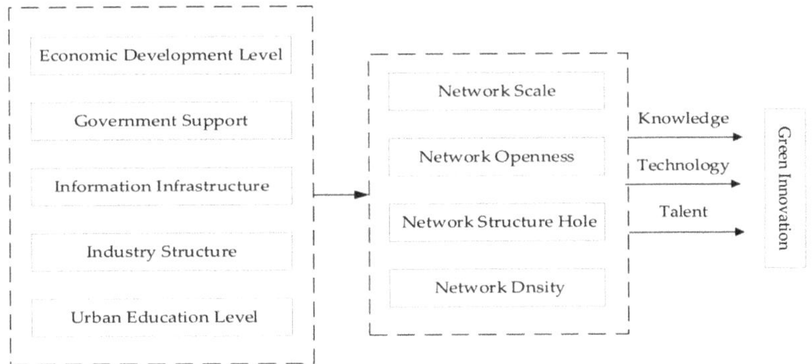

Figure 1. Influencing mechanisms of green innovation.

3.2.3. Information Infrastructure

Well-developed infrastructure is the primary condition for a region to attract innovative talent. A well-developed information network promotes knowledge flow and technology diffusion among innovation subjects. Moreover, convenient transportation and information networks provide essential conditions for expanding the scale of regional networks and enabling enterprises to engage in innovative activities.

3.2.4. Industry Structure

The industrial structure embedded in the innovation network is the reintegration and redistribution of resources. Innovative enterprises use innovation networks to enhance their competitiveness and optimize the structures of emerging industries. Based on undertaking traditional industries, urban agglomerations integrate into innovation networks, thus forming professional industrial layouts. The long-term, knowledge spillover in the innovation network leads to a nonlinear process from inferiority to continuous optimization of the industrial structure in innovation activities. Furthermore, with the improvement of resource utilization, innovation achievements are successfully transformed.

3.2.5. Urban Education Level

Urban education aims at cultivating high-quality and heterogeneous innovative talent. Importantly, only in the innovation network can heterogeneous talent realize resource complementarity and technology sharing, thus generating a synergistic effect among innovation subjects [25]. The embeddedness of network structure has an essential impact on innovation activities. Heterogeneous human capital, which is located in the hole of network structure, can obtain information resources by building relations with key innovation subjects and occupying information advantages. The more educated network members, the larger the network scale. Meanwhile, the more information resources embedded in the network, the more frequent communication and interaction conducive to constructing a complete network system.

4. Model Methodology and Variable Selection

4.1. Research Methodology

4.1.1. SBM Model with Non-Expected Output

Scholars mainly used the non-parametric DEA method in early studies on the measurement of green innovation efficiency. Since the traditional DEA model assumes that the output is the desired output, ignoring the influence of slack variables will significantly increase the measured efficiency values. At the same time, the indicators used in the experimental production process will produce lower efficiency. Therefore, the SBM model proposed by Tone [15] can effectively avoid the measurement bias caused by traditional DEA and measure the green innovation efficiency of the cities within the Great Changsha-Zhuzhou-Xiangtan City Group. The following is the constructed evaluation model of green innovation efficiency.

$$\mathrm{Min}\rho = \frac{1 - \frac{1}{m}\sum_{i=1}^{m}\frac{s_i^-}{x_{ik}}}{1 - \frac{1}{q_1+q_2}\left(\sum_{r=rk}^{q_1}\frac{s_r^+}{y_{rk}} + \sum_{t=1}^{q_2}\frac{s_t^-}{z_{rk}}\right)}$$

$$\text{s.t.}\begin{cases} \sum_{\substack{j=1 \\ j \neq k}}^{n} x_{ij}\lambda_j - s_i^- \leq x_{ik} \\ \sum_{\substack{j=1 \\ j \neq k}}^{n} y_{ij}\lambda_j - s_i^+ \leq y_{ik} \\ \sum_{\substack{j=1 \\ j \neq k}}^{n} z_{ij}\lambda_j - s_i^z \leq z_{rk} \\ \lambda, s^-, s^+ \geq 0 \end{cases} \tag{1}$$

In the formula, there are n decision units (DUM), which include the input, expected output, and unexpected output. The number of types is represented by m, q_1, q_2, respectively. The slack variables of input, expected output, and unexpected output are represented by s^-, s^+, s^z, respectively. In general, λ is the weight of the decision unit and ρ is the efficiency value of DMU. As known, the value range is between 0 and 1. Therefore, effective DUM is obtained through the SBM model of unexpected output.

4.1.2. Panel Tobit Model

The Tobit regression model, also known as a truncated model, was proposed by Tobit. The limited Tobit regression model can avoid the deviation of ordinary least square estimation results. The Tobit regression model is used to ensure unbiased and valid estimation results. After the measurement of green innovation efficiency is completed, the value range of GIE is between 0 and 1, and the basic model is as follows:

$$\begin{cases} Y_{it}^* = \alpha_0 + X_{it}\beta + \varepsilon_i \\ Y_{it} = Y_{it}^* \ if \ 1 \geq Y_{it}^* \geq 0 \\ Y_{it} = 0 \ if \ Y_{it}^* \leq 0 \\ Y_{it} = 1 \ if \ Y_{it}^* \geq 1 \end{cases} \tag{2}$$

In the formula, Y_{it}^* the explanatory variable vector, Y_i is the value of the observed dependent variable, β is the regression parameter vector, and ε_i represents the stochastic perturbation term. The panel Tobit regression model constructed based on influencing factors is as follows.

$$GIE = \alpha_0 + \beta_1 \ln GDP_{it} + \beta_2 \ln GOV_{it} + \beta_3 \ln INF_{it} + \beta_4 \ln INS_{it} + \beta_5 \ln EDU_{it} + \varepsilon_{it} \quad (3)$$

4.2. Variables and Data

4.2.1. Variable Selection

Input Variables

Green innovation efficiency is a complex system with multiple inputs and outputs. The Great Changsha-Zhuzhou-Xiangtan City Group constitutes a networked innovation system. Selecting corresponding indicators is based on green innovation input and green innovation output to estimate the green innovation efficiency of eight cities in the Great Changsha-Zhuzhou-Xiangtan City Group. According to the status quo of green innovation in the Great Changsha-Zhuzhou-Xiangtan City Group and the data availability, we constructed the evaluation index system of green innovation efficiency. The measurement of green innovation efficiency of cities in the Great Changsha-Zhuzhou-Xiangtan City Group should reflect the theme "Green" and the network effect of urban agglomeration innovation. Therefore, this paper integrates the green concept into the input-output factors and comprehensively considers the environmental pollution situation. Based on the network characteristics of urban agglomeration, the input indexes mainly include: First, using the number of universities, research institutions, and enterprises to represent network scale; second, network openness measures the ability of urban agglomeration to absorb external knowledge and information by foreign direct investment; third, the network structure hole selects the technology market transaction amount as the bridge of knowledge flow and technical cooperation among innovation subjects in the innovation network; fourth, the network density is represented by the external expenditure of Research and Development (R&D) expenditure in universities [23].

Output Variables

Output can be divided into expected output and unexpected output. Expected outputs mainly include patent applications and per capita GDP. Patent data can reflect the results of urban innovation output comprehensively; therefore, the number of patent applications is selected to represent. Per capita GDP demonstrates the level of regional economic growth. Generally, as unexpected outputs, the comprehensive environmental index is used to represent the environmental benefit output of the undesired output, and the environmental pollution comprehensive index is calculated by the entropy weight method of the three industrial wastes, which are the industrial wastewater discharge, the industrial sulfur dioxide discharge, and the industrial smoke emission respect, respectively. Specific indicators are shown in Table 1 below.

Table 1. Evaluation index system of green innovation efficiency.

Indicators	Categories	Composition	Specific Contents
Input	Network Structure Characteristics	Network Scale	The Number of Universities, Research Institutions, and Enterprises
		Network Openness	Foreign Direct Investment
		Network Structure Hole	Technology Market Transaction Amount
		Network Density	External R&D Expenditure of Universities
Output	Economic Benefits Output	Technology	Number of Technical Patent Applications
		Economic	Per Capita GDP
	Environmental Benefit Output	Environment	Environmental Pollution Index

Theoretically, in urban innovation activities, any factors related to innovation activities will impact innovation efficiency. According to the current situation of urban green development, economic development requires a high level of technical innovation. On the contrary, the economic development level is also bound to increase innovation input, forming a positive feedback effect. Therefore, the influencing factors of green innovation efficiency should include the level of economic development (*GDP*), which is represented by per capita GDP. Since communication infrastructure plays the most important role in innovation among the four infrastructures of the highway, railway, communication, and energy, it is expressed as the proportion of regional postal and telecommunication services in GDP (*INF*) [26]. Government support (*GOV*) is represented by the proportion of regional science and technology financial expenditure in local financial expenditure. Industrial structure (*INS*) uses the proportion of the added value of the tertiary industry in the gross regional product, which reflects the ratio of the tertiary industry in the national economy to measure. Urban education can provide high-quality human capital for innovation and become a significant factor influencing the regional innovation system. Therefore, this study selected urban education level (*EDU*) as the influencing factor and represented by the number of university students [27] (see Table 2 for details).

Table 2. Influencing factors and variables of green innovation efficiency.

Influencing Factors	Variable Abbreviation	Variable Definition
Economic Development Level	GDP	Per Capita GDP
Government Support	GOV	Regional Science and Technology Expenditure Accounted for the Proportion of Local Fiscal Expenditure
Information Infrastructure	INF	Regional Posts and Telecommunications Accounted for the Proportion of GDP
Industry Structure	INS	The Proportion of The Added Value of The Tertiary Industry in GDP
Urban Education Level	EDU	University Students

4.3. Data Sources

We included eight prefecture-level cities in the Great Changsha-Zhuzhou-Xiangtan City Group from 2012 to 2018 as the research objects, namely, Changsha, Zhuzhou, Xiangtan, Hengyang, Yueyang, Changde, Yiyang, and Loudi. The original data of the indicators in this paper came predominantly from the statistical bulletins of the corresponding prefecture-level cities in Hunan Statistical Yearbook, Hunan Science and Technology Statistical Yearbook, and China Urban Statistical Yearbook (2012–2018). Some missing values are complemented by linear interpolation.

5. Measurement and Characteristics of Green Innovation Efficiency of Great Changsha-Zhuzhou-Xiangtan City Group Based on Network

5.1. Measurement Results of Green Innovation Efficiency of the Great Changsha-Zhuzhou-Xiangtan City Group Based on Network

Using the constructed SBM model of unexpected output to calculate the green innovation efficiency of cities in the Great Changsha-Zhuzhou-Xiangtan City Group from 2012 to 2018, the results are shown in Table 3. Meanwhile, cloud model and system cluster analysis were used to analyze the normal distribution and segmentation of green innovation efficiency in the Great Changsha-Zhuzhou-Xiangtan City Group. The results are shown in Figures 2 and 3, respectively.

Table 3. Green innovation efficiency of urban agglomeration.

Region	2012	2013	2014	2015	2016	2017	2018	Mean	Ranking
Changsha	0.8153	1	1	1	1	1	0.8325	0.9497	1
Zhuzhou	1	0.7121	0.7709	0.8196	0.7895	0.9407	1	0.8864	3
Xiangtan	0.6488	0.6891	0.6624	0.7185	0.7687	0.7898	0.7953	0.7247	5
Hengyang	0.2335	0.4882	1	0.597	0.8911	1	0.5197	0.6756	6
Yueyang	1	1	1	1	1	0.5614	1	0.9373	2
Changde	0.4312	0.7856	0.8108	0.5851	0.6938	0.6854	0.6589	0.6644	7
Yiyang	0.4948	0.4064	0.3032	0.6408	1	0.4474	1	0.6132	8
Loudi	1	0.6047	1	1	1	1	0.3769	0.8545	4
Mean	0.7030	0.7108	0.8184	0.7951	0.8929	0.8031	0.7729		

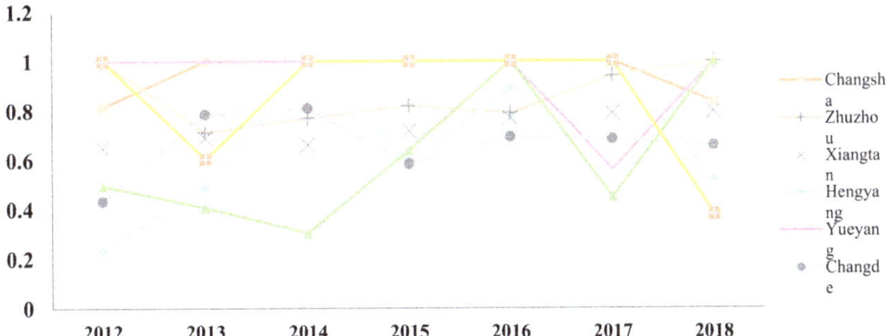

Figure 2. Distribution trend map of green innovation efficiency in the urban agglomeration.

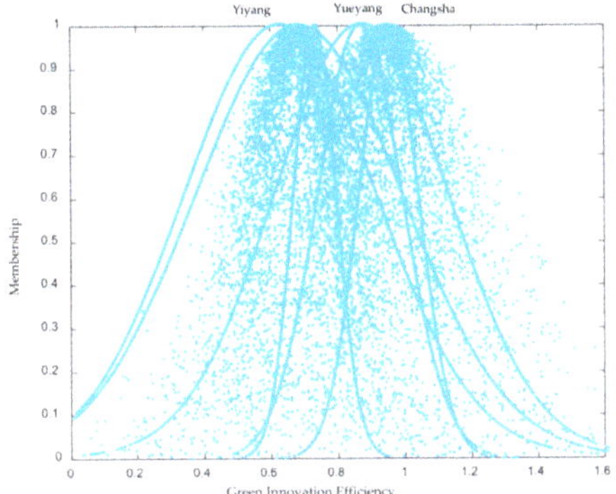

Figure 3. Normal cloud-membership degree distribution of green innovation efficiency in the Great Changsha-Zhuzhou-Xiangtan City Group.

5.2. Characteristics Analysis of Green Innovation Efficiency of the Great Changsha-Zhuzhou-Xiangtan City Group Based on Network

The green innovation efficiency of the Great Changsha-Zhuzhou-Xiangtan City Group was generally stable. In terms of the overall time series, the mean green innovation efficiency values of the Great Changsha-Zhuzhou-Xiangtan City Group from 2012 to 2018 were 0.7030, 0.7108, 0.8184, 0.7951, 0.8929, 0.8031, and 0.7729, respectively. The overall mean values of green innovation efficiency fluctuated at a medium level and showed a stable trend. This may be because Hunan province has focused on constructing a two-oriented society in the core area of urban agglomeration, improving the overall green development of the Great Changsha-Zhuzhou-Xiangtan City Group to a certain extent in recent years. At the same time, influenced by the international environment, China's economy has slowed down; with tremendous downward pressure, the transformation of technological innovation achievements into practical productivity in the Great Changsha-Zhuzhou-Xiangtan City Group is insufficient. In the meantime, rational allocation and utilization of innovative resources are hindered, so the overall efficiency is low. The distribution trend of green efficiency in urban agglomeration is shown in Figure 2. Therefore, the green innovation development of urban agglomeration cannot be achieved without a favorable innovation environment.

The green innovation efficiency of the Great Changsha-Zhuzhou-Xiangtan City Group presents the characteristics of "urban gradient" development. As shown in Figure 3, Changsha and Yueyang rank high in green innovation efficiency and have a maximum advantage for green innovation development. On the one hand, Changsha has an extensive network scale and high network openness. As a provincial capital city, Changsha is inextricably linked to its position in the network structure hole in the innovation system. As a network structure hole for strengthening regional cooperation, Yueyang is the "main battlefield" of Hunan's growth pole construction and an essential fulcrum for docking the Yangtze River Economic Belt construction. Other cities in the urban agglomeration—Zhuzhou and Xiangtan—have dominant industries concentrated in secondary industries, such as iron and steel industries. Therefore, extensive development may cause environmental pollution; their green innovation efficiency, therefore, is slightly behind Changsha. Yiyang has continuously lagged with its green innovation efficiency, which has restricted the balanced development of green innovation in the Great Changsha-Zhuzhou-Xiangtan City Group. According to the echo theory, the development of green innovation is limited by its economic scale, resulting in a lack of endogenous motivation and issues with green development [28]. Therefore, we can conclude that it can be divided into three gradients: (1) Changsha and Yueyang, (2) Yiyang, and (3) other cities.

The "mid-stream drive" difference of green innovation efficiency in the Great Changsha-Zhuzhou-Xiangtan City Group is noticeable. Figure 3 shows the apparent differences between cities such as Changsha and Yiyang and other cities in the urban agglomeration; however, other cities need further analysis based on Figures 4 and 5. We can see from Figure 4 that Xiangtan, Changde, and Hengyang are classified into one category, and the green innovation efficiency value is between 0.6 and 0.7. Combined with Figure 5, Xiangtan moves up by one place, Changde moves up by one place, and Hengyang moves up by two. The reasons are as follows. First, Xiangtan encourages independent innovation and strengthens R&D and project promotion by the government bonus policy. For example, the *Implementation Measures for Encouraging Scientific and Technological Innovation in Xiangtan High-tech Zone*, issued in 2016, enhanced the guiding role of government expenditure on green innovation and promoted the innovation efficiency of the region to maintain an excellent level. The Changde Office of Science and Technology issued some awards in *Changde Science and Technology Awards* in 2017 to stimulate the initiative of innovation subjects, accelerate the transformation of innovation achievements, and improve the innovation system. Hengyang proposed *Measures for The Administration of Subsidies for Science and Technology Services (trial)* for innovative services subsidies, enhancing the innovation capacity of relevant enterprises. Second, Zhuzhou and Loudi are the second types, with a green innovation efficiency ranging from 0.85 to 0.90. Zhuzhou moves up by one place, while Loudi moves down by four places. The reason may be that the limited economic scale is the main factor in developing green innovation. The small scale of a regional innovation network and insufficient innovation investment seriously hinder the improvement of innovation efficiency.

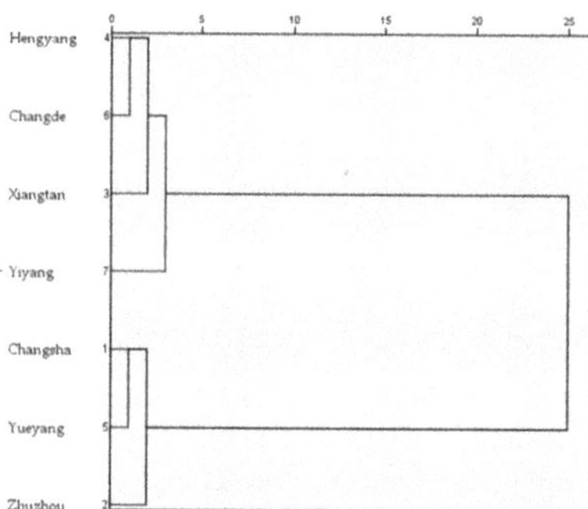

Figure 4. Cluster tree of green innovation efficiency in the Great Changsha-Zhuzhou-Xiangtan City Group.

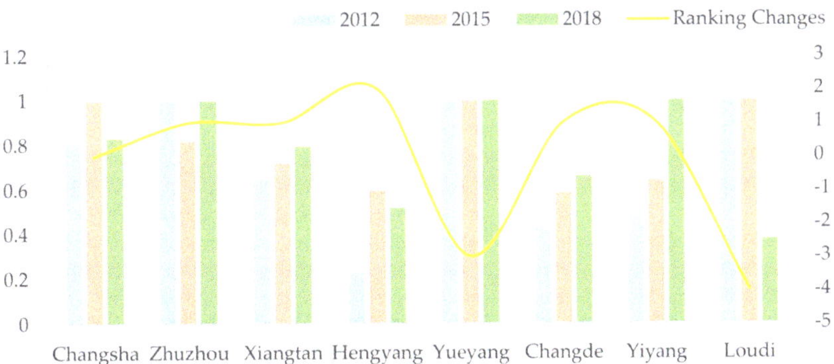

Figure 5. Trend of green innovation efficiency in the Great Changsha-Zhuzhou-Xiangtan City Group.

6. Research on The Influencing Factors of Green Innovation in the Great Changsha-Zhuzhou-Xiangtan City Group Based on Network

Based on the theoretical mechanism of the network and the green innovation efficiency measured by the unexpected output SBM model, the empirical analysis of influencing factors is conducted. Unit root tests were carried out prior to regression result analysis to ensure the rigor and correctness of regression results.

6.1. Unit Root Test

Combined with the SBM model of unexpected output, this paper calculates the green innovation efficiency data of the Great Changsha-Zhuzhou-Xiangtan City Group from 2012 to 2018. It conducts regression analysis with green innovation efficiency as the dependent variable. Firstly, the unit root test of panel data is carried out to enable a more rigorous and reliable empirical process and avoid the pseudo-regression caused by the non-stationary variables to ensure the effectiveness of the estimation results. The stationary unit root test mainly uses LLC, IPS, ADF-Fisher, and PP-Fisher methods to test the variables. The formula of the unit root test is as follows:

$$X_{it} = \rho_i X_{i,t-1} + \varepsilon_{it} \tag{4}$$

The unit root test can be divided into two categories: one category is the LLC test, which is the same regression coefficient. Additionally, it changes according to different cross-sections, namely, IPS test, ADF-Fisher test and PP-Fisher test. The first test results show that the sequence of a variable is unstable, so take the logarithm to carry on the stability test. As shown in Table 4, the results obtain the overall rejection of the original hypothesis of 'existence of unit root', and then carry out the panel Tobit random effect regression; the results are shown in Table 5.

6.2. Analysis of Regression Results of Influencing Factors

According to the regression results, the relevant influencing factors are as follows:

First: the economic development level. The economic development level regression results show a significantly positive impact on green innovation efficiency, and the impact coefficient is 0.9745. The level of regional economic development promotes the improvement of green innovation efficiency. A higher GDP encourages individuals to pursue green and low-carbon products and facilitates the optimal allocation of green innovation resources [29]. Moreover, it effectively avoids the redundancy of network scale, enables the effectiveness of the network structure as a whole, and improves the conversion rate of innovation achievements.

eration development, draw lessons from the experience of efficient city development, build an innovation cooperation center system, increase the scale of network construction, and gradually realize the transformation of innovation achievements. For cities with insufficient green innovation and development capacity in Yiyang, we should actively learn from the green development experience of Changsha, improve the construction of collaborative mechanisms for urban green innovation, and encourage technological innovation to benefit innovative development.

Second, it is essential to form an innovation network system with enterprises at its core, including higher-education institutions, research institutions, and other innovation subjects; expand the scale of the innovation network, strengthen the innovation links between subjects, further enhance the enthusiasm and initiative of independent innovation, and effectively improve the efficiency of green innovation.

Third, economic development mainly depends on technological innovation, accelerating technological innovation research and development. The transformation of the economic development model is essential in order to save resources, protect the environment, focus efforts on green innovation efficiency, actively learn from the developed areas of green innovation development, optimal green innovation resources allocation, and enhance the efficiency of green innovation.

Fourth, it is vital to vigorously improve the construction of education service systems and invest reasonably in education resources. We will foster outstanding innovative talent, guide creative and talented university graduates to form enterprises and realize the diffusion of technological knowledge. The purpose is to ensure the rational utilization of regional innovation resources, promote the upgrading of industrial institutions, and realize the development of green innovation.

Author Contributions: Conceptualization, L.W. and L.C.; methodology, L.W. and L.C.; software, L.W. and L.C.; formal analysis, L.W. and L.C.; resources, L.W. and L.C.; data curation, L.W. and L.C.; writing—original draft preparation, L.W.; writing—review and editing, L.W. and L.C.; supervision, L.W., W.Y. and L.C.; project administration, W.Y. and L.C.; funding acquisition, W.Y. All authors have read and agreed to the published version of the manuscript.

Funding: This work was funded by The National Social Science Fund of China (grant number 20BGL299).

Data Availability Statement: We are taking eight prefecture-level cities in the Great Changsha-Zhuzhou-Xiangtan City Group from 2012 to 2018 as the research objects, including Changsha, Zhuzhou, Xiangtan, Hengyang, Yueyang, Changde, Yiyang, and Loudi. The original data of the indicators in this paper mainly come from the statistical bulletins of the corresponding prefecture-level cities in *Hunan Statistical Yearbook, Hunan Science and Technology Statistical Yearbook, and China Urban Statistical Yearbook (2012–2018)*.

Acknowledgments: We would like to express our gratitude to all those who helped us during the writing of this article. Our deepest gratitude goes first and foremost to Yaping Hu, who are from the Hunan University of Science and Technology in China, for their constant encouragement and guidance.

Conflicts of Interest: The authors declare no conflict of interest.

References

1. Xu, Y.Q.; Zeng, G.; Wang, Q.Y. The evolution and optimization strategy of collaborative innovation network pattern in Yangtze River Delta urban agglomerations. *Econ. Geogr.* **2018**, *38*, 133–140.
2. Huallacháin, B.Ó.; Lee, D.-S. Urban centers and networks of co-invention in American biotechnology. *Ann. Reg. Sci.* **2014**, *52*, 799–823. [CrossRef]
3. Lee, D.-S. Towards Urban Resilience through Inter-City Networks of Co-Invention: A Case Study of U.S. Cities. *Sustainability* **2018**, *10*, 289. [CrossRef]
4. Sheng, Y.W.; Gou, Q.; Song, J.P. Research on network structure and innovation efficiency of urban agglomeration: A case study of Beijing-Tianjin-Hebei, Yangtze River Delta and Pearl River Delta. *Geogr. Sci.* **2020**, *40*, 1831–1839.
5. Ren, S.G.; Hu, C.Y.; Wang, L.W. An empirical study on the impact of network structure on regional innovation capability in China. *Syst. Eng.* **2011**, *29*, 50–55.

6. Wu, Z.C. Research on regional innovation performance based on stochastic frontier: From the perspective of innovation network Structure. *Technol. Econ.* **2020**, *39*, 120–131.
7. Li, X.Y. Regional science and technology innovation research based on network. *Sci. Manag. Res.* **2014**, *32*, 52–55.
8. Liu, J.T. The Canonical Correlation Analysis of innovation Network Structure Characteristics to innovation output. *Stat. Decis.* **2019**, *35*, 118–120.
9. Rennings, K. Redefining innovation—Eco-innovation research and the contribution from ecological economics. *Ecol. Econ.* **2000**, *32*, 319–332. [CrossRef]
10. Driessen, P.H.; Hillebrand, B.; Kok, R.; Verhallen, T.M.M. Green New Product Development: The Pivotal Role of Product Greenness. *IEEE Trans. Eng. Manag.* **2013**, *60*, 315–326. [CrossRef]
11. Kemp, R. Eco-innovation: Definition, Measurement and Open Research Issues. *Econ. Politica* **2010**, *60*, 397–420.
12. Li, D.; Zeng, T. Are China's s intensive pollution industries greening? An analysis based on green innovation. *J. Clean. Prod.* **2020**, *259*, 120901. [CrossRef]
13. Zeng, J.; Skare, M.; Lafont, J. The co-integration identification of green innovation efficiency in the Yangtze River Delta region. *Eur. J. Oper. Res.* **2021**, *134*, 252–262. [CrossRef]
14. Qian, L.; Wang, W.; Xiao, R. Research on the regional disparities of China's industrial enterprises green innovation efficiency from the perspective of shared inputs. China Population. *Resour. Environ.* **2016**, *26*, 149–157.
15. Tone, K. A slacks-based measure of efficiency in data envelopment analysis. *Eur. J. Oper. Res.* **2001**, *03*, 498–509. [CrossRef]
16. Ten, T.W.; Qu, C.Y.; Hu, S.L.; Zeng, G. Green innovation efficiency pattern differentiation and spatial correlation characteristics of Urban agglomeration in Yangtze River Delta. *J. E. China Norm. Univ.* **2019**, *5*, 107–117.
17. Zhang, W.; Li, H.L.; An, X.B. Discussion on the Theoretical Model and Thinking of Enhancing China's Green Innovation Ability by using FDI. *Manag. World* **2011**, *12*, 170–171.
18. Li, J.; Du, Y. Can Spatial effect of Environmental Regulation on Green Innovation Efficiency—Evidence from Prefectural-level Cities in China. *J. Clean. Prod.* **2020**, *286*, 125032. [CrossRef]
19. Zhou, M.; Wu, C.Q. The impact of government subsidies on some technological innovations. *Sci. Res. Manag.* **2017**, *38*, 574–580.
20. Yang, L.G. Evaluation of ecosystem security in urban agglomeration based on ecological footprint: A case study of Changsha-Zhuzhou-Xiangtan Urban agglomeration. *Study World Geogr.* **2009**, *18*, 74–82.
21. Gu, Z.L.; Ma, T.; Yuan, X.H.; Zhang, X.M.; Wang, X. Ecological protection and development of greenheart in Changsha-Zhuzhou-Xiangtan Urban agglomeration. *Resour. Environ. Yangtze Basin* **2010**, *19*, 1124–1131.
22. Li, W.B.; Li, C. Can "Two-oriented Society" Comprehensive Reform Zone promote green Development? *J. Financ. Econ.* **2018**, *44*, 24–37.
23. Wu, Z.C. Network structure, innovation infrastructure and regional innovation performance: An analysis based on network DEA multiplication Model. *J. Beijing Jiaotong Univ.* **2021**, *20*, 79–89.
24. Lin, F. Co-construction, Co-governance and Shared benefits: Regional integration from the perspective of innovative economy: A case study of Yangtze River Delta integration development. *West. BBS* **2020**, *30*, 68–77.
25. Li, Y.Z.; Yang, J.N.; Tian, X.F. Research on the relationship between incubation network embedment, entrepreneurial efficacy and entrepreneurial performance. *Sci. Sci. Sci. Technol. Manag.* **2016**, *37*, 169–180.
26. Lv, Y.W.; Xie, Y.X.; Lou, X. Research on spatial-temporal transition and convergence trend of regional green innovation efficiency in China. *J. Res. Quant. Tech. Econ.* **2020**, *37*, 78–97.
27. Li, J.Y.; Li, C.; Li, Z.Y. Urban green innovation efficiency evaluation and Its Influencing Factors analysis. *Stat. Decis.* **2017**, *20*, 116–120.
28. Dong, H.Z.; Li, X.; Zhang, R.J. Spatial and temporal characteristics and driving factors of green innovation efficiency in Guangdong-Hong Kong-Macao Greater Bay Area. *Econ. Geogr.* **2021**, *41*, 134–144.
29. Li, J.; Ma, X.F. Spatial and temporal differences and influencing factors of green innovation efficiency in Beijing-Tianjin-Hebei City. *Syst. Eng.* **2019**, *37*, 51–61.
30. Peng, R.Z.; Wang, L. Is service industry development green?—Based on total factor efficiency analysis of service industry environment. *Ind. Econ. Res.* **2016**, *4*, 18–28.
31. Peng, Y.W.; Zou, M.X. Comparative study on evaluation of regional green innovation efficiency in Hunan Province. *Econ. Forum* **2019**, *7*, 37–44.

 *land*

Article

Combining Traffic Microsimulation Modeling and Multi-Criteria Analysis for Sustainable Spatial-Traffic Planning

Irena Ištoka Otković [1,*], Barbara Karleuša [2], Aleksandra Deluka-Tibljaš [2], Sanja Šurdonja [2] and Mario Marušić [1]

[1] Faculty of Civil Engineering and Architecture Osijek, Josip Juraj Strossmayer University of Osijek, 31000 Osijek, Croatia; mariomarusic.10@gmail.com
[2] Faculty of Civil Engineering, University of Rijeka, 51000 Rijeka, Croatia; barbara.karleusa@uniri.hr (B.K.); aleksandra.deluka@uniri.hr (A.D.-T.); sanja.surdonja@gradri.uniri.hr (S.Š.)
* Correspondence: iirena@gfos.hr; Tel.: +385-91-224-0749

Abstract: Spatial and traffic planning is important in order to achieve a quality, safe, functional, and integrated urban environment. Different tools and expert models were developed that are aimed at a more objective view of the consequences of reconstruction in different spatial and temporal ranges while respecting selection criteria. In this paper we analyze the application of the multi-criteria analysis method when choosing sustainable traffic solutions in the center of a small town, in this case Belišće, Croatia. The goal of this paper is to examine the possibility of improving the methodology for selecting an optimal spatial–traffic solution by combining the quantifiable results of the traffic microsimulation and the method of multi-criteria optimization. Socially sensitive design should include psychological and social evaluation criteria that are included in this paper as qualitative spatial–urban criteria. In the optimization process, different stakeholder groups (experts, students, and citizens) were actively involved in evaluating the importance of selected criteria. The analysis of stakeholders' survey results showed statistically significant differences in criteria preference among three groups. The AHP (Analytic Hierarchy Process) multi-criteria analysis method was used; a total of five criteria groups (functional, safety, economic, environmental, and spatial–urban) were developed, which contain 21 criteria and 7 sub-criteria; and the weights of criteria groups were varied based on stakeholders' preferences. The application of the developed methodology enabled the selection of an optimal solution for the improvement of traffic conditions in a small city with the potential to also be applied to other types of traffic–spatial problems and assure sustainable traffic planning.

Keywords: sustainable spatial–traffic planning; microsimulation traffic modeling; AHP; multi-criteria analysis; sensitivity analysis; stakeholders' preferences; public participation

Citation: Ištoka Otković, I.; Karleuša, B.; Deluka-Tibljaš, A.; Šurdonja, S.; Marušić, M. Combining Traffic Microsimulation Modeling and Multi-Criteria Analysis for Sustainable Spatial-Traffic Planning. *Land* **2021**, *10*, 666. https://doi.org/10.3390/land10070666

Academic Editors: Iwona Cieślak, Andrzej Biłozor and Luca Salvati

Received: 14 June 2021
Accepted: 21 June 2021
Published: 24 June 2021

1. Introduction

Reconstruction of segments of the existing traffic network according to the principles of sustainable urban mobility includes multidisciplinarity and consideration of spatial interventions through holistic criteria. Sustainable traffic and spatial planning should equally meet urban mobility development goals and the needs of traffic users, as well as understand and preserve the natural environment [1].

The European Commission has adopted a Sustainable Urban Mobility Plan (SUMP), a document that sets the framework for planning and managing urban mobility with the goal of improving the urban quality of life by creating a safe, reliable, integrated, multimodal, efficient, and environmentally friendly traffic system [2].

In order to positively influence the choice of movement modalities and encourage active and more environmentally friendly forms of urban mobility, it is necessary to provide accessible, safe, comfortable, and attractive traffic areas intended for pedestrians and cyclists, and various forms of micromobility and, at the same time, respect necessary requirements for motorized traffic. Planning and construction of traffic infrastructure was traditionally aimed at meeting traffic demand, but the analyses clearly show that the planning approach is changing and that the design of traffic infrastructure is effectively used to generate the desired forms of mobility [3]. The central parts of cities were spatially defined in the distant past, and the continuous increase of the degree of motorization and of the size of the population impose the need to meet numerous and often conflicting objectives during reconstruction, and the choice of an optimal solution becomes a question of multidisciplinary cooperation and the application of a multi-criteria optimization [4].

Microsimulations are a frequently used tool in assessing the functional (traffic) characteristics of alternative transport infrastructure solutions [5,6]. The assessment of alternative solutions based on traffic and cost criteria does not provide a sufficiently broad approach for sustainable and socially sensitive spatial planning, so it is necessary to include other aspects of potential solutions.

The aim of this paper is to give a methodological contribution to the evaluation of alternative solutions of transport infrastructure, including cost, microsimulations as quantitative indicators of traffic conditions, and spatial–urban criteria as qualitative criteria in a broader context that allows for the application of multi-criteria analysis methods. The methodology for evaluating alternative solutions applied in this research is to combine the multi-criteria analysis methods, traffic micro-simulations, and a socially sensitive spatial planning approach to develop a model that unites a conventional approach to traffic planning based on traffic demand of all modes of traffic and a sustainable approach in planning based on ensuring accessibility of space with active public participation in the assessment of alternative solutions and selection of the optimal solution.

In this paper, the Analytic Hierarchy Process (AHP) [7] multi-criteria analysis method is applied to a case study, selecting the optimal solution for the reconstruction of traffic areas at the level of a conceptual solution for part of the traffic network in the center of the town of Belišće in Croatia. The local inhabitants, as well as the professional and non-professional public, were involved in the process of choosing the optimal solution through several steps in the procedure, and the traffic microsimulation technique was used to verify future traffic needs.

After the introduction, the Section 2 of this paper provides an overview of the literature from the research topic. Section 3, entitled "Study Area," describes the spatial and traffic characteristics of the selected case study, the central zone of the town of Belišće in Croatia. Section 4 explains in detail the methodological steps and their application in the selected case study—a combination of traffic microsimulation modeling and MCA. In the Section 5, the results of the application of the analyzed methodology are presented and discussed. The results of quantitative and qualitative criteria are analyzed, as well as different stakeholder preferences. Section 6 presents the concluding considerations arising from the application of the proposed innovative methodology for the assessment of alternative solutions for the reconstruction of transport infrastructure and selection of the optimal solution.

2. Literature Review

The reconstruction of traffic infrastructure in built and spatially defined urban zones according to the principles of sustainable urban mobility is a complex task. On the one hand, we have different transport users who have different mobility needs and modalities that include vehicle traffic demand, and on the other hand, there are traffic safety requirements, socially sensitive spatial design, environmental impact, and users and community costs that need to be reconciled in the limited available space. Active forms of mobility that are environmentally friendly have additional positive effects because they take up less space and are economically acceptable, so they are often the focus of designers when the reconstruction of transport infrastructure is planned in city centers.

Urban planners and analysts draw particular attention to the concept of urban "walkability", which connects urban design and traffic infrastructure with wider objectives such as public health, ecological and economic objectives, and social equality. The very notion of walkability is complex and is used to describe quite different kinds of phenomena. Pikora et al. [8] presented the development of a framework of potential environmental influences on walking and cycling. The framework includes four features: functional, safety, aesthetic, and destination. Ewing and Handy [9] highlighted the five criteria for the evaluation of traffic infrastructure intended for pedestrian movement: imageability, enclosure, human scale, transparency, and complexity. Forsyth [10] analyzed walkability from three different points of view: conditions (walkable environments, infrastructure, quality, safety), outcomes (making places lively and sociable), and better urban places (walkability provides a holistic solution to a variety of urban problems). According to Ruiz-Padillo et al. [11], the concept of walkability refers to the extent to which a neighborhood is walking-friendly and defines three key influential factors: public security, traffic safety, and pavement quality.

The traffic safety of pedestrians is an important influential factor in selecting the modality of movement, and numerous studies have focused on the analysis of pedestrian behavior, particularly in the risky segments of the traffic network, such as zones of potential conflict between pedestrians and vehicle–pedestrian crossings. Various studies have been performed and models developed for predicting the behavior of pedestrians [12–14] and vulnerable groups such as children [15] and the elderly [16], with the aim of identifying the influential parameters on pedestrian behavior in the traffic safety analysis. The results of different studies show that pedestrian movement is under significant influence of the cultural heritage of a specific environment and traffic and spatial conditions; therefore, such models are not universally applicable [14,17].

The presence and optimal placement of functionally mixed urban contents in walking distance are the prerequisite for the choice of the pedestrian movement modality. Dovey and Pafka [18] analyzed density, functional mix, and access networks as key factors of urban pedestrian mobility, but highlighted a key distinction between walkability and walking. These factors ensure spatial prerequisites that can be motivating or demotivating, but do not directly cause walking.

Su et al. [19] identified the most common indicators in existing indicator classification frameworks—connectivity, accessibility, suitability, serviceability, and perceptibility—and established an indicator classification system through expert panel evaluation for auditing street walkability in China. Blečić et al. [20] presented a survey of operational methods for walkability analysis and evaluation that are potentially useful as tools for sustainability-oriented urban design.

Zuniga-Teran et al. [21] identified gaps and strengths in the Leadership in Energy and Environmental Design for Neighborhood Development (LEED-ND) certification system, through the lens of the nine categories of walkability—connectivity, land use, density, traffic safety, surveillance, parking, experience, greenspace, and community. Within this evaluation system, parking is mentioned as an important element in the evaluation of spatial potentials and shows a clear link between parking and pedestrian infrastructure planning.

A personal car spends around 5% of the total time in movement [22], so the planning approach to stationary traffic for personal vehicles, which adapts to the needs of users, was proven to be ineffective in the time perspective. Traffic infrastructure intended for stationary traffic is efficiently used as an influential factor when choosing movement modalities [23], and in the lack of space in urban-defined central parts of cities, urban areas should be ceded to more efficient traffic modalities.

Zahabi et al. [24] analyzed the impact of surface purpose, public urban transport offers, parking capacities, and parking prices on the choice of urban mobility modalities. The results of the study showed that a USD 1 increase in the hourly parking price would imply an increase of 5% in the probability of using public transport and a 10% increase in the public transport fare would represent a 10% average reduction in the probability of using public transport.

Available and quality transport infrastructure is a prerequisite, but not necessarily a decisive factor in choosing the modality of urban mobility, as shown by the comparison between mobility habits in different cities [25]. Motivational factors for the choice of mobility modalities include environmental [1], economic level, climate, available public transport services [24], etc. A deeper analysis of the complex psychological and sociological factors influencing mobility motivation is explored within the sociology of mobility [26]. Mobility can be analyzed as capital that has significant social implications. The mobility capital (motility) affects social integration and can be considered an analytical tool that allows the analysis of the relationship between the dynamics of spatial and social mobility [27].

In small towns (up to 10,000 inhabitants) [28], active traffic is traditionally present; however, the construction of traffic areas intended for motor traffic, particularly for fast traffic, jeopardizes the safety of such a manner of movement while the areas intended for stationary traffic are often organized in very attractive urban locations. During spatial and traffic planning and reconstruction, small towns should be observed through spatial particularities of such urban spaces [29]; however, the potential for preserving and revitalizing healthy and environmentally friendly movement lies exactly in the smaller urban areas.

The first step of analysis when planning road infrastructure reconstruction is to identify existing traffic demand; however, it is mandatory to look at the reconstruction solutions through the criteria of future traffic demand. The application of microsimulation traffic modeling enables the analysis of alternative reconstruction solutions in different time and space ranges [30], simulating the behavior of all traffic participants. The advantage of microsimulation modeling is that it enables the analysis of a significant number of traffic and spatial indicators of alternative solutions in the planning phase, as well as the possibility to analyze the reconstruction impacts on a wider network coverage and with future traffic demand [31]. Microsimulations allow for the comparison of variable solutions through numerically expressed functional parameters—driving time, delays, queue parameters, etc., and dynamic parameters—speed of each individual vehicle, average speed, and other indicators related to traffic flow. Microsimulations have proven to be a valuable tool in traffic infrastructure analysis [5,6], but their potential has not been sufficiently exploited in a broader spatial–traffic analysis, as enabled by multi-criteria analysis methods.

Considering the numerous criteria based on which an optimal spatial and traffic solution should be selected, in this paper we analyze the possibility of using multi-criteria analysis combined with the use of VISSIM for traffic microsimulations in the procedure of preliminary analysis of possible developments of the traffic system. Multi-criteria analysis was applied to select the best alternative for the restoration of areas and the reconstruction of systems.

During spatial planning and design, the multi-criteria analysis method is used to optimize the selection of a solution for different types of problems, and the application to big-scale problems has shown its effectiveness [32,33]. Although a significant aspect of scientific interest is developing in the direction of selecting the appropriate optimization method for a particular observed problem, taking into account different MCA methods (including the fuzzy approach), and comparing their adequacy regarding the specifics of the problem analyzed (as suggested by [34,35]), when it comes to planning and designing traffic infrastructure in urban areas, the most commonly used procedure is the Analytic Hierarchy Process (AHP) multi-criteria analysis method [4,36]. Based on the review of multi-criteria analysis method applications in decision-making about transport infrastructure from 2003 to 2012 done by Deluka-Tibljaš et al. [4], in 70% of analyzed scientific papers the AHP method was applied. Other applied methods, in a much lower percentage, were ELECTRE (Elimination and (Et) Choice Translating Reality), PROMETHEE (Preference Ranking Organization Method for Enrichment Evaluations), SAW (Simple Additive Weighting), TOPSIS (Technique for Order Preference by Similarity to Ideal Solution), etc.

Broniewicz and Ogrodnik [37], in their review of global literature from 2010 to 2019 conducted on the subject of MCDM/MCDA (Multi-Criteria Decision Making/Multi-Criteria Decision Analysis) methods used in transportation infrastructure, proved that the most popular method used to solve multi-criteria decision problems in the field of transport is the AHP method, followed by TOPSIS, DEMATEL (Decision Making Trial and Evaluation Laboratory), and PROMETHEE and ELECTRE.

The AHP method enables the optimization of solutions already in the planning phase when alternative solutions, in general, are not elaborated in detail, because besides the usual multi-criteria assessment of every alternative regarding each criterion separately, it also provides the possibility of pair-wise comparison of alternatives in relation to a specific criterion [38].

Alemdar et al. [39] used microsimulation and MCA (AHP and TOPSIS) to select the optimal corridor design and analyzed different solutions for three intersections on the Vatan Street corridor in İstanbul, Turkey. The evaluation criteria in the study were vehicle delay, queue length, stopped delay, stops, travel time, vehicle safety, CO emission, fuel consumption, and construction cost. In the study done by Bayrak and Bayata [40], three intersection types were modeled and tested using VISSIM software in line with parameters: capacity and level of service, travel time, average delay, queue length, fuel consumption, and vehicle exhaust emissions. Safety and construction costs were added. A new model was then generated using the VISSIM results and the AHP results. Du et al. [41] used a combination of VISSIM microsimulation and AHP methods to create a model to evaluate access management. Alemdar et al. [42] applied AHP and VIKOR methods to optimize the position of pedestrian crossings on the road corridor in a case study in Erzurum, Turkey. Microsimulations in VISSIM were applied in the assessment of the impact of individual scenarios of pedestrian movements on the traffic conditions of the observed coverage.

A review of the literature shows that in recent times the results of microsimulations have been used as input data of MCA methods (most often AHP methods). In doing so, functional indicators are predominantly used, with the addition of cost and safety parameters.

Functional indicators obtained by applying microsimulations represent one of the five groups of criteria on the basis of which alternative solutions are evaluated in this paper. Each group of criteria brings one specific aspect of the evaluation of the solution, and their importance was analyzed by different stakeholders. The addition of spatial–urban criteria introduces a set of qualitative parameters that are psychologically and socially oriented and focused on the subjective experience of different stakeholders. Involving stakeholders in the evaluation of alternative solutions is not a completely new approach, but, like the application of microsimulations, it is insufficiently used in the broader context of the evaluation of alternative solutions. In this paper, the MCA (AHP method) and microsimulation methods are combined in order to achieve sustainable traffic goals in making decisions for a small city area, respecting relevant criteria for socially sensitive design but also local conditions by including local inhabitants as relevant stakeholders.

3. Study Area

The town of Belišće is an industrial town with around 10,000 inhabitants, is one of seven towns in Osijek-Baranja County, and covers a surface area of 68.75 km^2. The geographical and traffic position of Belišće makes this town the intersection of important regional road corridors.

The town of Belišće is associated with a 130-year-long industrial tradition of the production and processing of packaging paper, which plays an important role in the economic development of the region. Since 1998, a hydraulic press factory for the production of car tires has been operating in the inner city center of Belišće. The factory is predominantly export oriented and represents an important social and traffic pull factor. Industrial zones located near the town center (Figure 1) generate several major problems in the urban traffic network of the town of Belišće. The insufficient hierarchical structure of the urban network brings freight traffic to the very periphery of the secondary network, which is a significant functional and traffic safety problem.

In addition, increased traffic of trucks and heavy freight vehicles cause economic problems by means of accelerated deterioration of the pavement structure and environmental problems such as noise and air pollution. The observed problem should be seen in the context that Belišće does not offer public city passenger transport, except for a taxi service, and the choice of modalities is reduced to a private car or taxi car and active forms of mobility.

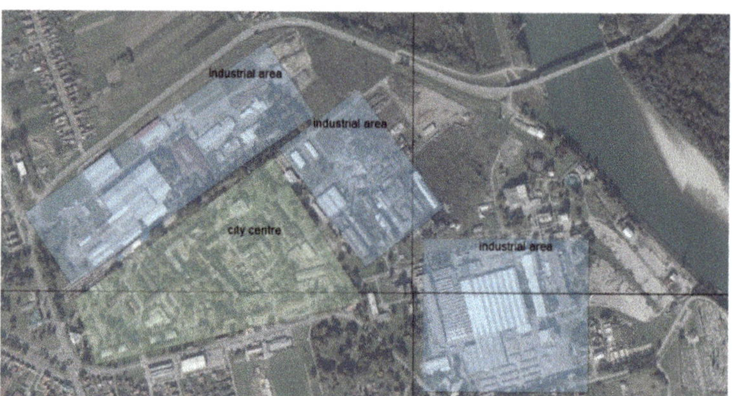

Figure 1. Area use layout—industrial zones (blue) and town center (green) [43].

The reconstruction zone is a central city zone, which, with 86 existing parking spaces in the observed coverage, cannot meet the requirements of stationary traffic. The city center is an administrative center with a number of public facilities (post office, police, health center, pharmacy, museum) that generate the need for a greater dynamic of stationary traffic (stopping and short-term parking). Within the considered segment of the traffic network, there is also a residential zone with long-term parking requirements. The vicinity of the industrial zones puts an additional pressure to the infrastructure intended for stationary traffic and a large number of workers who come to work cannot park their cars near the factories (Figure 2).

Figure 2. Stationary traffic problems in Belišće town center.

The wider analyzed coverage of the secondary network is shown in Figure 3 (left). The subject of reconstruction are the traffic areas of the secondary network, parking lots, and pedestrian paths shown in Figure 3 (right). The intersection on the access road marked in red is, according to the traffic count results, a critical point of the narrower secondary network reconstruction area, which is included in the model and analysis of functional characteristics of the alternative solutions.

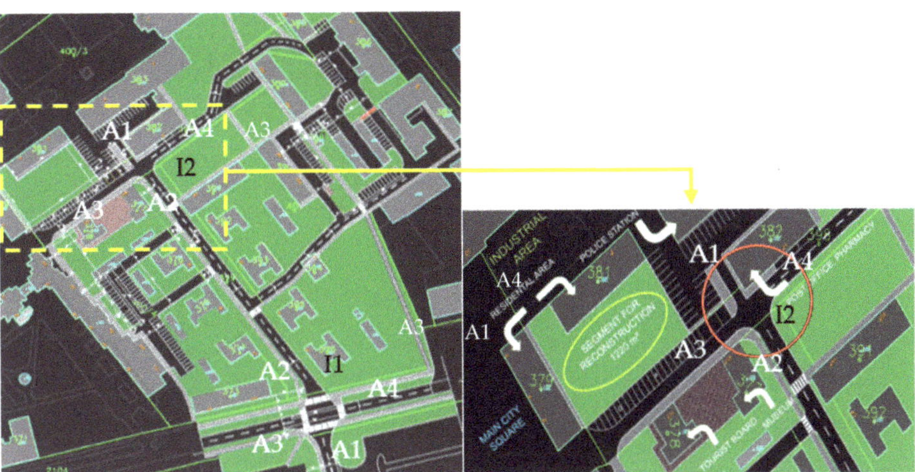

Figure 3. Observed secondary network coverage (**left**) and narrower area of reconstruction (**right**).

Traffic was recorded and counted several times in May 2019 under standard conditions—Tuesdays and Wednesdays at peak hour, from 2 to 3 pm, and the mean value of the total monitored motor traffic network coverage expressed in the equivalent units of a personal car (PCU) distributed on the traffic routes is shown in Table 1. The access points (A1–A4) for both intersections are marked in Figure 3 for the intersection located at the periphery of the reconstruction zone (I1) and for the intersection within the reconstruction zone (I2). In the mixed flow with motor vehicles, there were cyclists who were not counted in equivalent units, to highlight the frequency of this form of mobility (Table 1). The same table contains the existing traffic demand for pedestrian movements.

Table 1. Traffic count results.

		PCU/h			Cyclist/h			Ped/h	
		Straight	Right	Left	Straight	Right	Left	Route 1	Route 2
	Access 1	59	4	17	8	1	7	17	22
	Access 2	69	16	6	13	1	4	22	17
I1	Access 3	12	22	15	10	5	3	15	12
	Access 4	12	12	8	20	8	2	19	23
	Access 1	40	10	12	23	3	3	18	22
	Access 2	18	22	14	11	4	5	24	20
I2	Access 3	25	20	6	3	5	2	12	15
	Access 4	5	25	5	7	5	4	18	14

4. Materials and Methods

To select the optimal traffic solution for Belišće town center, the multi-criteria analysis (MCA) method was used in combination with the traffic micro-simulation method for preliminary screening of traffic development alternatives based on the expected traffic demand in the project period.

The MCA is applicable if a choice must be made between several solutions based on a larger number of criteria and different, both quantitative and qualitative, measures [44,45]. Although the problem of decision-making related to different infrastructure in urban areas is based largely on common principles, there are certain specifics for different types of infrastructure planning, so for a more detailed analysis, each infrastructure should be observed separately with the definition of specific criteria.

The selection of criteria in the multi-criteria analysis is a sensitive step in particular when quantitative and qualitative indicators are combined. A large number of criteria can be confusing for decision-makers and a small number of criteria does not provide an analysis of all relevant information [46]. The application of quantitative indicators contributes to the objectification of multi-criteria analysis, but the problem is that the selection of the optimal solution is done at the early project stages, when there is not enough detailed data, so the quantification of indicators is a challenging task. The criteria groups, criteria, and sub-criteria for the subject case study are explained in more detail in Section 4.1.

Significant support for decision-makers is provided by the tools of traffic modeling—traffic micro-simulations, which are used to analyze and compare alternative reconstruction solutions. Modeling results are functional traffic indicators such as travel time, delays, length of the vehicle queue, number of stops, delays related to stopping, level of service, dynamic characteristics, etc. Within the modeling results, it is possible to obtain environmental indicators (emissions of harmful gases, etc.), as well as economic indicators expressed through fuel consumption. Keeping in mind that this is a microsimulation model, said traffic indicators can be obtained for each specific entity within the simulation (personal vehicle, public transport vehicle, truck, cyclist, pedestrian) and also as mean values of indicators for individual traffic flows or in total for the observed network coverage. At each second of the simulation, the model calculates the spatial positions and dynamic characteristics of each entity and their interactions defined through the input parameters of traffic regulation, priority rules, and characteristics of the behavior of traffic participants. The modeling results are quantitative indicators, and the advantage of the model lies in its high-quality graphic interpretation, so it is possible to make a movie in 3D mode on the operationalization of the observed spatial solution, which enables a realistic insight of the spatial solution for decision-makers and the general public.

Due to the requirements of the analysis of alternative solutions (alternatives), the mean values of traffic indicators were analyzed, which provided a sufficiently realistic insight into the quality of the alternatives for the purposes of their comparison and evaluation. The calibration of the VISSIM model was not done specifically for Belišće, but the application of the VISSIM model on roundabouts in the city of Osijek, Croatia, made a difference between modeled and measured data in situ: around 10% for the uncalibrated and less than 5% for the calibrated model [47]. The town of Belišće is in the immediate vicinity of the city of Osijek, so the input parameters of the VISSIM model associated with driver behavior were shown to be applicable. The selected functional indicators for this analysis level were total delays, maximum queue length, number of stops, delays caused by stopping, and level of service (LOS). Functional indicators were analyzed for the existing traffic demand (determined by the traffic counting) and for the various scenarios of future (presumed) traffic demand, which includes an increase in the traffic load of vehicles, cyclists, and pedestrians. A larger offer of parking spaces, cycling paths, and a well-maintained pedestrian promenade are the pull-factors of the traffic load that were taken into account when evaluating the alternatives of reconstruction.

Traffic is stochastic by nature, so in order to get the modeling results to be as realistic as possible, 10 different driving scenarios of vehicle arrivals were analyzed, and the considered traffic indicators are the average values of functional indicators of each analyzed traffic scenario. The initial value of the random number generator (random seed) was a default value and the set increment was 10. The same 10 traffic scenarios were analyzed for the current state and for each alternative solution of reconstruction in order to ensure comparable simulation results.

Within this study, the microsimulation traffic models for the secondary road network for the existing situation (solution zero—A0) and for alternative reconstruction solutions (A1, A2, A3) were developed in VISSIM. The aim of the application of the microsimulation traffic models was to analyze alternative reconstruction solutions and to evaluate them based on the available simulation results according to the selected functional criteria, which were incorporated into the procedure of the multi-criteria analysis of alternatives.

In addition to functional indicators, the simulation results provided information regarding an environmental criterion—the emissions of harmful gases, and an economic criterion—user costs through fuel consumption, which was included in the evaluation through the defined criteria. The results of the microsimulation were sufficiently accurate as relative indicators for the purpose of comparing alternatives, but we could not use them as absolute values, so more sensitive models with more spatial input data [48] should be used to model air pollution.

Multi-criteria optimization methods are applied to ill-structured problems, including those related to planning and designing traffic infrastructure. Ill-structured problems are those with very complex objectives, often vaguely formulated, with many uncertainties, and the nature of the observed problem gradually changes during the process of problem solving [49]. The results of ill-structured problems are different dimensions' criteria for the evaluation of solutions and variable constraints.

The alternative solutions in the case study were analyzed using the AHP multi-criteria analysis methodology [7,44,46,49].

AHP is a priority method applicable to problems that can be represented by a hierarchical structure [7]. The top of the hierarchy is represented with the goal, one level lower are criteria, with more levels of (sub)criteria possible, while the lowest level is represented by alternatives.

The basic methodological steps applied in this research are shown in Figure 4.

The AHP method is based on estimating relative priorities (weights) of criteria and alternatives on which a pair-wise comparison matrix for criteria and pair-wise comparison matrices for alternatives (one matrix for each criterion) are generated. The pair-wise comparison of alternatives regarding their importance with respect to each criterion or pair-wise comparison of criteria respect to the goal is done using the pair-wise comparison scale shown in Table 2.

Table 2. AHP pair-wise comparison scale [7].

Intensity of Weight, Importance, Preference	Definition
1	Equal importance (no preference)
3	Moderate importance (moderate preference)
5	Strong importance (strong preference)
7	Very strong importance (very strong preference)
9	Extreme importance (extreme preference)
2, 4, 6, 8	Intermediate values

The result of the AHP method is the overall priority vector that defines the priority (weight/importance) of each alternative with respect to the goal so the ranking of alternatives can be made.

The advantage of this multi-criteria method is that it can be used when just the pair-wise comparison of alternatives according to each criterion and the pair-wise comparison of criteria towards the goal are known, but also when the alternatives are exactly valued in regard to each criterion separately—that is, if the importance of every single criterion is exactly defined, but it also gives the possibility to combine these two approaches in the same analysis. This was applied in our research.

In this case study, the application of the AHP methodology was based on a total of 5 criteria groups: functional criteria, traffic safety criteria, economic criteria, environmental criteria, and spatial–urban criteria. Each alternative was evaluated according to each criterion and sub-criterion defined within the specified criterion group. Within each of these groups, part of the (sub)criteria were defined by measurable values except in a group of spatial–urban criteria assessed based on the collected subjective evaluations of criteria given by different categories of stakeholders—local population, experts, and students—whereas economic criteria of construction and maintenance costs were analyzed by alternative pair-wise comparison.

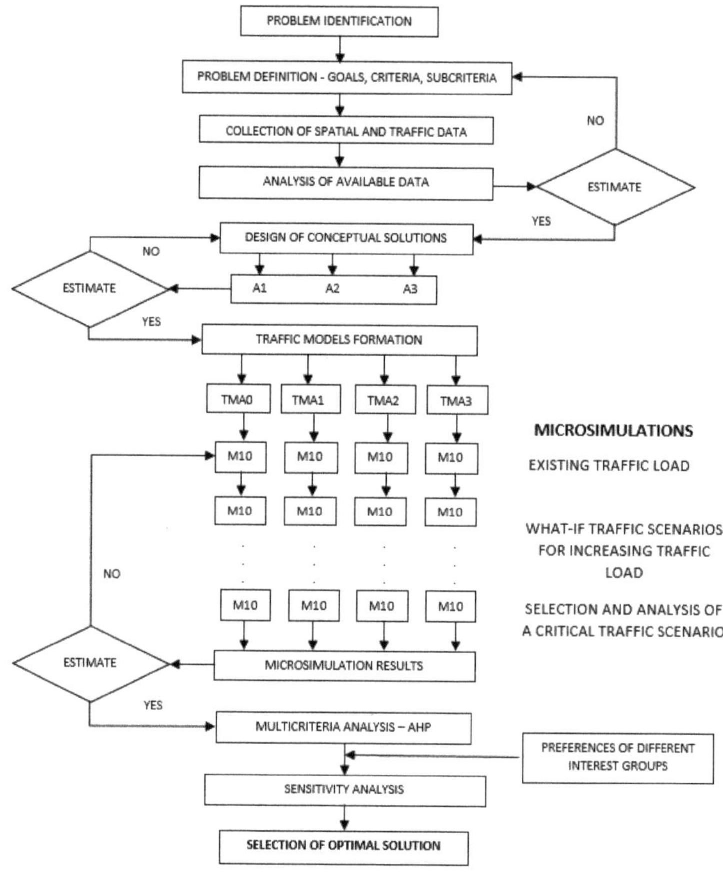

Figure 4. The flow diagram of the basic methodology steps.

4.1. Problem Definition

In the process of spatial and traffic planning, the definition of solutions starts with a clear identification and definition of the problem by defining the basic elements needed for the implementation of the procedure: expected outputs—goals, required input data, definition of expected limitations, and criteria according to which the alternative solutions will be evaluated.

The goal of the project was to meet the traffic demand of different traffic users in Belišće town center, according to the principles of sustainable urban mobility. The data collected from field research and from the existing documentation related to the coverage area served as a basis for the development of alternative solutions [50].

The selected criteria for the assessment of the alternatives using MCA are briefly explained below:

4.1.1. Functional Criteria—F

The functional criteria were used to evaluate the extent to which a particular alternative solution meets the traffic multimodal requirements of certain categories of users, as visible from the selected criteria and sub-criteria. The requirements and quality of traffic flow needed to be analyzed for the current and future traffic load, which, by analyzing different scenarios for increasing traffic load, would be evaluated as critical. Traffic requirements of cycling and pedestrian traffic as well as of stationary traffic were analyzed. The traffic conditions and the interaction of multimodal traffic in conflict zones was evaluated through five of the functional criteria for future traffic demand.

F1—functional traffic criteria of motor vehicles and integrated flows and traffic interactions for a critical traffic scenario of future traffic demand, obtained as a result of the application of traffic microsimulations in VISSIM.

F11—the maximum queue length (m) is the longest line that appears within the traffic simulation and the traffic conditions of the peak load are simulated for 3600 s, i.e., 1 h.
F12—total mean delays per vehicle (sec/veh) are time losses caused by all influential parameters, such as traffic load, traffic structure, type of conflict flows, traffic regulation, reaction time of traffic participants, dynamic conditions of each entity (driving speed, acceleration, deceleration, pedestrian speed), safety clearance, the influence of infrastructure elements, etc.
F13—the average number of stops of each vehicle (number) in the traffic flow caused by traffic conditions, traffic regulation, conflict flows, parking/unparking, etc.
F14—the average delays caused by stopping per vehicle (sec/veh) are a measure of the complexity of individual traffic situations and interactions, because there may be traffic scenarios in which there are more short stops, or traffic scenarios in which there are fewer stops, but traffic circumstances are complex and stops last longer.
F15—the level of service (LOS) demonstrated categorically from A to F is a qualitative indicator of traffic conditions and is ranked in six levels, where the conditions of traffic flow of level A are the best and consistent with the movement of vehicles in free flow, and level F practically means standing or very slow forced movement in a line of vehicles. The basis for evaluation of the level of service is the user-oriented parameter expressed through the mean delays, unlike the previously used theoretical criteria of reserve capacity.

F2—functional traffic criterion of bicycle traffic expressed through the length of bicycle paths (m).
F3—functional traffic criterion of pedestrian traffic expressed through the length of pedestrian infrastructure (m).
F4—functional traffic criterion for stationary traffic expressed through the number of parking spaces.

F2, F3, and F4 are defined in the project for alternative solutions (Section 4.2).

4.1.2. Safety Criteria—S

Direct traffic safety indicators such as the number of traffic accidents and the number of severe traffic accidents (with a dead or seriously injured person) could be analyzed only for the existing traffic solution (A0), so the safety criteria were analyzed through indirect indicators.

S1—speed (km/h) is correlated with the number of traffic accidents and is highly correlated with outcomes, i.e., the severity of traffic accidents, especially in the vehicle–pedestrian interaction. The increase in speed from 30 km/h to 50 km/h increases the likelihood of fatal and severe outcomes for pedestrians from the range of 5–22% to the range of 45–85% [51]. The mean speed is obtained by applying the traffic microsimulations in VISSIM.
S2—the degree of segregation (expressed through the number of separated traffic flows) is an indicator of how many traffic flows have separate areas for movement. The pedestrian flows are the last to be integrated into the common traffic area, and this must be hierarchi-

cally (secondary network, access street), safety-wise (vehicle speeds adjusted to pedestrian walking speed), and functionally (low traffic load) justified.

S3—the number of potential conflict points (number) of opposing vehicle–vehicle traffic flows.

S4—the number of potential conflict points (number) of opposing vehicle–pedestrian traffic flows.

S2, S3, and S4 are assessed from the project for alternative solutions (Section 4.2).

4.1.3. Economic Criteria—EC

The economic criteria are usually assessed through the cost of construction, maintenance, the value of the facility at the beginning and at the end of the planned period, the direct and indirect costs of users, etc. Due to the level of project documentation, the cost of construction and maintenance costs could not be expressed in numerical terms, but were analyzed by comparing the solution pairs as one of the options, i.e., benefits of the AHP methodology.

EC1—construction cost–pair-wise comparison.

  EC11—reconstructed area in m^2.
  EC12—use of modern technologies (camera/displays with data about the number of available parking spaces)

EC2—maintenance cost–pair-wise comparison.

EC3—fuel consumption (US gal lqd) for a critical traffic scenario of future demand, obtained as a result of micro-simulations in VISSIM.

4.1.4. Environmental Criteria—EN

Environmental criteria were assessed through the quantity of exhaust gases in grams for a critical scenario of future traffic demand, and were derived from the results of micro-simulations in VISSIM. The reported numerical indicators could not be used as an absolute, but rather as relative indicators for analysis and comparison of alternative solutions.

EN1—carbon monoxide (CO) emission in grams.

EN2—nitrogen oxide emission (NOx) in grams.

EN3—volatile organic compounds (VOC) in grams.

4.1.5. Spatial–Urban Criteria—SU

The spatial–urban criteria are qualitative indicators expressed through the selected criteria describing the attractiveness of the solution and the spatial potential. These qualitative indicators include a subjective experience, and in this study, we analyzed subjective evaluations of performance of alternative solutions according to the criteria expressed in surveys by different target groups.

SU1—walkability potential and spatial motivation for pedestrian movement.

SU2—potential and spatial motivation for cycling.

SU3—attractiveness.

SU4—potential for social interactions.

SU5—assessment of a sense of comfort.

SU6—assessment of a sense of safety for the most vulnerable traffic groups.

SU7—parking policy—adequate attitude toward a stationary traffic solution (how much space we agree to spend on parking lots).

The importance of the criteria for evaluation of alternative solutions, i.e., the weight coefficients of individual criteria, reflect the preferences of the decision-makers and have an impact on the optimization process and on the final outcome of the alternative solution selection. Within this study, three groups of stakeholders were involved in the survey. They analyzed the relevance of criteria and gave their evaluation of alternative solutions in relation to qualitative criteria (SU). The similarities and differences of the relevance of the criteria between different stakeholder groups were statistically analyzed.

4.2. Case Study Description—Alternative Reconstruction Solutions/Alternatives

As input parameters for the design of a possible manner of reconstruction of the traffic network, data on movement for all types of traffic and data on existing traffic infrastructure as well as on all land uses in the wider coverage zone (Section 3) were collected.

Three alternative solutions (A1, A2, A3) for the reconstruction of traffic areas in the center of Belišće were developed and further analyzed, and were conceptually different according to the solution of all relevant elements of traffic demand—pedestrian traffic, cycling traffic, motor vehicle traffic, and stationary traffic. All three alternatives were designed to preserve the three large chestnut trees that are a symbol of the town and are within the reconstruction coverage. The comparison of alternatives [50] is given in Table 3.

Table 3. Comparison of proposed alternatives.

Alternative A1: PEDESTRIAN STREET
Construction of a parking lot and of a pedestrian promenade and repurposing of the road into access to the parking lot using modern technological solutions—cameras in parking lots and a display with the number of free parking spaces on each access road in the wider coverage of the secondary network. Cycling traffic is in the mixed flow together with vehicles, but due to low speeds and elimination of the vehicles that are entering the parking zone inefficiently, the traffic conditions are better.

Reconstructed area: 1630 m^2
Parking places: 109 (20 new)
New pedestrian paths: 130 m
New bicycle paths: 0 m
Intersection: three-leg

Alternative A2: SHARED SPACE
Concept with full integration of traffic flows on a common surface designed to meet the needs of pedestrian and cycling movements, with fewer parking spaces than the existing solution, in order to influence the selection of active modalities of urban mobility and demotivate the choice of personal cars as the primary modality. An addition to the solution is the construction of a network of bicycle paths in the coverage area that provides greater safety to bicycle flows.

Table 3. *Cont.*

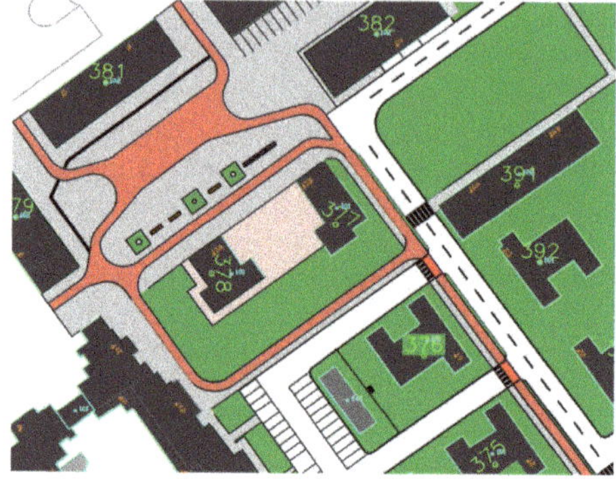

Reconstructed area: 3880 m^2
Parking places: 48 (38 fewer)
New pedestrian paths: 160 m
New bicycle paths: 535 m
Intersection: four-leg

Alternative A3: TRAFFIC-CALMING ZONE

Reconstruction of the existing collector road leading to the inner city center, in a reduced speed zone ("30 zone"), with 31 new longitudinal parking spaces. The existing traffic areas for pedestrians and cyclists, along with the areas intended for stationary traffic (82 parking spaces), remain the same.

Reconstructed area: 1150 m^2
Parking places: 116 (31 new)
New pedestrian paths: 145 m
New bicycle paths: 210 m
Intersection: four-leg

4.3. Formation of Traffic Models

For the analysis of functional indicators, microsimulation traffic models for the existing situation (A0) and for the considered alternative solutions (A1, A2, A3) were formed. Traffic load and traffic distribution data for existing traffic demand were entered into the models, after which a number of what-if scenarios of traffic load increase were analyzed in order to verify whether alternative reconstruction solutions could meet functional traffic demand requirements in the future. The input data of the model are the dynamic characteristics of the vehicles, speed, acceleration, deceleration, type, and engine power of the individual vehicles, as well as characteristics associated with driver behavior, such as reaction time, frequency, and length of scanning traffic situations, etc. The model approaches the stochastic nature of the traffic flow by using a random number generator and distributions, and the entered data, for example for the speed of individual vehicle categories, are used in the model as the median of the speed distribution. Of the 10 different traffic scenarios of the arrival of vehicles within the same traffic load, the mean values of functional and other traffic indicators were analyzed.

5. Results and Discussion

5.1. Microsimulation Results

The formed microsimulation traffic models for the existing traffic solution and all alternative reconstruction solutions enabled the analysis of different traffic loads and traffic structures. The existing traffic requirements were the first scenario analyzed, but the future traffic requirements were adopted as a relevant traffic load. The projection of the future traffic load is based on economic growth and development of industrial production, for which there is a significant market demand and which would also generate an increase in traffic demand. Scenarios for increasing traffic demand of 50, 75, and 100% for all modes of traffic and in combination were analyzed and a critical scenario was found to be a 100% increase in motor vehicle traffic, without a significant impact on functional indicators of different increases in other modes of traffic. A comparison of functional indicators for a critical traffic scenario for the future traffic load is shown in the diagram in Figure 5. The maximum queue length is expressed in meters; the total delays of vehicles are reported as the mean value in seconds per vehicle. The average number of stops is expressed as the number of stops per vehicle and the delays caused by the stops are expressed in seconds per vehicle. Level of service is a qualitative indicator of the operating conditions of the traffic that is user-oriented and related to delays.

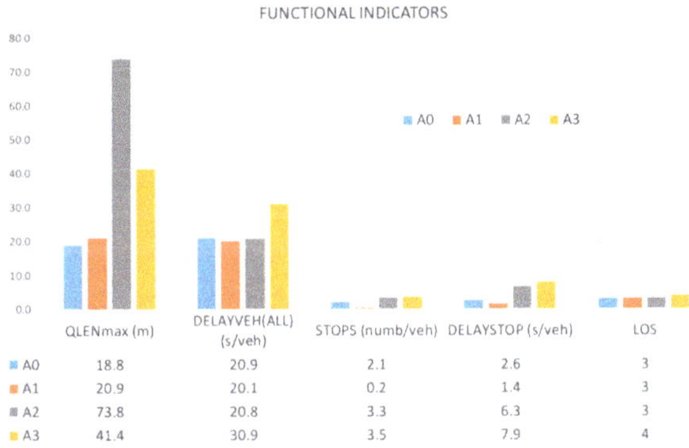

	QLENmax (m)	DELAYVEH(ALL) (s/veh)	STOPS (numb/veh)	DELAYSTOP (s/veh)	LOS
A0	18.8	20.9	2.1	2.6	3
A1	20.9	20.1	0.2	1.4	3
A2	73.8	20.8	3.3	6.3	3
A3	41.4	30.9	3.5	7.9	4

Figure 5. Comparison of functional indicators for critical traffic load. A0—existing solution, A1–A3—alternative reconstruction solutions.

The simulation results show that all alternative solutions could meet the presumed increase in traffic with a satisfactory level 3 (or C) of service, with the exception of A3, which showed a greater sensitivity to traffic increase due to frequent parking and unparking maneuvers causing higher time losses and level 4 (or D) of service. Service level D is considered an acceptable level in the conditions of the secondary urban traffic network. The alternative solution with shared space (A2) proved to be sensitive to an increase in traffic according to the criterion of the maximum queue, but this solution is designed to demotivate the use of personal vehicles as the primary form of mobility, as per the spatial concept and the number of available parking spaces. However, in order for the alternative solution to be comparable, it was analyzed for the same increase in traffic and pointless entry into the share-space zone, without the possibility of parking the vehicle.

Figure 6 shows the simulation results of carbon monoxide (CO), nitrogen oxides (NOx), and volatile organic compounds (VOC) for the existing traffic load Figure 6a and for future traffic demand in a critical scenario Figure 6b.

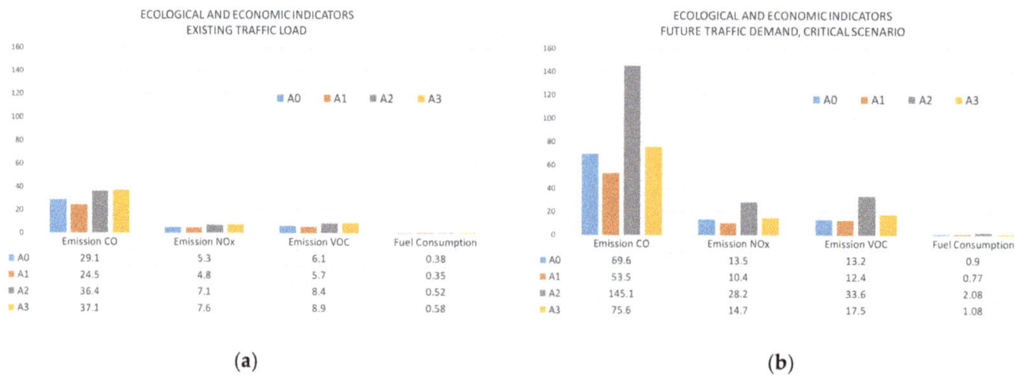

(**a**) (**b**)

Figure 6. Comparison of gas emissions and fuel consumption for the existing traffic load (**a**) and for future traffic demand (**b**).

The results clearly show the difference between the alternative reconstruction solutions in conditions of future demand, which justifies the use of microsimulation traffic modeling in the evaluation of alternative solutions.

5.2. Results of the Analysis of Qualitative Spatial–Urban Criteria

An online survey involving 120 participants formed of different groups of stakeholders was used to assess the quantitative spatial–urban criteria. They were presented with the current condition of the traffic areas of the town of Belišće in the coverage area of a secondary traffic network and three alternative reconstruction solutions. Three groups of 40 participants were surveyed, including experts, students, and citizens. The experts were mostly members of the academic community, and the students included in the survey were equally divided among students in their final years of study in civil engineering and architecture and urbanism. Citizens who participated in the survey were selected by a random sample of residents of Belišće. All respondents were over 20 years of age, and all surveyed students were between 20 and 30 years of age. The gender and age structure of the respondents is shown in Table 4.

Table 4. Gender and age structure of respondents.

Respondents	Distribution by Gender (%)		Distribution by Age (%)		
	Female	Male	<40	40–60	>60
Experts	48	52	35	57	8
Citizens	50	50	45	45	10
Students	38	62			

Seven spatial–urban criteria were selected that are applicable to a specific problem and arose from the analysis of research and urban studies [8,9,20].

Respondents subjectively evaluated the existing situation and each reconstruction solution with scores from 1 to 5 according to seven spatial–urban criteria (Table 5). Score 1 means that the solution does not meet the spatial criteria at all, and score 5 means that the solution fully meets the spatial criteria.

The subjective evaluation of all groups of respondents was that according to the spatial–urban criterion, the existing solution is the worst, and the best solution is the alternative with a pedestrian promenade and separate parking (A1) according to all criteria except the cycling potential. The shared space alternative (A2) received high scores for cycling potential, walkability, attractiveness, and social interactions, but experts and students gave it a poor score for parking policy. What is interesting is that the citizens gave this solution a good score for the parking policy and recognized the potential of this solution in promoting active forms of mobility while limiting parking in the zone. Experts were skeptical about this solution, because the successful application of such a solution is related to a shift in the paradigm of personal mobility in people's minds, and the experience shows that it is a longer process than the implementation of the solution. Ultimately, it often happens that such solutions in the local environment do not contribute to the motivation of active forms of mobility to the expected extent, but to the relocation of stationary traffic problems to another, close location. It is interesting that the solution of longitudinal parking in the collector street (A3), which offers the largest increase in the number of parking spaces, was not the best rated solution according to the criteria of parking policy in any group of respondents. The fact that alternative A3 was the worst rated solution according to the criterion of feeling of traffic safety in all three groups of respondents speaks in favor of the fact that respondents rated longitudinal parking in the road profile as a functionally and safety-wise worse solution than the others. Alternative solution A3 was also rated the worst by the criterion of social interactions, because it does not enrich the space with areas that people would use for mutual encounters, although it effectively increases the number of parking spaces in the coverage area.

Table 5. Subjective evaluation of the alternative solution according to spatial–urban criteria.

	Evaluation Criteria	A0	A1	A2	A3
Experts	Walkability	2.2	4.3	3.9	3.1
	Cycling	1.9	3.4	4.0	3.4
	Attractiveness	1.9	4.1	3.9	2.9
	Social interactions	2.1	4.2	4.1	3.1
	Pleasure	2.0	4.3	3.9	2.8
	Sense of safety	2.2	4.2	3.7	3.3
	Parking policy	2.2	4.3	2.8	3.6
	MEAN SCORE	2.1	4.1	3.8	3.2
Students	Walkability	2.4	4.6	3.9	3.3
	Cycling	2.0	3.5	4.5	3.8
	Attractiveness	2.0	4.5	4.0	3.2
	Social interactions	2.2	4.7	4.2	3.2
	Pleasure	2.2	4.6	3.8	3.3
	Sense of safety	2.1	4.5	3.4	3.2
	Parking policy	2.2	4.5	2.9	4.0
	MEAN SCORE	2.1	4.1	3.8	3.2
Citizens	Walkability	2.2	4.5	4.0	3.4
	Cycling	2.1	3.8	4.2	3.9
	Attractiveness	2.1	4.5	4.0	3.5
	Social interactions	2.2	4.6	4.0	3.7
	Pleasure	2.3	4.4	4.0	3.5
	Sense of safety	2.0	4.0	3.7	3.4
	Parking policy	1.8	4.2	3.6	3.8
	MEAN SCORE	2.1	4.3	3.9	3.6
	MEAN OVERALL SCORE	2.1	4.3	3.8	3.4

5.3. Application of the AHP Method

Table 6 shows criteria groups, criteria, and sub-criteria (explained in detail in Section 4.1) with numerical values derived from design solutions or obtained as a result of microsimulation traffic modeling, and for spatial–urban criteria they are obtained as a result of the subjective evaluation of respondents, with scores from 1 to 5.

Table 6. Criteria and sub-criteria.

Criterion	Sub-Criterion	Target	Units	A0	A1	A2	A3
F—FUNCTIONAL CRITERIA							
	F11—Queue$_{max}$	min	m	18.8	20.9	73.8	41.4
F1—Functional	F12—Delays $_{veh(all)}$	min	sec/veh	20.9	20.1	20.8	30.9
indicators/critical	F13—Stops	min	number/veh	2.1	0.2	3.3	3.5
scenario	F14—Delays$_{stops}$	min	sec/veh	2.6	1.4	6.3	7.9
	F15—Level of service	min	rating	C(3)	C(3)	C(3)	D(4)
F2—Parking—number of spaces		max	number	86	109	48	116
F3—Cyclists—length of bike paths		max	m	-	-	535	210
F4—Pedestrians-length of pedestrian paths		max	m	570	700	730	715
S—SAFETY CRITERIA							
S1—Speed		min	km/h	40	40	20	30
S2—Segregation of traffic flows		max	number	2	2	0	3
S3—Number of conflict points veh/veh		min	number	75	25	42	105
S4—Number of conflicting points pedes/veh		min	number	16	14	30	16
EC—ECONOMIC CRITERIA							
EC1—Construction	EC11—Reconstruction of the area	Pair-wise comparison					
	EC12—Advanced technology	Pair-wise comparison					
EC2—Maintenance		Pair-wise comparison					
EC3—Fuel consumption		min	US gal lqd	0.90	0.77	2.08	1.08
EN—ENVIRONMENTAL CRITERIA—EXHAUST GASES							
EN1—CO		min	grams	69.6	53.5	145.1	75.6
EN2—NOx		min	grams	13.5	10.4	28.2	14.7
EN3—VOC		min	grams	13.2	12.4	33.6	17.5
SU—SPATIAL–URBAN CRITERIA							
SU1—Walkability [1]		max	score	2.3	4.5	3.9	3.3
SU2—Cycling [2]		max	score	2.0	3.6	4.2	3.7
Su3—Attractiveness of the solution		max	score	2.0	4.4	4.0	3.2
SU4—Social interaction		max	score	2.2	4.5	4.1	3.3
SU5—Comfort score		max	score	2.2	4.4	3.9	3.2
Su6—Safety score		max	score	2.1	4.2	3.6	3.3
SU7—Parking policy [3]		max	score	2.1	4.3	3.1	3.8

A0—existing solution, A1–A3—alternative reconstruction solutions. [1] Potential and motivation for walking. [2] Potential and motivation for cycling. [3] Does not apply to the number of parking spaces, but rather to adequate/efficient planning of space intended for stationary traffic.

Economic criteria, construction, and maintenance costs due to the level of project documentation were analyzed by alternatives pair-wise comparison based on the reconstruction area and the application of advanced technological solutions according to the data from Table 6.

The pair-wise comparison of alternative solutions in regard to economic criteria, construction, and maintenance cost is shown in Table 7 using the AHP pair-wise comparison scale shown in Table 2.

Table 7. Pair-wise comparison for construction and maintenance costs.

	EC11—Construction (Area)				EC12—Construction (Technology)				EC2—Maintenance			
	A0	A1	A2	A3	A0	A1	A2	A3	A0	A1	A2	A3
A0		5	9	3		9	1	1		3	6	2
A1			3	−2			−9	−9			3	−2
A2				−4				1				−4
A3	In=	0.01			In=	0.0			In=	0.01		

The consistency of pair-wise comparison matrices of alternatives, criteria, and also of overall priority matrix should be analyzed by calculating the inconsistency index. The inconsistency index should be lower than 0.1, in which case the evaluations are consistent [7]. This was the case of pair-wise comparison of alternatives for EC11, EC12, and EC2 (Table 7).

The hierarchy of criteria groups, criteria, and sub-criteria is shown in Figure 7.

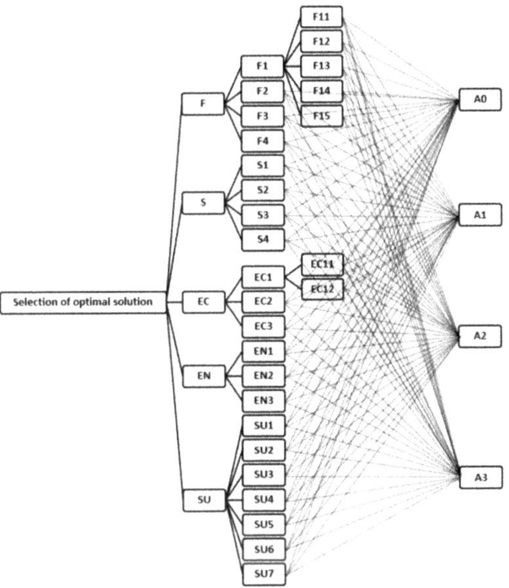

Figure 7. The AHP hierarchy of the problem (top left is the goal; on the bottom right are the alternatives).

The Analysis of Preferences

The evaluation of the importance of five selected criteria groups applied in the AHP method was included in a survey in which three groups of respondents participated (experts, students, citizens), and the total database contains the results of the evaluation of 120 respondents. Table 8 shows the criteria scores as evaluated by individual groups of respondents, but also the scores obtained on the entire database. Criteria scores ranked from 1 to 10, with score 1 meaning the criteria are not at all relevant, score 5 meaning the criteria are neither relevant nor irrelevant, and score 10 meaning the criteria are very relevant.

Table 8. Mean values of marks and rank of criteria (criteria groups).

	N	Functional Criteria	Safety Criteria	Economic Criteria	Ecological Criteria	Spatial Urban
Experts	40	9.55	9.53	7.20	8.13	8.08
Students	40	9.63	9.58	7.68	8.55	8.43
Citizens	40	8.65	9.33	6.80	8.45	9.05
Total	120	9.28	9.48	7.23	8.38	8.58
Rank		2	1	5	4	3

Table 8 shows that none of the criteria were evaluated as irrelevant, and all mean scores of criteria groups had values greater than 7. By analyzing the mean values of each criteria group, the highest mean score is assigned to the safety criteria in total and by the citizens. Experts and students gave a slightly higher score to the functional criteria, but the citizens put the functional criteria in third place. Economic criteria were assigned with the lowest scores by all groups, and the lowest mean score was assigned to them by the citizens.

The environmental criteria were evaluated by experts and students as third by relevance of the criteria, but citizens thought they were in fourth place. Spatial–urban criteria were the second most important to citizens, after the safety criteria, and other groups rated these criteria as the fourth most relevant.

The basic statistical indicators of the database for each group of respondents are shown in Table 9.

Table 9. Basic statistical indicators.

Criteria Group	Groups of Respondents	N	Mean	StDev	Median	Min	Max
Functional criteria	Experts	40	9.55	0.71	10	8	10
	Students	40	9.63	0.67	10	7	10
	Citizens	40	8.65	1.25	9	6	10
Safety criteria	Experts	40	9.53	0.70	10	7	10
	Students	40	9.58	0.93	10	6	10
	Citizens	40	9.33	1.05	10	6	10
Economic criteria	Experts	40	7.20	1.51	7	4	10
	Students	40	7.68	0.89	8	5	9
	Citizens	40	6.80	1.86	7	1	10
Ecological criteria	Experts	40	8.13	1.73	8.5	3	10
	Students	40	8.55	1.20	9	5	10
	Citizens	40	8.45	1.72	9	1	10
Spatial urban criteria	Experts	40	8.25	1.55	8	4	10
	Students	40	8.43	1.24	8.5	5	10
	Citizens	40	9.05	0.876	9	7	10

According to the results from Table 9, the evaluations of economic and environmental criteria had the largest standard deviations and the largest range.

According to the Anderson–Darling test for all groups of respondents and all evaluation criteria, the data did not follow a normal distribution. Non-parametric Bonett and Levene tests were used to evaluate the preferences of individual groups of respondents and to analyze the relationship between variance and standard deviations [52,53]. The null hypothesis is that there is no statistically significant difference between groups of respondents $(\sigma(G1)/\sigma(G2) = 1)$ and set significance level $\alpha = 0.05$. The results are shown in Table 10.

Table 10. Comparison of preferences of individual groups of respondents—statistical analysis.

Criteria Group	Test	Experts/Students		Experts/Citizens		Students/Citizens	
		Statist. Test	*p*-Value	Statist. Test	*p*-Value	Statist. Test	*p*-Value
Functional criteria	Bonett	0.07	0.795	15.43	**0.00**	14.48	**0.00**
	Levene	0.24	0.625	17.58	**0.00**	23.44	**0.00**
Safety criteria	Bonett	0.64	0.425	1.63	0.202	0.14	0.706
	Levene	0.17	0.685	2.66	0.107	1.27	0.263
Economic criteria	Bonett	9.18	**0.002**	1.14	0.286	8.38	**0.004**
	Levene	5.77	**0.019**	0.93	0.338	9.94	**0.002**
Ecological criteria	Bonett	2.71	0.099	0.00	0.994	1.04	0.307
	Levene	2.75	0.101	0.20	0.656	0.99	0.322
Spatial–urban criteria	Bonett	1.32	0.250	7.13	**0.008**	3.32	0.068
	Levene	1.31	0.256	7.99	**0.006**	3.64	0.060

Statistically significant differences between the preferences of individual groups of respondents are given in Table 10 (bold). There were statistically significant differences in preferences for functional criteria group between experts and citizens and students and citizens, but there was no statistically significant difference in evaluations between experts and students, which was the expected result. The result clearly shows that experts considered functional criteria to be as equally important as safety criteria, whereas citizens assigned them a slightly lower score. The economic criteria were best evaluated by students and their score was statistically significantly different from the other groups of respondents. There was no statistically significant difference between experts and citizens in the ranking of the economic criteria group. All groups of respondents gave high scores to the safety criteria. There was no statistically significant difference in preferences between individual groups of respondents for safety and environmental criteria. For spatial–urban criteria, there was a statistically significant difference among preferences between experts and citizens, and a comparison of evaluations of other groups did not show statistical significance in the difference in scores. This result clearly shows that the spatial potential of the solution that contains social and psychological components of the evaluation was important to the citizens and that they gave it more importance than the profession that preferred functional criteria.

The inclusion of preferences in the AHP method was done through the analysis of five different scenarios with different weighting coefficients, thus leading to a sensitivity analysis. The weights were assigned only to criteria groups, not to lower levels of criteria. A description of individual scenarios is shown in Table 11.

Table 11. Assignment of different preferences—analysis of 5 scenarios.

Scenario 1	All criteria groups' weights are equal.
Scenario 2	Weights are assigned to criteria groups according to the ranking of all respondents (the entire database).
Scenario 3	Weights are assigned to criteria groups according to the experts' ranking.
Scenario 4	Weights are assigned to criteria groups according to the students' ranking.
Scenario 5	Weights are assigned to criteria groups according to the citizens' ranking.

The ranking results according to the above scenarios are shown in Table 12.

Table 12. The results of ranking according to the analyzed scenarios.

Rank	Scenario 1	Scenario 2	Scenario 3	Scenario 4	Scenario 5
1	A1 (0.325)	A1 (0.327)	A1 (0.326)	A1 (0.326)	A1 (0.330)
2	A3 (0.273)	A3 (0.273)	A3 (0.273)	A3 (0.273)	A3 (0.273)
3	A0 (0.227)	A0 (0.220)	A0 (0.221)	A0 (0.222)	A0 (0.217)
4	A2 (0.125)	A2 (0.180)	A2 (0.180)	A2 (0.179)	A2 (0.181)
Inconst:	0.03	0.06	0.06	0.06	0.06

Results of the AHP analysis for Scenarios 1 and 2 are shown in Figures 8 and 9.

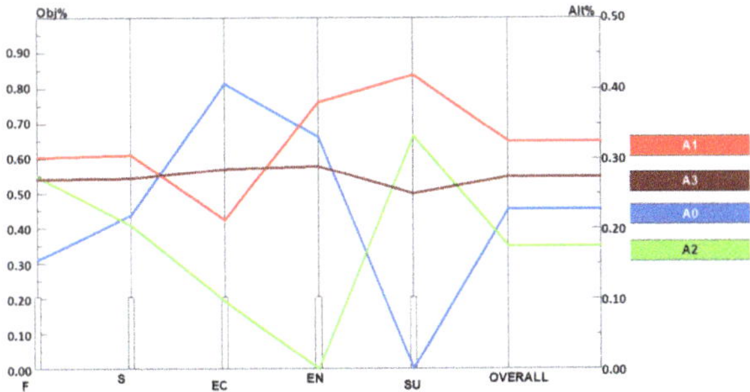

Figure 8. The AHP analysis for Scenario 1.

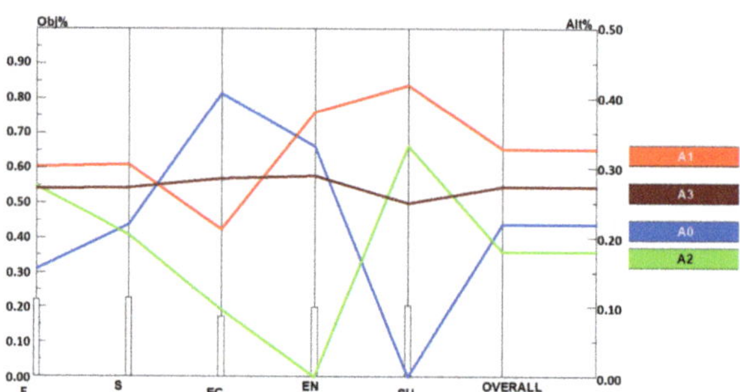

Figure 9. The AHP analysis for Scenario 2.

The results of the AHP analysis for the preferences of individual groups of respondents are shown in Figure 10a for experts, Figure 10b for students, and Figure 10c for citizens of Belišće.

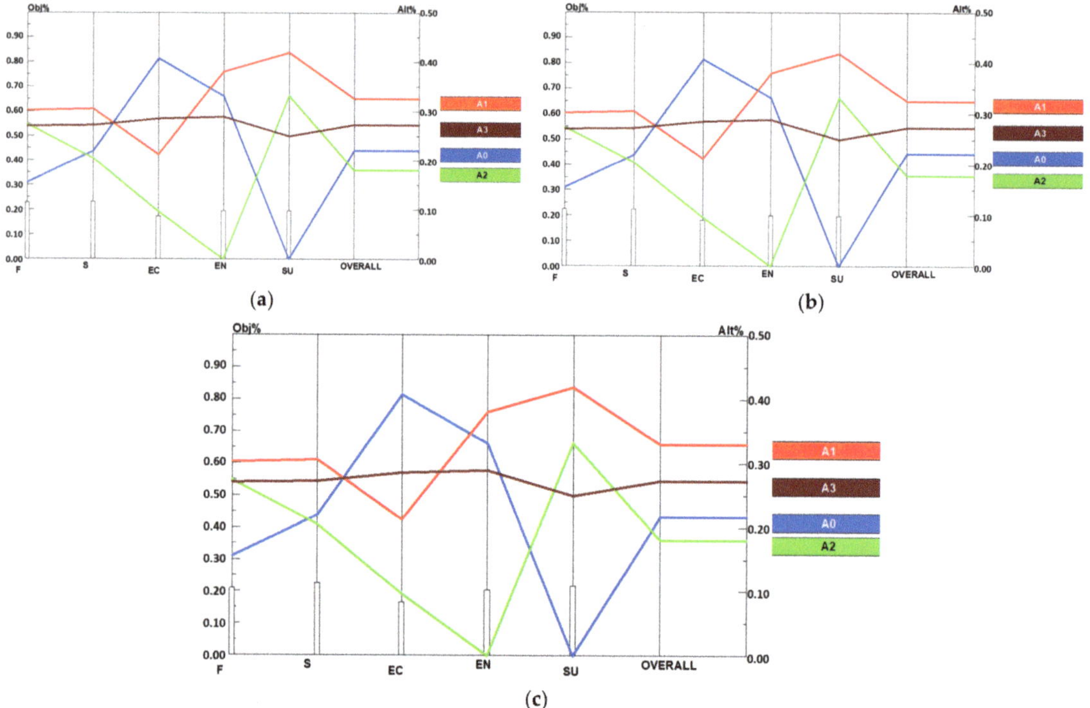

Figure 10. The AHP analysis for Scenarios 3 (**a**), Scenarios 4 (**b**) and Scenarios 5 (**c**).

Regardless of the selected scenario and preferences, A1 stands out as the best alternative solution for reconstruction, followed by A3, A0, and A2 as the worst. The overall priority vector values for alternatives and selected scenarios ranged for A1 from 0.325 to 0.330, for A0 from 0.217 to 0.227, and for A2 from 0.125 to 0.181, while for A3 the value was the same at 0.273. It can be noticed that the introduction of preferences (weights of criteria groups) in Scenario 2 resulted in minor changes in the overall priority vector values for A0 and A2, whereas for A1 this difference was extremely small. The difference between Scenarios 2, 3, and 4 was also negligible, whereas for Scenario 5 A1 had the highest overall priority vector values, and the difference between the overall priority vector values for A0 and A2 was the lowest.

In order to avoid the rank reversal phenomenon, the "ideal mode" AHP was used to rank alternatives according to all scenarios [54]. Then the ranking results were tested by decomposing the initial problem into sub-problems. First the alternatives were compared two at a time and then based on removing one alternative at a time from the whole group [55,56]. The results show that in all cases the ranking remained the same.

All groups pointed out functional criteria (experts, students) or traffic safety criteria (citizens) as the most important two groups of criteria, which, taking into account different scenarios, is best reconciled by alternative A1. Alternative A1 is based on the well-known concept of the pedestrian zone, which contributes to the quality of pedestrian traffic and space in general. At the same time, the solution for motor traffic (including stationary) does not jeopardize the standard of motor traffic because it offers even more parking spaces than those that already exist. A more innovative concept of the A2 alternative—shared space, is, perhaps as expected, a less desirable solution because it raises the issue of traffic safety in the area shared by all road users, and also significantly reduces the number of parking spaces in the zone. This is something to which the tenants living in this zone were particularly sensitive. Alternative A3—calm traffic, as estimated, does not make visible improvements to existing spatial–traffic conditions, as it continues to favor motor traffic by increasing the total number of parking spaces in the road profile. Longitudinal parking in the road profile was rated by all groups of respondents as worse in terms of the feeling of safety, and even worse than the shared space (A2) alternative solution.

6. Conclusions

Key decisions about traffic infrastructure reconstruction, such as choosing an alternative solution that has direct implications on the quality of the planned reconstruction, are made in the early design stages when the level of project detail is such that decisions are traditionally based on the evaluation and experience of the designers. In order to prevent subjectivity in selecting the optimal solution, the application of microsimulation traffic modeling is a logical choice, as it allows a numerically based analysis of traffic conditions for planned alternative reconstruction solutions for different spatial and time-related ranges. The traditional approach of selecting the best solution, along with the analysis of traffic conditions, is based on the cost estimation of solution implementation. This approach is especially present when it comes to spatial and traffic interventions in smaller urban environments. On the other hand, sustainable and socially sensitive traffic planning must take into account a wider range of criteria as well as the subjective criteria of different stakeholders involved in the planning process, and equally important of future users—citizens—so this pointed in the direction of MCA application.

In this paper, we presented the results of the application of the AHP multi-criteria analysis method, which is the most common method in traffic analysis, to the selection of the optimal solution for the reconstruction of a segment of the secondary traffic network in the center of the small town of Belišće in Croatia.

Traffic microsimulations were used to define a critical scenario for future traffic demand before applying the MCA. However, in order to choose among three alternative solutions, other than functional, we defined four more criteria groups with a greater number of criteria on lower levels. The AHP methodology use enabled us to combine different manners of assessing the criteria in the next step: numerically expressed sub-criteria obtained from microsimulation; pair-wise comparison of alternatives in relation to a specific criterion, which was used to evaluate economic criteria, since at this stage it was not possible to exactly define the costs of construction and maintenance; and assessment of preferences of the criteria of the various stakeholders involved in the process, which was used to evaluate the spatial–urban criteria.

The AHP analysis was performed according to the preferences of each of the groups involved—experts, citizens, and students. The sensitivity analysis for selecting the optimal reconstruction solution for all combinations of the weight coefficients of the criteria groups ultimately gave the same ranking order of alternative reconstruction solutions, although statistically significant differences were shown for the relevance of the evaluation criteria themselves for different groups of respondents. The ranking was also analyzed for the rank reversal phenomenon, which did not occur in this case.

Experts and students highlighted functional criteria as the most important criteria, and citizens highlighted traffic safety criteria as most important, which was expected. The next most important criteria for citizens were spatial–urban criteria, which they ranked with a very high weight, which indicates their interest in the manner in which they will use their everyday space in the future and the justification for their active involvement in this analysis with the aim of information and education.

Compared to existing studies, this research brings a methodological contribution through a holistic approach to the evaluation of alternative solutions, which in addition to quantitative introduces socially oriented qualitative parameters and involves stakeholders in the decision-making process.

The described methodology of the application of MCA in combination with traffic micro-simulations in spatial–traffic planning, although applied to problems of larger spatial coverage, proved to be justified when applied in a smaller urban settlement. Alternative A1, which was chosen as optimal, best reconciles the requirements of future traffic demand with environmental impacts and the expectations of future users related to traffic safety and the quality of space.

The methodology for selecting the optimal reconstruction solution presented in this paper is applicable to different segments of the urban traffic network of larger and smaller cities and should be validated in future research.

A potential challenge for the future is the application of the developed methodology to a wider urban area with more complex traffic and spatial situations. A potential issue is whether the large area should be analyzed as one zone or divided into smaller network segments. As for the traffic microsimulation, one larger zone would give a better understanding of the future functionality of the network and the implication of the reconstruction to the whole area. The partialization of large spatial coverage is justified in terms of traffic, due to the temporary regulation of traffic, and also in terms of construction, due to the dynamic plan of reconstruction work.

In larger urban areas, real potential users should be taken into account and a representative sample of citizens should be included in the evaluation of alternative solutions. In the continuation of this research, the influence of different weight factors on the lower-level criteria on the final result of the selection of the optimal alternative should be analyzed, and the needs of vulnerable traffic users, such as children and the elderly and people with disabilities, should be analyzed in the evaluation of alternative reconstruction solutions, too.

The selected MCA that was applied in this research was AHP, so further research should take into account other MCA methods, including the fuzzy approach, and compare their adequacy regarding the specifics of the problem analyzed.

Author Contributions: Conceptualization, I.I.O., B.K., A.D.-T., S.Š. and M.M.; methodology, I.I.O., B.K. and A.D.-T.; software, I.I.O., B.K. and M.M.; validation, B.K., A.D.-T. and S.Š.; formal analysis, I.I.O., B.K. and M.M.; investigation, I.I.O., B.K., A.D.-T., S.Š. and M.M.; data curation, I.I.O., B.K. and M.M.; writing—original draft preparation, I.I.O., B.K. and A.D.-T.; writing—review and editing, A.D.-T. and S.Š.; visualization, I.I.O., B.K. and M.M.; supervision, I.I.O., B.K. and A.D.-T. All authors have read and agreed to the published version of the manuscript.

Funding: This research was co-funded by the University of Rijeka projects: "Transport infrastructure in the function of sustainable mobility" (uniri-tehnic-18-143) and "Implementation of innovative methodologies, approaches and tools for sustainable river basin management" (uniri-tehnic-18-129).

Institutional Review Board Statement: The study was conducted according to the guidelines of the Declaration of Helsinki, and approved by the Faculty of Civil Engineering and Architecture Osijek Ethics Committee, Approval number: 003-06/21-05/04 from 5 February 2021.

Informed Consent Statement: Informed consent was obtained from all subjects involved in the study.

Data Availability Statement: Not applicable.

Conflicts of Interest: The authors declare no conflict of interest.

References

1. Yli-Pelkonen, V.; Kohl, J. The role of local ecological knowledge in sustainable urban planning: Perspectives from Finland. *Sustain. Sci. Pract. Policy* **2005**, *1*, 3–14. [CrossRef]
2. Wefering, F.; Rupprecht, S.; Bührmann, S.; Böhler-Baedeker, S.; Granberg, M.; Vilkuna, J.; Saarinen, S.; Backhaus, W.; Laubenheimer, M.; Lindenau, M.; et al. *Developing and Implementing a Sustainable Urban Mobility Plan Guidelines—Developing and Implementing a Sustainable Urban Mobility Plan*; European Commission: Brussels, Belgium, 2014.
3. Wen, L.; Kenworthy, J.; Guo, X.; Marinova, D. Solving traffic congestion through street renaissance: A perspective from dense Asian cities. *Urban Sci.* **2019**, *3*, 18. [CrossRef]
4. Deluka-Tibljaš, A.; Karleuša, B.; Dragičević, N. Review of multicriteria-analysis methods application in decision making about transport infrastructure. *Građevinar* **2013**, *65*, 619–631.
5. Giuffrè, T.; Trubia, S.; Canale, A.; Persaud, B. Using microsimulation to evaluate safety and operational implications of newer roundabout layouts for European road networks. *Sustainability* **2017**, *9*, 2084. [CrossRef]
6. Tollazzi, T.; Jovanović, G.; Renčelj, M. New type of roundabout: Dual One-lane roundabouts on two levels with right-hand turning bypasses–"Target Roundabout". *Promet Traffic Transp.* **2013**, *25*, 475–481. [CrossRef]
7. Saaty, T.L. *The Analytic Hierarchy Process*, 2nd ed.; RWS Publications: Pittsburg, PA, USA, 1996.
8. Pikora, T.; Giles-Corti, B.; Bull, F.; Jamrozik, K.; Donovan, R. Developing a framework for assessment of the environmental determinants of walking and cycling. *Soc. Sci. Med.* **2003**, *56*, 1693–1703. [CrossRef]
9. Ewing, R.; Handy, S. Measuring the unmeasurable: Urban design qualities related to walkability. *J. Urban Des.* **2009**, *14*, 65–84. [CrossRef]
10. Forsyth, A. What is a walkable place? The walkability debate in urban design. *Urban Des. Int.* **2015**, *20*, 274–292. [CrossRef]
11. Ruiz-Padillo, A.; Pasqual, F.M.; Larranaga-Uriarte, A.M.; Cybis, H.B.B. Application of multi-criteria decision analysis methods for assessing walkability: A case study in Porto Alegre, Brazil. *Transp. Res. Part D Transp. Environ.* **2018**, *63*, 855–871. [CrossRef]
12. Zeng, W.; Chen, P.; Nakamura, H.; Iryo-Asano, M. Application of social force model to pedestrian behavior analysis at signalized crosswalk. *Transp. Res. Part Emerg. Technol.* **2014**, *40*, 143–159. [CrossRef]
13. Zhang, S.; Abdel-Aty, M.; Yuan, J.; Li, P. Prediction of Pedestrian Crossing Intentions at Intersections Based on Long Short-Term Memory Recurrent Neural Network. *Transp. Res. Rec.* **2020**, *2674*, 57–65. [CrossRef]
14. Gruden, C.; Ištoka Otković, I.; Šraml, M. Neural Networks applied to microsimulation: A prediction model for pedestrian crossing time. *Sustainability* **2020**, *12*, 5355. [CrossRef]
15. Ištoka-Otković, I.; Deluka-Tibljaš, A.; Šurdonja, S.; Campisi, T. Development of models for children—Pedestrian crossing speed at signalized crosswalks. *Sustainability* **2021**, *13*, 777. [CrossRef]
16. Neider, M.B.; Gaspar, J.G.; Mccarley, J.S.; Crowell, J.A.; Kaczmarski, H.; Kramer, A.F. Walking & talking: Dual-task effects on street crossing behavior in older adults. *Psychol. Aging* **2011**, *26*, 260–268. [CrossRef]
17. Deluka-Tibljaš, A.; Ištoka Otković, I.; Campisi, T.; Šurdonja, S. Comparative analyses of parameters influencing children pedestrian behavior in conflict zones of urban intersections. *Safety* **2021**, *7*, 5. [CrossRef]
18. Dovey, K.; Pafka, E. What is walkability? The urban DMA. *Urban Stud.* **2019**, *57*, 93–108. [CrossRef]
19. Su, S.; Zhou, H.; Xu, M.; Ru, H.; Wang, W.; Weng, M. Auditing street walkability and associated social inequalities for planning implications. *J. Transp. Geogr.* **2019**, *74*, 62–76. [CrossRef]
20. Blečić, I.; Congiu, T.; Fancello, G.; Trunfio, G.A. Planning and design support tools for walkability: A guide for urban analysts. *Sustainability* **2020**, *12*, 4405. [CrossRef]
21. Zuniga-Teran, A.A.; Orr, B.J.; Gimblett, R.H.; Chalfoun, N.V.; Going, S.B.; Guertin, D.P.; Marsh, S.E. Designing healthy communities: A walkability analysis of LEED-ND. *Front. Archit. Res.* **2016**, 433–452. [CrossRef]
22. Button, K. The political economy of parking charges in "first" and "second-best" worlds. *Transp. Policy* **2006**, *13*, 470–478. [CrossRef]
23. Kimpton, A.; Pojani, D.; Ryan, C.; Ouyang, L.; Sipe, N.; Corcoran, J. Contemporary parking policy, practice, and outcomes in three large Australian cities. *Prog. Plan.* **2020**, *100506*, 100506. [CrossRef]
24. Zahabi, S.A.H.; Miranda-Moreno, L.F.; Patterson, Z.; Barla, P. Evaluating the effects of land use and strategies for parking and transit supply on mode choice of downtown commuters. *J. Transp. Land Use* **2012**, *5*, 103–119. [CrossRef]
25. Kaufmann, V. Modal practices: From the rationales behind car & public transport use to coherent transport policies. Case studies in France & Switzerland. *World Transp. Policy Pract.* **2000**, *6*, 8–17.
26. Kaufmann, V. *Re-Thinking Mobility: Contemporary Sociology*; Ashgate Pub Ltd.: Surrey, UK, 2002.
27. Kaufmann, V.; Bergman, M.M.; Joye, D. Motility: Mobility as capital. *Int. J. Urban Reg. Res.* **2004**, *28*, 745–756. [CrossRef]
28. Zimmermann, R. Prijedlog određivanja srednjih gradova u Hrvatskoj. *Društvena Istraživanja* **1999**, *1*, 21–43.
29. Kehagia, F. Transforming small towns by remedial street design. *Transp. Res. Proc.* **2017**, *24*, 507–514. [CrossRef]

30. Fang, F.C.; Elefteriadou, L. Some guidelines for selecting microsimulation models for interchange traffic operational analysis. *J. Transp. Eng.* **2005**, *131*, 535–543. [CrossRef]
31. Flötteröd, G.; Chen, Y.; Nagel, K. Behavioral calibration and analysis of a large-scale travel microsimulation. *Netw. Spat. Econ.* **2012**, *12*, 481–502. [CrossRef]
32. Tsolaki-Fiaka, S.; Bathrellos, G.D.; Skilodimou, H.D. Multi-criteria decision analysis for an abandoned quarry in the Evros Region (NE Greece). *Land* **2018**, *7*, 43. [CrossRef]
33. Shahumyan, H.; Williams, B.; Petrov, L.; Foley, W. Regional development scenario evaluation through land use modelling and opportunity mapping. *Land* **2014**, *3*, 1180. [CrossRef]
34. Sałabun, W.; Wątróbski, J.; Shekhovtsov, A. Are MCDA methods benchmarkable? A comparative study of TOPSIS, VIKOR, COPRAS, and PROMETHEE II methods. *Symmetry* **2020**, *12*, 1549. [CrossRef]
35. Rehman, A.U.; Shekhovtsov, A.; Rehman, N.; Faizi, S.; Sałabun, W. On the analytic hierarchy process structure in group decision-making using incomplete fuzzy information with applications. *Symmetry* **2021**, *13*, 609. [CrossRef]
36. Yannis, G.; Kopsacheili, A.; Dragomanovits, A.; Petraki, V. State-of-the-art review on multi-criteria decision-making in the transport sector. *J. Traffic Transp. Eng. Engl. Ed.* **2020**, *7*, 413–431. [CrossRef]
37. Broniewicz, E.; Ogrodnik, K. Multi-criteria analysis of transport infrastructure projects. *Transp. Res. Part D Transp. Environ.* **2020**, *83*, 102351. [CrossRef]
38. Deluka-Tibljaš, A.; Karleuša, B.; Benac, Č. AHP methodology application in garage-parking facility location selection. *Promet Traffic Transp.* **2011**, *23*, 303–313. [CrossRef]
39. Alemdar, K.D.; Tortum, A.; Kaya, Ö.; Atalay, A. Interdisciplinary evaluation of intersection performances—A microsimulation-based MCDA. *Sustainability* **2021**, *13*, 1859. [CrossRef]
40. Bayrak, O.Ü.; Bayata, H.F. Multi-criteria decision-based safety evaluation using microsimulation. *Proc. Inst. Civ. Eng. Transp.* **2020**, *173*, 345–357. [CrossRef]
41. Du, Y.; Zhao, C.; Zhang, X.; Sun, L. Microscopic simulation evaluation method on access traffic operation. *Simul. Model. Pract. Theory* **2015**, *53*, 139–148. [CrossRef]
42. Alemdar, K.D.; Kaya, Ö.; Çodur, M.Y. A GIS and microsimulation-based MCDA approach for evaluation of pedestrian crossings. *Accid. Anal. Prev.* **2020**, *148*, 105771. [CrossRef]
43. Geoportal of the National Spatial Data Infrastructure (City of Belišće). Available online: https://geoportal.dgu.hr/ (accessed on 20 April 2021).
44. Opricović, S. *Višekriterijumska Optimizacija*; Naučna knjiga: Beograd, Serbia, 1986.
45. Triantaphyllou, E. *Multi-Criteria Decision Making Methods: A Comparative Study*; Springer: Boston, MA, USA, 2000. [CrossRef]
46. Morales, F., Jr.; de Vries, W.T. Establishment of land use suitability mapping criteria using analytic hierarchy process (AHP) with practitioners and beneficiaries. *Land* **2021**, *10*, 235. [CrossRef]
47. Ištoka-Otković, I.; Tollazzi, T.; Šraml, M. Calibration of microsimulation traffic model using neural network approach. *Expert. Syst. Appl.* **2013**, *40*, 5965–5974. [CrossRef]
48. Abou-Senna, H.; Radwan, E.; Westerlund, K.; Cooper, C.D. Using a traffic simulation model (VISSIM) with an emissions model (MOVES) to predict emissions from vehicles on a limited-access highway. *J. Air Waste Manag. Assoc.* **2013**, *63*, 819–831. [CrossRef] [PubMed]
49. Đorđević, B. *Vodoprivredni Sistemi*; Naučna knjiga: Beograd, Serbia, 1990.
50. Marušić, M. Assessment of Urban Transport Network Segment Reconstruction. Master's Thesis, Civil Engineering, Faculty of Civil Engineering and Architecture Osijek, Josip Juraj Strossmayer University of Osijek, Osijek, Croatia, 2019.
51. Han, Y.; Yang, J.; Mizuno, K.; Matsui, Y. Effects of vehicle impact velocity, vehicle front-end shapes on pedestrian injury risk. *Traffic Inj. Prev.* **2012**, *13*, 507–518. [CrossRef] [PubMed]
52. Montgomery, D.C. *Applied Statistics and Probability for Engineers*; John Wiley & Sons, Inc.: Hoboken, NJ, USA, 2003.
53. Bonett, D.G.; Seier, E. Confidence interval for a coefficient of dispersion in nonnormal distributions. *Biom. J.* **2006**, *48*, 144–148. [CrossRef]
54. Saaty, T.L.; Vargas, L.G. *Models, Methods, Concepts and Applications of the Analytic Hierarchy Process*; Kluwer Academic Publishers: Boston, MA, USA, 2001.
55. Triantaphyllou, E. Two new cases of rank reversals when the AHP and some of its additive variants are used that do not occur with the multiplicative AHP. *J. Multi-Criteria Decis. Anal.* **2001**, *10*, 11–25. [CrossRef]
56. Wang, X.; Triantaphyllou, E. Ranking irregularities when evaluating alternatives by using some ELECTRE methods. *Omega* **2008**, *36*, 45–63. [CrossRef]

 *land*

Article

Transport Accessibility in a Suburban Zone and Its Influence on the Local Real Estate Market: A Case Study of the Olsztyn Functional Urban Area (Poland)

Agnieszka Szczepańska 

Department of Socio-Economic Geography, Institute of Spatial Management and Geography, Faculty of Geoengineering, University of Warmia and Mazury in Olsztyn, Prawocheńskiego 15, 10-724 Olsztyn, Poland; aszczep@uwm.edu.pl

Abstract: The development of real estate markets in the vicinity of cities is linked with suburbanization processes. The migration of the population to suburban areas contributes to the growth of the residential property market (houses, apartments and construction plots). To minimize commuting costs, property buyers opt for locations that are situated close to the urban core. This article analyzes construction plots on the local real estate market in the Olsztyn Functional Urban Area, in terms of their temporal accessibility and demographic changes. Spatial variations in population distribution were analyzed with the use of the Gini index and geostatistical interpolation techniques. Spearman's rank correlation coefficient was calculated to determine the relationships between the analyzed variables. The study revealed differences in the spatial distribution of the population and real estate transactions as well as strong correlations between average transaction price, number of transactions, commuting time and population. The highest number of transactions were observed in cadastral districts situated in the direct vicinity of Olsztyn's administrative boundaries and the major transportation routes due to their high temporal accessibility.

Keywords: transport accessibility; real estate market; population; concentration

Citation: Szczepańska, A. Transport Accessibility in a Suburban Zone and Its Influence on the Local Real Estate Market: A Case Study of the Olsztyn Functional Urban Area (Poland). *Land* **2021**, *10*, 465. https://doi.org/10.3390/land10050465

Academic Editor: Luca Salvati

Received: 11 April 2021
Accepted: 23 April 2021
Published: 29 April 2021

Publisher's Note: MDPI stays neutral with regard to jurisdictional claims in published maps and institutional affiliations.

1. Introduction

Suburbanization leads to the sprawl of residential areas in the vicinity of an urban core. This process affects the morphology of suburban areas, their functions and the development of suburban infrastructure. The migration of the population to suburban areas contributes to the formation of settlement subsystems around the city center. This process stimulates the growth of local real estate markets, in particular, the number of transactions involving construction plots. City residents migrating to suburbs base their purchasing decisions on a property's location relative to the urban core. Real estate in the city is swapped for much cheaper and more attractively located properties in a suburban zone. Lower prices and a cleaner environment are the main reasons why city dwellers seek property in suburban areas.

Commuting time strongly influences the decisions made by the buyers of suburban property [1,2]. In suburban areas, the share of space intended for residential, commercial and industrial purposes are increasing [3]. The urbanization of rural areas surrounding the city is closely linked with distance to the functional urban core [3] and the availability of transport infrastructure [4–7]. Feedback mechanisms between accessibility, land-use and travel behavior were observed [8]. Buyers search for property in areas with well-developed road networks. Dwellers who migrate to suburbs continue to maintain close links with the city, and commuting entails high costs [9,10]. The cost and time of travel to work [11,12], school or commercial outlets is lower in suburban areas that are easily accessed from the urban core [13,14]. Daily commuting is the consequence of population shift from cities to suburban areas, which increases the demand for locations [15,16] with well-developed transport infrastructure [17–19]. Property buyers hope to maximize benefits and minimize costs by searching for a compromise between the price and quality of property and its distance from the urban core. They strive to minimize both the physical and temporal distance between the place of residence and the city. The above increases the demand for locations characterized by shorter commuting time and lower transportation costs [20–22].

Commuting increases road traffic because passenger vehicles are the main mode of transport for suburban residents [23] since public transport options are limited and do not always cover local transport needs. Daily commuting costs are a significant item in household budgets, but they are compensated by lower property prices in suburban areas. It has been said that the expenditures associated with daily travel are a different form of mortgage carrying costs. Time is also a factor that has specific economic value because it is a limited resource; therefore, commuting time, rather than distance, is the key determinant of transport accessibility [24]. Research studies conducted at various levels have also demonstrated that transport accessibility can influence property prices [25–27]. The balance between transport accessibility and the remaining attributes of real estate in rural areas (specific locations) have also been examined in suburban areas in Poland [28,29]. Therefore, the intensity and the spatial extent of suburbanization are inversely proportional to the demand for real estate, and the demand for and supply of construction plots influences real estate prices and the development of real estate markets in the suburban zone [30]. The demand for construction plots shifts the boundaries of the urban real estate market outwards [31–33]. The intensity and trends of suburbanization are reflected on the suburban real estate market and the accompanying changes in land use, population structure and investments. These processes are mutually interdependent, and they contribute to the formation of the urbanization cycle (Figure 1).

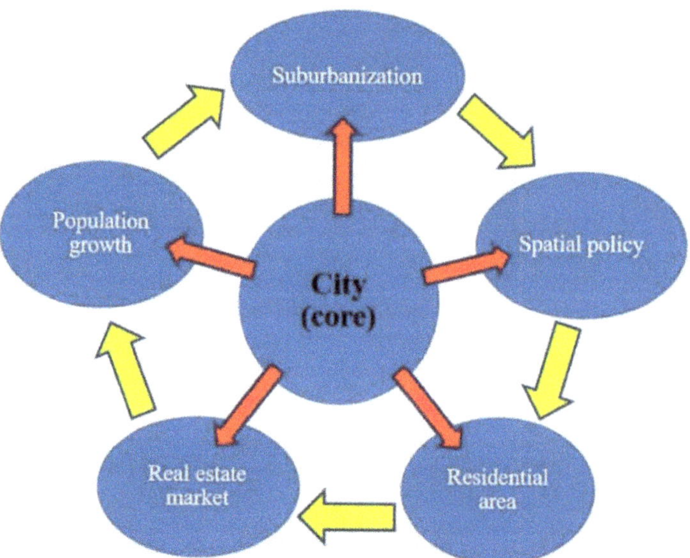

Figure 1. The urbanization cycle. Source: [34].

The suburban real estate market is characterized by a high and growing number of transactions involving vacant land. The supply of and the demand for construction plots continues to increase, which increases property prices. The prices of residential property are influenced by numerous factors, but the prices of suburban property are determined mainly by the size of the town or village and its distance from the urban core. Land prices generally decrease with an increase in distance from the city [35–37], and the development of transport infrastructure drives urbanization and home prices [36,38–40]. This observation is consistent with the new urban economics concept, which postulates that the prices of residential property are influenced by distance from the city core (land prices decrease as the distance from the city increases).

Urban expansion is thus directly linked to the growth of the real estate market, and real estate prices are determined by numerous factors. Deng et al. [41] proposed the following formula to illustrate differences in urban expansion as a function of economic and geophysical factors:

$$\text{Urban land area} = f\ (\text{income growth, demographic shifts, agricultural land value, transportation costs, changes in the structure of the economy, other location and geophysical variables}) \tag{1}$$

These factors also influence the real estate market [42,43]. Transportation costs are one of such factors, and the changes in and the reach of functional urban areas are linked with the real estate market [41,44]. Therefore, temporal accessibility is one of the key factors that influence the choice of location and, consequently, the real estate market. Transportation costs, expressed both in time and money, significantly influence property value, and buyers strive to maximize benefits in the process of selecting the optimal location.

Similar processes and interactions are observed in the former Soviet bloc countries of Central-Eastern Europe. In the socialist era, cities were relatively homogeneous functional urban areas with clearly delimited centers, whereas suburban zones were not well developed. The political transformations of 1989 led to changes inside urban areas and their surroundings. Central planning was abandoned, and free market principles were introduced. These changes affected the housing market and introduced a new urban order, which gave rise to capitalist-type cities and urban sprawl. Growing levels of social and economic inequality induced new types of migration, including urban to suburban migration. Rural areas surrounding cities developed rapidly, and suburbanization became one of the most spectacular phenomena in the former Soviet bloc countries [45,46]: Estonia [47], Slovakia [48], Bulgaria [49], Romania [50] and the Czech Republic [51].

The introduction of ground rents and property rights as a result of political and economic transformations also triggered suburbanization processes in Poland [52–55], and real estate purchases in suburban areas increased due to growing affluence. The growth of real estate markets in suburban areas accelerated after the political transformations of 1989 [56,57], especially around big cities [58–60]. Since 1990, spatial planning policies and progressing suburbanization have led to significant changes in land-use structure (increase in residential area), infrastructure development, population and the number of transactions involving residential property. The type and scope of these changes have varied intensity and territorial scope, and they are largely determined by an urban area's status in the hierarchy of cities. The migration of the urban population to suburbia proceeded most rapidly in the largest cities, followed by medium-sized towns.

Suburbanization is a relatively new phenomenon in Poland, and it largely follows the trends observed in other countries. However, this process has certain unique characteristics in Poland [53,61]:

- it proceeds far more rapidly than in Western Europe,
- it is not effectively controlled by planning authorities,
- technical and social infrastructure in suburban areas are extremely inadequate,
- rural municipalities in suburban areas are formally independent of cities and are not a part of their administrative structure,
- the distance from the urban core plays a key role in property purchase decisions, and buyers tend to opt for larger construction plots,
- new homes are built along the existing transport routes,
- the original structure of suburban villages is completely transformed by new development, mostly single-family homes with typical urban design,
- large-scale residential developers have a limited interest in suburban areas, and
- suburban development is homogeneous and linked with the family life cycle.

In Poland, most suburbanization processes are uncontrolled due to the passive stance of urban planning authorities and the relaxation of planning regulations. The above leads to spatial chaos, low economic effectiveness of new settlements and a shortage of vital services. The low spatial accessibility of suburban settlements increases development costs and public expenditures, including transport costs and the time needed to introduce complementary services that are essential for the functioning of territorial and social systems [62–64]. Transport accessibility to the urban core is determined by the daily demand for commuter services, and accessibility tends to decrease over distance from the city [65–67]. Transport accessibility significantly impacts purchase decisions and consumer behavior on property markets. The choices made by real estate buyers can be analyzed based on the information about the location and prices of traded property.

Olsztyn, the city in north-eastern Poland, has a rapidly developing suburban zone. This medium-sized city is the capital of the Warmińsko-Mazurskie Voivodeship, and it occupies a high rank in the hierarchy of Polish cities. The following research hypotheses were tested in the study:

- Suburbanization processes affect the local real estate market. Analyses of local property markets can generate valuable information about the territorial reach and dynamics of suburbanization.
- Distinctive features of Polish suburbanization processes are observed in the vicinity of Olsztyn. Distance from the urban core and commuting time play a key role in property purchase decisions; new homes are built along transport routes, which influences the prices of construction plots; single-family homes are the predominant type of new construction in suburban areas, which affects the prices and availability of land sold for residential development.

The aim of the study was to determine whether suburbanization processes in the vicinity of Olsztyn conform to trends that are characteristic of Poland, and to define the territorial reach and dynamics of suburbanization based on local market data. The areas most affected by suburbanization were identified, and the demand for and supply of construction plots were analyzed. In this article, the number of transactions involving construction plots on the suburban real estate market was analyzed based on commuting time and distance from the city core. The observed processes were linked with demographic changes, which are the key determinant of suburbanization. The links between temporal accessibility expressed by commuting time, the prices on the local real estate market and demographic processes were identified, and their spatial distribution was described based on data for 2007–2018.

2. Materials and Methods

2.1. Study Area

The studied area was the Olsztyn Functional Urban Area (FUA) composed of six municipalities that are directly adjacent to the city: Barczewo, Dywity, Gietrzwałd, Jonkowo, Purda and Stawiguda (Figure 2). Olsztyn is the capital city of the Warmińsko-Mazurskie Voivodeship in north-eastern Poland and is an important regional hub. Olsztyn had a population of around 170,000 in 2018. The population of the suburban zone was estimated at 44,000.

The residents of the Olsztyn FUA were surveyed in 2014, and the results were published in a report entitled "Spatial relations in public and private transport in the Olsztyn Functional Urban Area in the context of urban mobility" [68]. The study revealed that private cars were the preferred mode of transportation for suburban residents commuting to the city. Passenger cars are widely available [69,70], and they are the main means of transport [71] between suburban zones and the urban core. The surveyed respondents relied on private means of transport to commute to work, school and commercial outlets, and the choice of private cars was dictated mainly by shorter commuting time. As a result, the number of passenger cars registered in the county of Olsztyn increased from 48,013 in 2009 to 71,549 in 2018 (data from the Central Statistical Office). The study also revealed considerable differences in access to residential areas in the Olsztyn FUA, and temporal accessibility was highest in towns and villages located along the main transportation routes. Nearly a third of the towns and villages in the Olsztyn FUA were characterized by low transport accessibility.

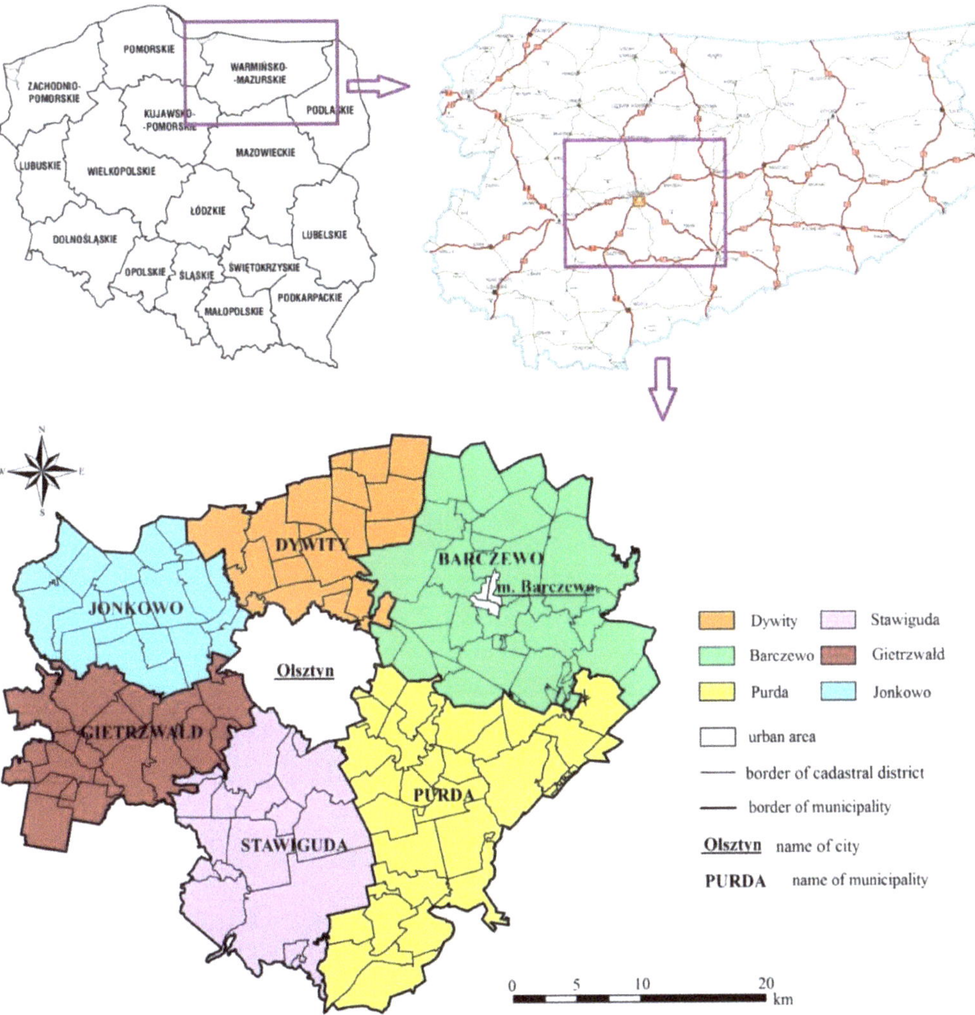

Figure 2. Study area. Source: own elaboration.

Public transport services include bus lines operated by the municipalities (public bus lines reach only 11 out of 138 towns and villages, some of them only on a seasonal basis) and, to a limited extent, by railway lines. Bus lines are also operated by commercial carriers, but they only target the most highly populated towns and villages. The vast majority of suburban settlements have to rely on private means of transport which are characterized by the shortest commuting time and the highest travel comfort [72–74].

Integrated Territorial Projects (ITP) are one of the priority goals of the Strategy of the Olsztyn FUA, and they were initiated to improve the quality of transportation in that area. This development tool was introduced to implement territorial strategies in an integrated manner. The initiative enables local authorities to launch joint projects that are co-financed by the European Regional Development Fund and the Social Fund as part of the European Union's Regional Operational Programs.

2.2. Methods

Temporal accessibility is evaluated based on physical distance as well as travel time. Travel time in the existing road network of the Olsztyn FUA was determined with the use of the Google Maps application. Travel time was measured from the administrative center of Olsztyn (City Hall) to different locations (cadastral districts) in the suburban zone. The analysis was carried out for travel by passenger car on the same date and at the same time (Wednesday, averaged result for 5 and 19 September 2018, 6 p.m.). Commuting times were interpolated for the entire studied area. The spatial distribution of the unit prices of construction plots in 2007 and 2018 were interpolated in the next stage of the analysis.

The information about the prices of construction plots (Register of Real Estate Prices and Values) and their location (base maps in the Land and Building Register, data in dxf format for cadastral plots in Barczewo, Dywity, Gietrzwałd, Jonkowo, Purda and Stawiguda) were obtained from the County Center for Geodetic and Cartographic Documentation in Olsztyn.

In both cases (temporal accessibility and prices of construction plots), data were interpolated by kriging. This method is widely applied to interpolate the spatial distribution of data, in particular, prices [75–81]. Kriging is the main method for estimating continuous surfaces based on the measured points. The estimated value is calculated as the weighted average of n measured points surrounding the estimated location. The kriging estimate (Formula (2)) is a realization of the random function $Z(si)$:

$$Z^*(S_0) = \sum_{i=1}^{n} w_i Z(s_i),\tag{2}$$

where w_i denotes kriging weights, $Z^*(S_0)$ is the value at the interpolated point and $Z(S_0)$ is the value at the measured point.

Kriging is used to define the range and anisotropy of spatial correlations and to identify directional changes in spatial phenomena. Data were interpolated in ArcGIS software.

Spearman's rank correlation coefficient was calculated (Formula (3)) to describe the relationships between the number of transactions, average unit price, population and temporal accessibility. This ranking method is not highly sensitive to outliers.

$$r_s = \frac{\frac{1}{6}\left(n^3 - n\right) - \left(\sum_{i=1}^{n} d_i^2\right) - T_z - T_y}{\sqrt{\left(\frac{1}{6}(n^3 - n) - 2T_z\right)\left(\frac{1}{6}(n^3 - n) - 2T_y\right)}},\tag{3}$$

where r_s is Spearman's rank correlation coefficient, $d_i = R_{x_i} - R_{y_i}$ is the difference between the ith rank of variable x and the ith rank of variable y, T_z and T_y are the coefficients of tied ranks expressed as $T = \frac{1}{12}\sum_j(t_j^3 - t_j)$ and t_j is the number of observations with the jth rank in the analyzed dataset.

Spearman's rank correlation coefficient was calculated, and scatter plots were generated in the Statistica program.

The Gini index, which measures concentration relative to uniform distribution in the overall set of values of the analyzed variable, was calculated to determine the inequality in the spatial distribution of the analyzed variables (Formula (4)).

$$G_i = \frac{\sum_{i=1}^{n}(2i - n - 1)y_i}{n^2\bar{y}},\tag{4}$$

where G_i is the Gini index, y_i is the value of the ith observation, $\bar{y}$ is the average value of all observations and n is the total number of observations.

The Gini index was computed with the use of an online calculator available at [82].

3. Results

The construction of the Olsztyn ring road commenced during the analyzed period. Planning and design work began in 2005. The contracts for the construction of the first and second segment of the ring road were signed in 2015 and 2016, and the first segment was commissioned for use in December 2018. The construction of the Olsztyn–Olsztynek express road intersecting Stawiguda municipality took place in the same period, and it improved transport accessibility in Stawiguda which already had the most highly developed road infrastructure in the Olsztyn suburban zone. The ring road considerably improved transport conditions, in particular, by diverting heavy transit traffic away from downtown Olsztyn. Despite the above, the ring road had only a minor impact on the temporal accessibility of suburban settlements.

The travel time for the entire studied area (between downtown Olsztyn and all settlements in the Olsztyn urban functional area) was interpolated based on the calculated commuting times (Figure 3). The location of the construction plots sold between 2007 and 2018 were marked on the temporal accessibility map.

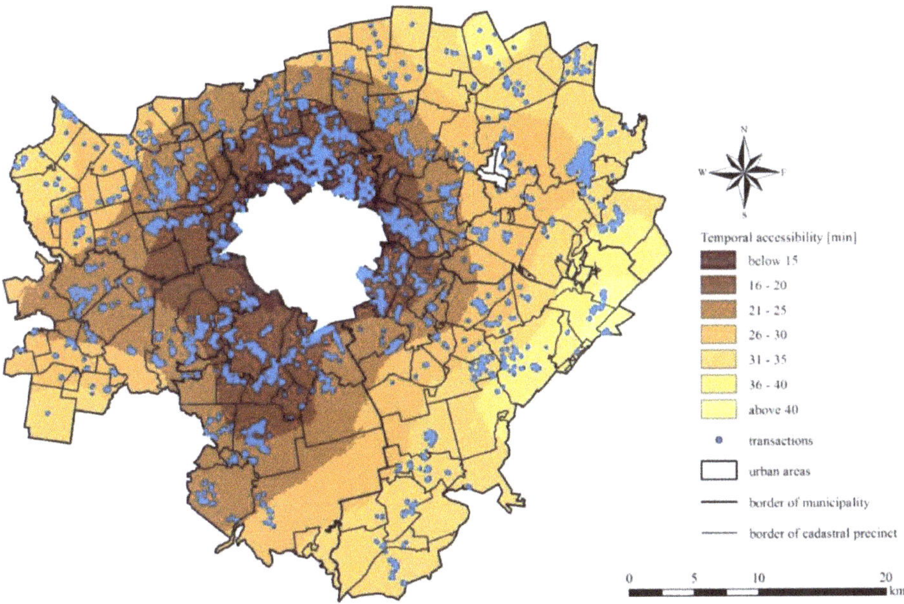

Figure 3. Temporal accessibility and spatial distribution of transactions in the suburban zone of Olsztyn in 2007–2018. Source: own elaboration.

The spatial distribution of temporal accessibility revealed local variations, mostly in the southern and south-western parts of the Olsztyn FUA. These areas are intersected by major roads, including the road linking Olsztyn with the Polish capital of Warsaw. The number of transactions change with an increase in travel time. These results indicate that the density of settlements decreased exponentially with an increase in distance from the urban core, which is consistent with urban development theories [83]. The map in Figure 2 also illustrates the extent to which the quality of transportation routes influences temporal accessibility. The number of transactions concluded in the cadastral districts of the Olsztyn FUA are presented in Figure 4.

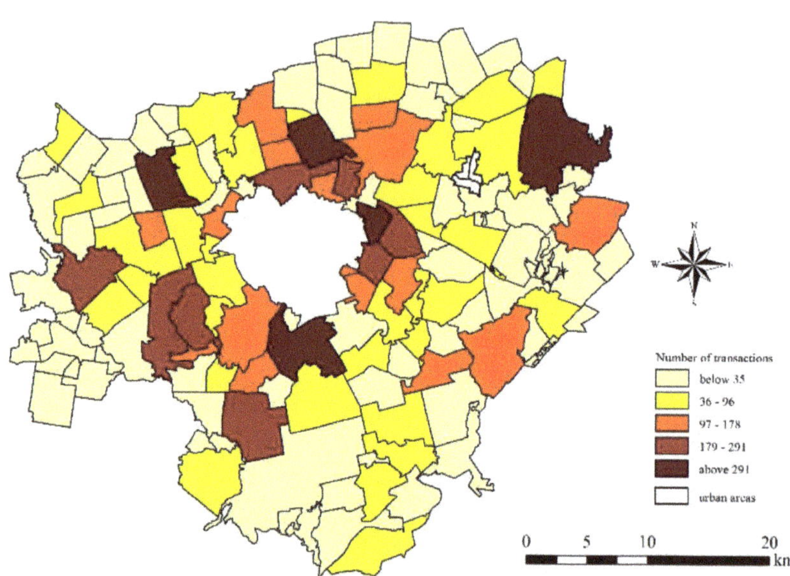

Figure 4. Number of transactions concluded in the cadastral districts of the Olsztyn FUA between 2007 and 2018.

The distance to the urban core and the number of real estate transactions are bound by an inversely proportional relationship. The number of transactions decreases with an increase in distance from Olsztyn's administrative boundaries (Figures 2 and 3). The highest number of transactions were concluded in cadastral districts situated in the direct proximity of Olsztyn and along the express road to Warsaw (south-west of Olsztyn). The outlier in the north-eastern part of the studied area is a cadastral district which remains under the influence of another city, Mrągowo.

The distribution of the unit prices of construction plots was also analyzed to determine the spatial pattern of real estate transactions on the local market. This is an important consideration because the choice of location is linked with prices, and prices influence decision-making [84]. The spatial distribution of the unit prices of construction plots was interpolated (Figure 5a,b) to identify trends in the development of the suburban zone and prime locations (based on prices).

The results of the interpolation point to an increase in the prices of construction plots between 2007 and 2018 and reveals prime locations with the highest temporal accessibility. The prices of construction plots were highest along the administrative boundaries of Olsztyn and in Stawiguda municipality, followed by the municipality of Gietrzwałd. These areas were characterized by the highest temporal accessibility.

The choices made by real estate buyers influence land use. New residential areas are created, which leads to changes in population [85]. Temporal accessibility induces demographic changes, and differences in commuting time influence population density. In the studied area, changes in population were strongly linked with access to transport infrastructure. As demonstrated in Figure 6, population decreased with an increase in commuting time. More people settle in areas with better accessibility.

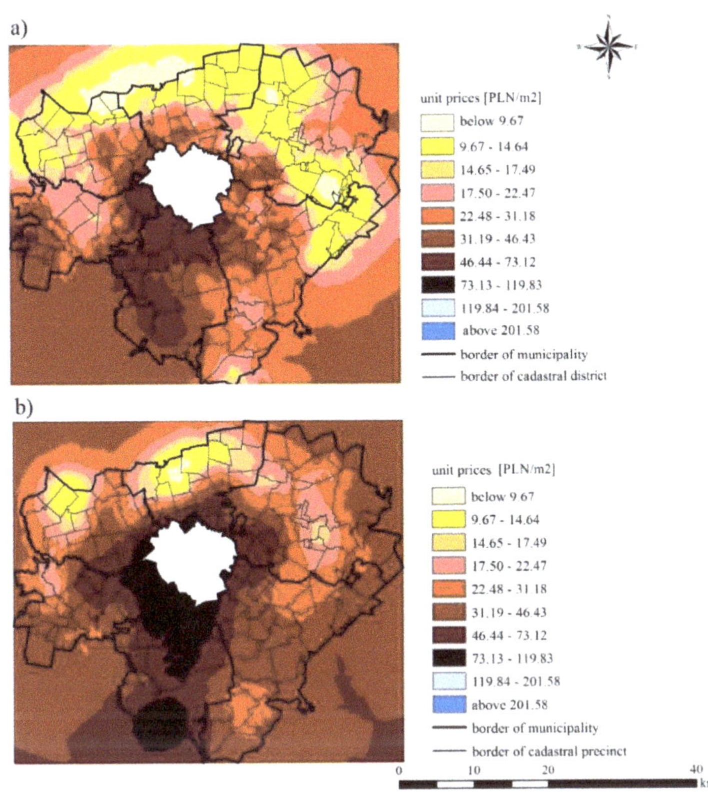

Figure 5. Interpolation of the unit prices of construction plots in the Olsztyn FUA in 2007 (**a**) and 2018 (**b**). Source: Own elaboration based on the Register of Real Estate Prices and Values kept by the County Center for Geodetic Documentation and Maps in Olsztyn.

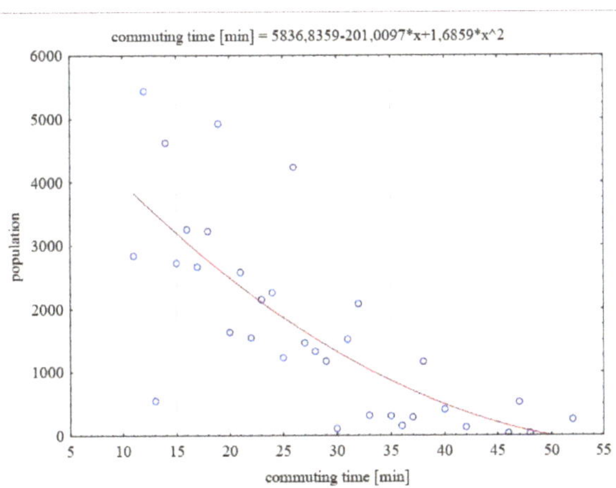

Figure 6. Relationship between population and commuting time in 2018. Source: own elaboration.

Similar population trends were noted in cadastral districts. Relative population growth was highest in the vicinity of Olsztyn and in districts with the highest temporal accessibility (Figure 7). At the same time, the population of the city of Olsztyn decreased by about 2%.

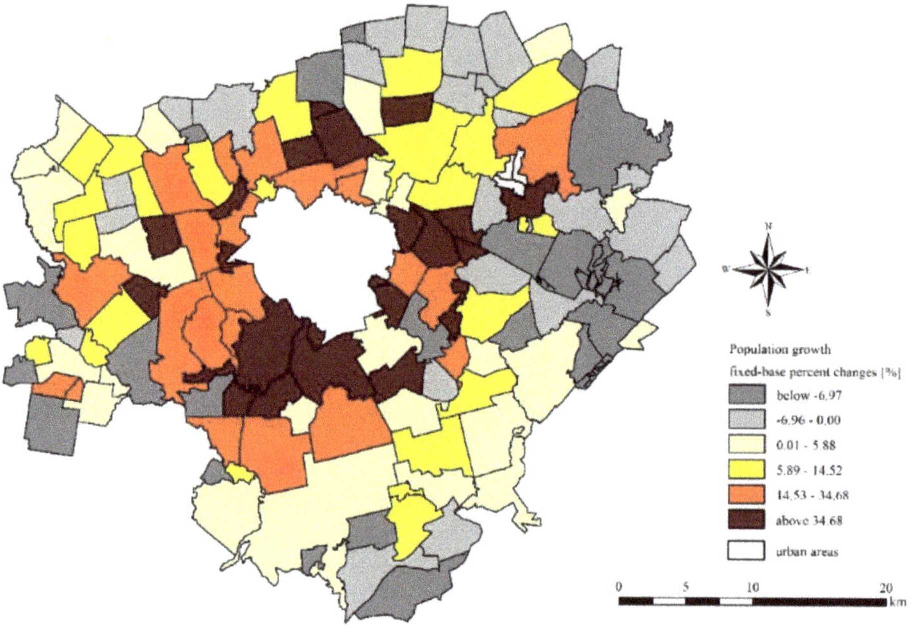

Figure 7. Relative population growth in the Olsztyn FUA in 2007–2018. Source: own elaboration.

The distribution of transactions, population and unit prices determines the homogeneity of the processes observed in the suburban zone. Therefore, the Gini index was calculated for the above variables (Table 1).

Table 1. Gini index of variables in the suburban zone of Olsztyn.

Gini Index	Population		Population Growth	Average Price		Number of Transactions	Number of Transactions	Number of Transactions	Commuting Time
	2007	2018	2017–2018	2007	2018	2007	2018	2007–2018	2018
Olsztyn FUA	0.4934	0.5372	0.8233	0.4298	0.4364	0.5853	0.6110	0.6274	0.4584

Source: own elaboration.

The calculated values of the Gini index revealed growing inequality in population distribution in the Olsztyn FUA (decrease in population uniformity). The number of transactions and the average unit prices in cadastral districts were also characterized by high values of the Gini index, which points to variations in the spatial distribution of these variables. The values of the Gini index also revealed considerable differences in the temporal accessibility of the analyzed districts, which suggests strong links between these variables.

The relationships between the analyzed variables were evaluated by calculating Spearman's rank correlation coefficients (Table 2).

Table 2. Spearman's rank correlation coefficients for the analyzed variables.

Variable	Spearman's Rank Correlation Coefficients, $p < 0.05$								
	Travel Time 2018	Population Growth 2007–2018	Population 2007	Population 2018	Number of Transactions 2007–2018	Number of Transactions 2007	Number of Transactions 2018	Average Price 2007	Average Price 2018
Travel time 2018	1.0000	−0.6683	x	−0.5230	−0.5388	x	−0.3851	x	−0.5799
Population growth 2007–2018	x	1.0000	0.3553	0.5950	0.5897	0.6117	0.2574	0.2842	0.5616
Population 2007	x	x	1.0000	0.8860	0,6468	0.5650	0.3687	0.1706	0.1411
Population 2018	x	x	x	1.0000	0.7236	0.6663	0.3961	0.2425	0.3118
Number of transactions 2007–2018	x	x	x	x	1.0000	0.2825	0.2575	0.2496	0.3750
Number of transactions 2007	x	x	x	x	x	1.0000	0.4435	0.2241	0.3157
Number of transactions 2018	x	x	x	x	x	x	1.0000	0.1412	0.1714
Average price 2007	x	x	x	x	x	x	x	1.0000	0.7015
Average price 2018	x	x	x	x	x	x	x	x	1.0000

Source: own elaboration.

The calculated values of the Spearman's rank correlation coefficient point to moderate and strong correlations between the analyzed variables (>0.5000). The strongest relationships were noted between the number of transactions and population, between travel time and population growth, and between travel time and average price. Travel time was negatively correlated with the remaining variables.

The results validate the assumption that the local real estate market was directly linked to urbanization processes expressed by demographic data. The analyses also confirmed the presence of cause-and-effect relationships between the evaluated variables.

The correlations between commuting time vs. the number of transactions and average unit prices were illustrated in scatter plots (Figure 8a,b). The scatter plots also present the influence of temporal accessibility on the local real estate market.

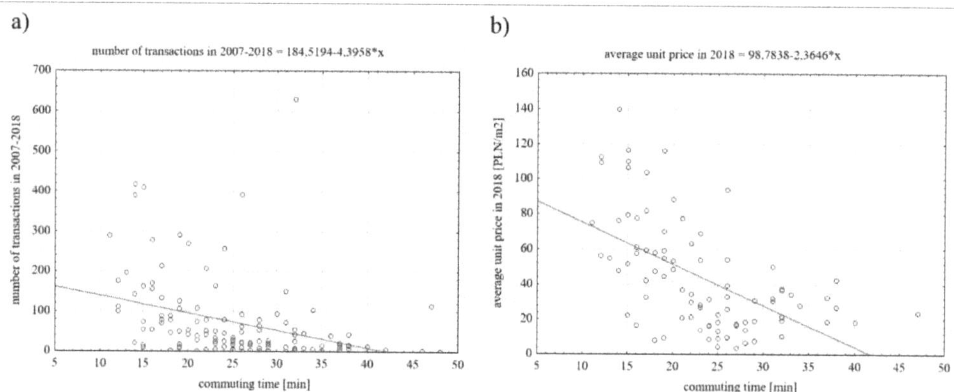

Figure 8. Correlations between (**a**) commuting time and the number of transactions, (**b**) commuting time and average unit prices. Source: own elaboration.

4. Discussion

The study demonstrated that suburbanization proceeds rapidly in the areas surrounding Olsztyn, but the spatial distribution of suburban settlements is irregular and characterized by significant local variations. The present findings validate the hypothesis that suburbanization influences the number of transactions on local real estate markets, and that transaction volumes reflect the dynamics and territorial reach of suburbanization. The results of this study also indicate that access to the transportation network significantly influences local real estate markets in suburban areas. Similar observations were made by other authors, who demonstrated that transport accessibility influences prices and transactions on the real estate market [86,87]. The correlations noted in this and other studies were similar, regardless of the size of the urban core [88,89].

Transport accessibility also influences the choice of locations on the real estate market and, consequently, population growth [90]. This feedback mechanism stimulates the growth of the real estate market. Buyers have a greater interest in residential areas with well-developed technical infrastructure and public facilities. Most areas that meet the above requirements are located in the direct vicinity of cities, and their greatest advantage is the proximity of the urban core. However, suburban areas are often deficient in social infrastructure, and many of them feature only basic retail outlets. Local residents have to commute to Olsztyn to satisfy other life needs. In contrast, rural municipalities situated further away from Olsztyn rely on local potential and attributes to develop local infrastructure. These municipalities have schools, healthcare facilities, local government offices and shops. As a result, basic life needs are met locally, and the residents do not have to travel to Olsztyn.

Herding behavior can also play an important role in real estate choices. Herding behaviors disrupt rational decision making because individuals follow other people and imitate group behaviors. Buyers who search for suburban property often follow trends without realizing that the most popular locations are characterized by high levels of urbanization and do not differ considerably from cities (in particular, locations adjacent to city limits). In an attempt to escape the hustle and bustle of urban life, many buyers ultimately pay a high price for locations that resemble the city.

The above factors lead to spatial disruptions. The analyzed variables were not evenly distributed and contributed to disproportions between the evaluated cadastral districts. A comparison of the Gini index for the first and last year of the analyzed period revealed growing differences in population distribution, the number of transactions and average prices, which contributed to disproportions between cadastral districts. Due to similar transport accessibility, cadastral districts adjacent to Olsztyn are characterized by the highest population, the highest number of property transactions and the highest prices of construction plots. However, an analysis of the areas located south-east of Olsztyn revealed considerable deficiencies in transport planning policies both at the local (municipal) and supralocal level. These observations are validated by the spatial distribution of the analyzed variables in the maps presented in Section 3. High values of the Gini index point to uneven distribution, which suggests that the Olsztyn FUA is characterized by considerable fragmentation of the analyzed phenomena. The above generates spatial chaos, and it compromises effective land management. The intensity and spatial concentration of the investigated processes were determined by distance from the city core and the proximity of the main transportation routes, which influences transport accessibility of the studied cadastral districts. Land prices, the number of real estate transactions and changes in the population growth rate decrease with an increase in distance from the urban core. Similar trends were reported by other authors [83,91].

5. Conclusions

Olsztyn is an example of a city where transport accessibility has an influence on the development of the suburban area and the local real estate market. Similar processes are observed in all large cities in Poland, particularly in areas situated in the vicinity of regional capitals and along the main transport corridors, for example, Białystok [31], Toruń-Bydgoszcz [92,93], Kraków [94] and Gdańsk-Gdynia-Sopot [95], because suburban dwellers rely mainly on private transport, whereas other forms of transport play a secondary role [63,96]. The results of this study validate the observations made by Śleszyński [97], who noted that accessibility expressed in monetary terms does not discount the improvement in temporal accessibility resulting from the construction of new roads.

This article presents the results of the first stage of research. The eastern, southern and western fragments of the Olsztyn ring road were commissioned for use in July 2019. The ring road and the accompanying access roads will considerably improve transport accessibility in the Olsztyn FUA. Faster travel and the higher throughput of traffic entering the city will increase temporal accessibility between cadastral districts in the evaluated suburban zone. The new road system will facilitate transport between the urban core and suburban districts, and it will provide suburban commuters with easier access to the city. These changes could stimulate new residential development projects in the suburban zone. Similar observations were made by other authors [98,99], including in Polish cities [100,101].

The new ring road plays a very important role in the 2016 Strategy of the Olsztyn Functional Urban Area. The new route will facilitate transit between Olsztyn and suburban districts, and the residents of the Olsztyn FUA will no longer have to cross the city center when traveling between suburban municipalities. The ring road will eliminate traffic bottlenecks, increase the throughput of the existing transportation routes and make travel between Olsztyn and suburban municipalities more comfortable. These changes could affect local real estate markets in the suburban zone. In the second stage of the research, the real estate markets in the Olsztyn FUA will be evaluated two years after the launch of the ring road.

The added value of the present research is that data were analyzed at the level of cadastral districts, whereas most studies of the type are conducted at the level of municipalities. The adopted approach supports more accurate assessments of the observed changes, which is particularly important in medium-sized (and smaller) cities and towns that exert a weaker effect on the surrounding areas. In contrast, analyses conducted at the level of municipalities produce generalized and less precise observations.

Funding: This research received no external funding.

Conflicts of Interest: The authors declare no conflict of interest. The funders had no role in the design of the study; in the collection, analyses, or interpretation of data; in the writing of the manuscript; or in the decision to publish the results.

References

1. Zhang, W.; Guhathakurta, S. Residential Location Choice in the Era of Shared Autonomous Vehicles. *J. Plan. Educ. Res.* **2018**, *41*, 0739456X18776062. [CrossRef]
2. Filion, P. Enduring Features of the North American Suburb: Built Form, Automobile Orientation, Suburban Culture and Political Mobilization. *Urban Plan.* **2018**, *3*, 4–14. [CrossRef]
3. Tokarczyk-Dorociak, K.; Kazak, J.; Szewranski, S. The Impact of a Large City on Land Use in Suburban Area: The Case of Wrocław (Poland). *J. Ecol. Eng.* **2018**, *19*, 89–98. [CrossRef]
4. Aljoufie, M.; Brussel, M.; Zuidgeest, M.; Van Maarseveen, M. Urban growth and transport infrastructure interaction in Jeddah between 1980 and 2007. *Int. J. Appl. Earth Obs. Geoinf.* **2013**, *21*, 493–505. [CrossRef]
5. Murphy, E. Urban spatial location advantage: The dual of the transportation problem and its implications for land-use and transport planning. *Transp. Res. Part A Policy Pract.* **2012**, *46*, 91–101. [CrossRef]
6. García-Palomares, J.C. Urban sprawl and travel to work: The case of the metropolitan area of Madrid. *J. Transp. Geogr.* **2010**, *18*, 197–213. [CrossRef]
7. Maat, C. *Built Environment and Car Travel: Analyses of Interdependencies*; IOS Press BV: Amsterdam, The Netherlands, 2009; Volume 29.
8. Geurs, K.T.; van Wee, B. Accessibility evaluation of land-use and transport strategies: Review and research directions. *J. Transp. Geogr.* **2004**, *12*, 127–140. [CrossRef]
9. Jones, P.; Lucas, K. The social consequences of transport decision-making: Clarifying concepts, synthesising knowledge and assessing implications. *J. Transp. Geogr.* **2012**, *21*, 4–16. [CrossRef]
10. Vega, A.; Reynolds-Feighan, A. A methodological framework for the study of residential location and travel-to-work mode choice under central and suburban employment destination patterns. *Transp. Res. Part A Policy Pract.* **2009**, *43*, 401–419. [CrossRef]
11. Zolnik, E.J. The Effect of Sprawl on Private-Vehicle Commuting Outcomes. *Environ. Plan. A Econ. Space* **2011**, *43*, 1875–1893. [CrossRef]
12. Karst, T.; Van Eck, J.R.R. Evaluation of Accessibility Impacts of Land-Use Scenarios: The Implications of Job Competition, Land-Use, and Infrastructure Developments for the Netherlands. *Environ. Plan. B Plan. Des.* **2003**, *30*, 69–87. [CrossRef]
13. Lee, Y.; Washington, S.; Frank, L.D. Examination of relationships between urban form, household activities, and time allocation in the Atlanta Metropolitan Region. *Transp. Res. Part A Policy Pract.* **2009**, *43*, 360–373. [CrossRef]
14. Crane, R. The Influence of Urban Form on Travel: An Interpretive Review. *J. Plan. Lit.* **2000**, *15*, 3–23. [CrossRef]
15. Delclòs-Alió, X.; Miralles-Guasch, C. Suburban travelers pressed for time: Exploring the temporal implications of metropolitan commuting in Barcelona. *J. Transp. Geogr.* **2017**, *65*, 165–174. [CrossRef]
16. Kortum, K.; Paleti, R.; Bhat, C.R.; Pendyala, R.M. Joint Model of Residential Relocation Choice and Underlying Causal Factors. *Transp. Res. Rec. J. Transp. Res. Board* **2012**, *2303*, 28–37. [CrossRef]
17. Dubé, J.; Thériault, M.; Rosiers, F.D. Commuter rail accessibility and house values: The case of the Montreal South Shore, Canada, 1992–2009. *Transp. Res. Part A Policy Pract.* **2013**, *54*, 49–66. [CrossRef]
18. Jaume, M.T. Sub-Centres and mobility patterns: A study on Social and Environmental Costs due to Commuting in the Barcelona Metropolitan Region. In Proceedings of the 59th Annual North American Meetings of the Regional Science Association International, Ottawa, ON, Canada, 7–10 November 2012.
19. Poelmans, L.; Van Rompaey, A. Complexity and performance of urban expansion models. *Comput. Environ. Urban Syst.* **2010**, *34*, 17–27. [CrossRef]
20. Szczepanska, A.; Senetra, A. Migrations of city dwellers to suburban areas—The example of the city of Olsztyn. *Bull. Geogr. Socio-Econ. Ser.* **2012**, *18*, 117–124. [CrossRef]
21. Nurlaela, S.; Curtis, C. Modeling Household Residential Location Choice and Travel Behavior and Its Relationship with Public Transport Accessibility. *Procedia Soc. Behav. Sci.* **2012**, *54*, 56–64. [CrossRef]
22. Molin, E.J.E.; Timmermans, H.J.P. Accessibility considerations in residential choice decisions: Accumulated evidence from the Benelux. In Proceedings of the 82nd Annual Meeting of the Transportation Research Boards, Washington, DC, USA, 12–16 January 2003.

23. Maat, K.; Timmermans, H.J. Influence of the residential and work environment on car use in dual-earner households. *Transp. Res. Part A Policy Pract.* **2009**, *43*, 654–664. [CrossRef]

24. Rodrigue, J.P. *The Geography of Transport. Systems*; Routledge: New York, NY, USA, 2020.

25. Niu, F.; Wang, F.; Chen, M. Modelling urban spatial impacts of land-use/ transport policies. *J. Geogr. Sci.* **2019**, *29*, 197–212. [CrossRef]

26. Jun, M.J.; Kim, H.J. The effects of Seoul's suburban beltway on accessibility, residential development, and housing rents: A transport–land use simulation approach. *Ann. Reg. Sci.* **2016**, *56*, 565–589. [CrossRef]

27. Zhuge, C.; Shao, C.; Gao, J.; Dong, C.; Zhang, H. Agent-based joint model of residential location choice and real estate price for land use and transport model. *Comput. Environ. Urban Syst.* **2016**, *57*, 93–105. [CrossRef]

28. Więcław-Michniewska, J.; Czado, J. Rynek nieruchomości w podkrakowskich gminach na przykładzie działek budowlanych i rolnych. In *Człowiek i Rolnictwo*; Górka, Z., Zborowski, A., Eds.; Uniwersytet Jagielloński: Kraków, Poland, 2009; pp. 239–249.

29. Hermann, B.; Świdurska, B. Analiza cen transakcyjnych nieruchomości zurbanizowanych w powiecie obornickim w kontekście przemian funkcjonalno-przestrzennych. *Biul. Stow. Rzeczozn. Mająt. Woj. Wielkop.* **2017**, *47*, 107–116.

30. Plantinga, A.J.; Lubowski, R.N.; Stavins, R.N. The effects of potential land development on agricultural land prices. *J. Urban Econ.* **2002**, *52*, 561–581. [CrossRef]

31. Harasimowicz, A. Procesy urbanizacji w strefie zewnętrznej Białegostoku. *Optimum. Econ. Stud.* **2019**, *97*, 125–150. [CrossRef]

32. Musiał-Malagó, M. Stan i kierunki rozwoju infrastruktury transportowej w Polsce. *Zesz. Nauk. Uniw. Ekon. Krak.* **2013**, *914*, 97–111.

33. Sýkora, L.; Bouzarovski, S. Multiple Transformations. *Urban Stud.* **2011**, *49*, 43–60. [CrossRef]

34. Szczepańska, A. *Modelowanie Zmian Przestrzennych Wywołanych Decyzjami Lokalizacyjnymi z Zastosowaniem Sieci Bayesa i Statystyk Morana na Przykładzie Strefy Podmiejskiej Olsztyna*; Wydawnictwo Uniwersytetu Warmińsko-Mazurskiego w Olsztynie: Olsztyn, Poland, 2017.

35. Jacob, B.; McMillen, D. Border Effects in Suburban Land Use. *Natl. Tax J.* **2015**, *68*, 855–873. [CrossRef]

36. Zheng, S.; Kahn, M.E. Land and residential property markets in a booming economy: New evidence from Beijing. *J. Urban Econ.* **2008**, *63*, 743–757. [CrossRef]

37. Martínez, F.J.; Araya, C.A. A Note on Trip Benefits in Spatial Interaction Models. *J. Reg. Sci.* **2000**, *40*, 789–796. [CrossRef]

38. Rahadi, R.A.; Wiryono, S.K.; Koesrindartoto, D.P.; Syamwil, I.B. Factors influencing the price of housing in Indonesia. *Int. J. Hous. Mark. Anal.* **2015**, *8*, 169–188. [CrossRef]

39. Holcombe, R.G.; Williams, D.W. Urban Sprawl and Transportation Externalities. *Rev. Reg. Stud.* **2010**, *40*, 257–273. [CrossRef]

40. Colwell, P.F.; Munneke, H.J. The Structure of Urban Land Prices. *J. Urban Econ.* **1997**, *41*, 321–336. [CrossRef]

41. Deng, X.; Huang, J.; Rozelle, S.; Uchida, E. Growth, population and industrialization, and urban land expansion of China. *J. Urban Econ.* **2008**, *63*, 96–115. [CrossRef]

42. Kayzer, D.; Szczepański, P.; Zbierska, A.; Zydroń, A. Czynniki wpływające na wartość nieruchomości niezabudowanych przeznaczonych na cele budowlane. *Stud. Pr. WNEiZ* **2015**, *42*, 257–267. [CrossRef]

43. Gawron, H. Wpływ cech fizycznych działek na ceny gruntów budowlanych w aglomeracji miejskiej (na przykładzie aglomeracji poznańskiej). *Stud. Mater. TNN* **2012**, *20*, 47–57.

44. McGrath, D.T. More evidence on the spatial scale of cities. *J. Urban Econ.* **2005**, *58*, 1–10. [CrossRef]

45. Kovács, Z. Cities from state-socialism to global capitalism: An introduction. *GeoJournal* **1999**, *49*, 1–6. [CrossRef]

46. Stanilov, K.; Sýkora, L. *Confronting Suburbanization*; Wiley: Hoboken, NJ, USA, 2014; pp. 1–32.

47. Tammaru, T.; Kulu, H.; Kask, I. Urbanization, Suburbanization, and Counterurbanization in Estonia. *Eurasian Geogr. Econ.* **2004**, *45*, 212–229. [CrossRef]

48. Tóth, V. Urban development of Bratislava: Suburbanization in years 1995–2009. *J. Geogr.* **2012**, *7*, 115–126.

49. Slaev, A.D.; Nedović-Budić, Z.; Krunic, N.; Petric, J.; Daskalova, D. Suburbanization and sprawl in post-socialist Belgrade and Sofia. *Eur. Plan. Stud.* **2018**, *26*, 1389–1412. [CrossRef]

50. Dumitrache, L.; Zamfir, D.; Nae, M.M.; Simion, G.; Stoica, V. The Urban Nexus: Contradictions and Dilemmas of (Post)Communist (Sub)Urbanization in Romania. *Hum. Geogr. J. Stud. Res. Hum. Geogr.* **2016**, *10*, 38–50. [CrossRef]

51. Šašek, M.; Hlaváček, P.; Holub, J. Suburbanization processes of large cities in the Czech Republic in terms of migration. *Bibl. Reg.* **2019**, *19*, 209–225. [CrossRef]

52. Gałka, J.; Warych-Juras, A. Regionalne uwarunkowania suburbanizacji w Polsce. *Stud. Miej.* **2011**, *3*, 147–157.

53. Heffner, K. Proces suburbanizacji a polityka miejska w Polsce. In *Miasto–Region–Gospodarka w Badaniach Geograficznych*; Marszał, T., Ed.; Wydawnictwo Uniwersytetu Łódzkiego: Łódź, Poland, 2016; pp. 75–110.

54. Gałka, J.; Warych-Juras, A. Suburbanization and migration in Polish metropolitan areas during political transition. *Acta Geogr. Slov.* **2018**, *58*. [CrossRef]

55. Szmytkie, R. Rozrost terytorialny dużych miast w Polsce. *Prz. Geogr.* **2020**, *92*, 499–520. [CrossRef]

56. Kałkowski, L. Polski rynek nieruchomościbilans otwarcia po wejściu Polski do Unii Europejskiej. *Probl. Rozw. Miast* **2007**, *4*, 12–45.

57. Gołąbeska, E. Współczesne trendy na rynku nieruchomości mieszkaniowych. Gospodarowanie przestrzenią w warunkach rozwoju zrównoważonego. In *Gospodarowanie Przestrzenią w Warunkach Rozwoju Zrównoważonego*; Broniewicz, E., Ed.; Oficyna Wydawnicza Politechniki Białostockiej: Białystok, Poland, 2017; pp. 85–106.

58. Tanaś, J. Differentiation of Local Housing Markets in the Poznań Suburban Area. *Real Estate Manag. Valuat.* **2013**, *21*, 88–98. [CrossRef]
59. Trojanek, M. The local property markets for single-family housing in Poznan agglomeration in the period 1995–2010. *Int. J. Acad. Res.* **2013**, *5*, 211–219. [CrossRef]
60. Luchter, B. Wpływ transformacji gospodarczej na zmiany użytkowania ziemi w strefie zewnętrznej Krakowa. *Zesz. Nauk. Akad. Ekon. Krak.* **2003**, *617*, 177–191.
61. Wagner, W. 3. The Failure of Planning in a Fragmented Property Market: Poland's Model of Suburbanization. In *Old Europe, New Suburbanization?* University of Toronto Press: Toronto, ON, Canada, 2017; pp. 41–65.
62. Maćkiewicz, B.; Andrzejewski, A.; Kacprzak, E. Koszty bezładu przestrzennego dla rynku nieruchomości. *Studia KPZK* **2018**, *182*, 280–316.
63. Śleszyński, P. Społeczno-ekonomiczne skutki chaosu przestrzennego dla osadnictwa i struktury funkcjonalnej terenów. *Studia KPZK* **2018**, *182*, 29–80.
64. Zybała, A. Polityka przestrzenna i jej rezultaty w warunkach rosnącej złożoności jej problemów. *Stud. z Polityki Publicz* **2019**, *22*, 103–122. [CrossRef]
65. Śleszyński, P. Metodyczne problemy wyznaczania obszarów urbanizacji. *Prz. Urban.* **2015**, *9*, 60–61.
66. Rosik, P.; Pomianowski, W.; Kołoś, A.; Guzik, R.; Goliszek, S.; Stępniak, M.; Komornicki, T. Dostępność gmin transportem autobusowym. *Pr. Kom. Geogr. Komun. PTG* **2018**, *21*, 54–64. [CrossRef]
67. Borowska-Stefańska, M.; Wiśniewski, S. The use of network analysis in the process of delimitation as exemplifi ed by the administrative division of Poland. *Geodesy Cartogr.* **2017**, *66*, 155–170. [CrossRef]
68. UNIREGIO. *Spatial Relations in Public and Private Transport in the Olsztyn Functional Urban Area in the Context of Urban Mobility*; UNIREGIO: Kraków, Poland, 2014.
69. Ding, C.; Wang, D.; Liu, C.; Zhang, Y.; Yang, J. Exploring the influence of built environment on travel mode choice considering the mediating effects of car ownership and travel distance. *Transp. Res. Part A Policy Pract.* **2017**, *100*, 65–80. [CrossRef]
70. Maat, K.; Timmermans, H.J. A causal model relating urban form with daily travel distance through activity/travel decisions. *Transp. Plan. Technol.* **2009**, *32*, 115–134. [CrossRef]
71. Banet, K.; Rogala, S. Znaczenie współpracy miast i gmin w kreowaniu efektywnych systemów transportowych w odpowiedzi na zjawisko urban sprawl / Importance of cooperation between cities and communes in creating effective transportation systems as an answer to negative effects of urban sprawl. *Pr. Nauk. Uniw. Ekon. Wroc.* **2016**, *443*, 23–32. [CrossRef]
72. *Mobility Plan of the Urban Functional Area until 2025*; REFUNDA Sp. z o.o.: Wrocław, Poland, 2017.
73. *Strategy for the Development of Public Transport in Olsztyn by 2027*; Public Transport Consulting: Olsztyn, Poland, 2015.
74. Guzik, R.; Kołoś, A.; Biernacki, W.; Działek, J.; Gwosdz, K.; Kocaj, A.; Panecka, M.; Wiedermann, K. *Relacje Przestrzenne Komunikacji Zbiorowej i Indywidualnej Miejskiego Obszaru Funkcjonalnego Olsztyna w kKntekście Mobilności Miejskiej. Raport Końcowy Wraz z Raportem Metodologicznym*; UNIREGIO Centrum Studiów Regionalnych: Kraków, Poalnd, 2014.
75. Cellmer, R.; Trojanek, R. Towards Increasing Residential Market Transparency: Mapping Local Housing Prices and Dynamics. *ISPRS Int. J. Geo-Inf.* **2019**, *9*, 2. [CrossRef]
76. Calka, B. Estimating Residential Property Values on the Basis of Clustering and Geostatistics. *Geoscience* **2019**, *9*, 143. [CrossRef]
77. Kuntz, M.; Helbich, M. Geostatistical mapping of real estate prices: An empirical comparison of kriging and cokriging. *Int. J. Geogr. Inf. Sci.* **2014**, *28*, 1904–1921. [CrossRef]
78. Montero, J.; Larraz, B. Interpolation Methods for Geographical Data: Housing and Commercial Establishment Markets. *J. Real Estate Res.* **2011**, *33*, 233–244. [CrossRef]
79. Li, L.; Revesz, P. Interpolation methods for spatio-temporal geographic data. *Comput. Environ. Urban Syst.* **2004**, *28*, 201–227. [CrossRef]
80. Pagourtzi, E.; Assimakopoulos, V.; Hatzichristos, T.; French, N. Real estate appraisal: A review of valuation methods. *J. Prop. Invest. Financ.* **2003**, *21*, 383–401. [CrossRef]
81. McCluskey, W.J.; Deddis, W.G.; Lamont, I.G.; Borst, R.A. The application of surface generated interpolation models for the prediction of residential property values. *J. Prop. Invest. Financ.* **2000**, *18*, 162–176. [CrossRef]
82. Wessa, P. Concentration and Inequality (v1.0.1) in Free Statistics Software (v1.2.1), Office for Research Development and Education. 2016. Available online: http://www.wessa.net/rwasp_concentration.wasp/. (accessed on 10 January 2021).
83. Cheng, J.; Masser, I. Urban growth pattern modeling: A case study of Wuhan city, PR China. *Landsc. Urban Plan.* **2003**, *62*, 199–217. [CrossRef]
84. Hurtubia, R.; Nguyen, M.H.; Glerum, A.; Bierlaire, M. Integrating psychometric indicators in latent class choice models. *Transp. Res. Part A Policy Pract.* **2014**, *64*, 135–146. [CrossRef]
85. Sener, I.N.; Pendyala, R.M.; Bhat, C.R. Accommodating spatial correlation across choice alternatives in discrete choice models: An application to modeling residential location choice behavior. *J. Transp. Geogr.* **2011**, *19*, 294–303. [CrossRef]
86. Osland, L.; Thorsen, I. Effects on Housing Prices of Urban Attraction and Labor-Market Accessibility. *Environ. Plan. A Econ. Space* **2008**, *40*, 2490–2509. [CrossRef]
87. Smersh, G.T.; Smith, M.T. Accessibility Changes and Urban House Price Appreciation: A Constrained Optimization Approach to Determining Distance Effects. *J. Hous. Econ.* **2000**, *9*, 187–196. [CrossRef]

88. Siejka, M. Rozbudowa układu komunikacyjnego i jej wpływ na poziom cen i aktywność lokalnego rynku nieruchomości, na przykładzie obiektu Świniarsko w powiecie nowosądeckim. *Infrastrukt. Ekol. Teren. Wiej.* **2012**, *2*, 87–96.
89. Adair, A.; McGreal, S.; Smyth, A.; Cooper, J.; Ryley, T. House Prices and Accessibility: The Testing of Relationships within the Belfast Urban Area. *Hous. Stud.* **2000**, *15*, 699–716. [CrossRef]
90. Kotavaara, O.; Pukkinen, M.; Antikainen, H.; Rusanen, J. Role of Accessibility and Socio-Economic Variables in Modelling Population Change at Varying Scale. *J. Geogr. Inf. Syst.* **2014**, *6*, 386–403. [CrossRef]
91. Borzacchiello, M.T.; Nijkamp, P.; Koomen, E. Accessibility and urban development: A grid-based comparative statistical analysis of Dutch cities. *Environ. Plan. B Plan. Des.* **2010**, *37*, 148–169. [CrossRef]
92. Idczak, P.; Mrozik, K. Suburbanizacja w wybranych gminach wiejskich Bydgosko-Toruńskiego Obszaru Metropolitalnego. *Stud. Pr. WNEiZ* **2015**, *42*, 183–194. [CrossRef]
93. Piątkowski, K. Suburbanizacja jako wyzwanie dla rozwoju transportu w miejskim obszarze funkcjonalnym Bydgoszczy i Torunia. *Rozw. Reg. Polit. Reg.* **2019**, *47*, 141–154. [CrossRef]
94. Kurek, S.; Gałka, J.; Wójtowicz, M. *Wpływ Suburbanizacji na Przemiany Wybranych Struktur Demograficznych i Powiązań Funkcjonalno-Przestrzennych w Krakowskim Obszarze Metropolitalnym*; Wydawnictwa Instytutu Geografii Uniwersytetu Pedagogicznego: Krakowie, Poland, 2015.
95. Masik, G. Suburbanizacja demograficzna i przestrzenna na Obszarze Metropolitalnym Gdańsk-Gdynia-Sopot: Demographic and spatial suburbanization in the Gdańsk-Gdynia-Sopot Metropolitan Area. *Stud. Obsz. Wiej.* **2018**, *50*, 155–170. [CrossRef]
96. Palak, M. *Nowe Oblicza Przedmieść. Socjologiczne Studium Suburbanizacji w Polsce na Przykładzie Rzeszowa*; Wydawnictwo Uniwersytetu Rzeszowskiego: Rzeszów, Poland, 2016.
97. Śleszyński, P. Dostępność ekonomiczna miast wojewódzkich w świetle kosztów dojazdu samochodem osobowym. *Pr. Kom. Geogr. Komun. PTG* **2017**, *20*, 7–18. [CrossRef]
98. Thomas, T.; Tutert, B. Route choice behavior in a radial structured urban network: Do people choose the orbital or the route through the city center? *J. Transp. Geogr.* **2015**, *48*, 85–95. [CrossRef]
99. Boarnet, M.G.; Chalermpong, S. New highways, house prices, and Urban development: A case study of toll roads in orange county, Ca. *Hous. Policy Debate* **2001**, *12*, 575–605. [CrossRef]
100. Kudłacz, M. Zewnątrzmiejska dostępność komunikacyjna w kontekście budowania atrakcyjności lokalizacyjnej metropolii w Polsce na przykładzie województwa małopolskiego. *Zesz. Nauk. Uniw. Èkon. Krak.* **2014**, *935*, 59–77. [CrossRef]
101. Rudnicki, A. Zrównoważona mobilność a rozwój przestrzenny miasta. *Czas. Tech. Archit.* **2010**, *107*, 57–74.

Article

The Role and Importance of a Footbridge Suspended over a Highway in the Opinion of Its Users—Trabzon (Turkey)

Maria Hełdak [1,*], Sultan Sevinc Kurt Konakoglu [2], Banu Cicek Kurdoglu [3], Hande Goksal [3], Bogdan Przybyła [4] and Jan K. Kazak [1]

[1] Institute of Spatial Management, Wrocław University of Environmental and Life Sciences, ul. Grunwaldzka 55, 50-357 Wrocław, Poland; jan.kazak@upwr.edu.pl

[2] Department of Urban Design and Landscape Architecture, Amasya University, 05100 Amasya, Turkey; sultansevinc.kurt@amasya.edu.tr

[3] Department of Landscape Architecture, Karadeniz Technical University, 61080 Trabzon, Turkey; banukurdoglu@ktu.edu.tr (B.C.K.); hande.altug@hotmail.com (H.G.)

[4] Department of Mechanics of Structers and Urban Engineering, Wrocław University of Science and Technology, 50-370 Wrocław, Poland; bogdan.przybyla@pwr.edu.pl

* Correspondence: maria.heldak@upwr.edu.pl

Abstract: In the urban landscape, footbridges appeared along with the development of urbanization through the implementation of more complex spatial structures. The introduction of transport transit to cities or sometimes urban gravity towards the important communication routes imposed their construction in order to ensure pedestrians' safety and smooth flow of traffic. The aim of the study is to determine how an overpass in the city of Trabzon is used by people of different ages, the security and motivation problem, the possibility of a footbridge, and how an overpass is perceived as a crossing over a highway. The study addresses the problem of safety and motivation related to the use of an overpass by people of different ages and is focused on the perception of an overpass as a crossing over a highway. The overpass connects the northern part of the city with the parking lot and the seaside boulevards in Trabzon. It has been constructed over the road no. D010, also known as the Black Sea Coastal Highway. In total, 124 members of the urban population who used the overpass participated in a questionnaire, which included multiple-choice and open-ended questions. The 'Semantic Differential Scale' was also used to evaluate the results. The research revealed that the respondents who chose the road through the overpass for safety in most cases used it relatively rarely (59%). People who used the footbridge every day or frequently marked safety as the reason in 39% of cases only. This means that as pedestrians use the overpass more often, they begin to notice other functional features of the footbridge. The results of the study showed that the overpass was most frequently used for the purpose of going down to the coast (76.0%), although 51.2% reported rarely using the overpass, which showed it was used very little by pedestrians.

Keywords: footbridge; urbanization; functions of pedestrian bridges; Trabzon

Citation: Hełdak, M.; Kurt Konakoglu, S.S.; Kurdoglu, B.C.; Goksal, H.; Przybyła, B.; Kazak, J.K. The Role and Importance of a Footbridge Suspended over a Highway in the Opinion of Its Users—Trabzon (Turkey). *Land* **2021**, *10*, 340. https://doi.org/10.3390/land10040340

Academic Editor: Iwona Cieślak

Received: 5 March 2021
Accepted: 22 March 2021
Published: 27 March 2021

1. Introduction

An urban space as a whole consists of structures and open areas where all urban activities are perceived by the inhabitants of the city. In other words, these spaces are where the activities in the lives of the urban dwellers take place. They include places for housing, working, entertainment, transportation and recreation [1].

According to Hasan and Napiah [2], for centuries, people have tried to walk over environments that contained a variety of terrains such as mountains, hills, valleys and rivers. With the advancement of civilization and urbanization of cities, which resulted in more complexes and civilized environments of habitation, and also with the development of cities having more complex communication systems, attempts were made to solve the problem of pedestrians' safety.

A footbridge was one of the objects that appeared in space along with the development of urbanization. The reason for implementing this type of solution was, firstly, ensuring the safety of pedestrians and, secondly, maintaining uninterrupted vehicle traffic. Therefore, the contemporary definition of a footbridge indicates that it is a vertical separation device used to separate pedestrians from road traffic without risking an accident [3].

Urban inhabitants must use urban spaces in all aspects of their lives. However, only the presence of appropriate and adequate technical, functional, aesthetic and behavioral reinforcement elements can enable people to use urban spaces in the best possible way. Reinforcing elements not only meet the needs of the people but also have psychological effects on them [4].

The most active forms of transportation are walking and bicycling, which have the lowest impact on the environment and improve the physical health of pedestrians and bicyclists. The most common problem impeding the preference for walking and bicycling is traffic safety [5–7].

Pedestrians are most at risk when they are crossing the road. This represents a significant proportion of all fatalities among pedestrians, amounting to 50% in non-built-up areas and 75% in built-up areas. The most frequent reason for this kind of accident is failure to give way [8,9].

Pedestrians who are crossing are considered to be one source of congestion [10]. According to Binti Kadzim, a pedestrian bridge is the best solution for pedestrians to cross the road. A pedestrian bridge is only one way to increase road network capacity [11].

Bridges connect destinations in communities and provide access to emergency and essential services. Bridges that lack pedestrian and bicycle accommodations can force substantial detours or sever routes entirely, discouraging or eliminating the option to walk and bike for transportation. Those who do travel on bridges without proper accommodations may increase their risk of being involved in a crash [12,13].

Pedestrian and bicycle bridges will provide that pedestrian and bicycle users can cross the road safely and go to school or businesses safely. Therefore, where it is needed, the design and the material used in the design are very important. Pedestrian and bicycle bridges should be both "functional" and "aesthetic" [14].

In both urban and landscape design studies, there is a need for a variety of structural elements that meet the physical needs of the population and are necessary for their understanding, safety and comfort in terms of health and a clean environment. These elements are defined as urban elements, and they should be compatible with other elements and improve the visual quality of the landscape design as a whole when used correctly [15]. Among these elements are pedestrian overpasses that provide pedestrian access, i.e., pedestrian bridges. These overpasses or bridges have the especially important goal of providing pedestrians with a means to travel within or between the areas that we call urban spaces. Although the main purpose of pedestrian bridges is to facilitate transport and increase pedestrian safety, this function alone is not enough for users. Bridges must also have an aesthetic value. They should also accommodate different types of activities for pedestrians. This condition determines the value of an urban pedestrian bridge [16–19].

Tunnels and pedestrian bridges should be used as collision-free crossings when [20] pedestrian routes intersect with higher class roads, pedestrian routes intersect with the roads of G (main road) or GP (major trunk road) class with heavy traffic, and the location of an at-grade pedestrian crossing may pose a serious hazard to pedestrians or cause long delays for vehicles and pedestrians.

According to studies conducted by Jamroz et al. [20], Ivan et al. [21], Congiu et al. [22] and Arroya et al. [23], most pedestrians and bicyclists were at risk of accidents. In today's cities, pedestrian zones are being implemented to provide comfortable and safe passage for pedestrians.

Pedestrian overpasses are structures that allow pedestrians to carry out their movements easily and conveniently. For this reason, some of them are built over vehicle traffic roads, some over parks and gardens and some over waterways. Their construction materials can be wood, stone, brick, reinforced concrete, steel or metal, and hybrid solutions are also used, in which the structural elements are made of different materials working together. They can be in the form of beam, arch, cantilever, suspension or truss bridges [24–28].

These structures pass over roads that are open to vehicle traffic and, in addition to their functional purpose of facilitating and directing the continuous flow of uninterrupted pedestrian movement and ensuring pedestrian safety, these passageways function as urban reinforcement elements that, with their aesthetic aspects, can affect the city's appearance. Therefore, the design criteria used for features and fixtures in urban spaces should also be applied to pedestrian overpasses.

Various examples of functional solutions for pedestrian bridges over roads can be listed as well as the attempts of classifying such constructions can be taken up considering their secondary function—apart from the basic one, i.e., ensuring safety for pedestrian, pedestrians and cycling traffic in cities, or providing the possibility of reaching dangerous and attractive landscape spots located in open spaces. In the urbanized area (city) the following can be identified:

-   pedestrian bridges over roads and among them very spectacular solutions such as, e.g., "Ponte Segunda Circular Bridge" in Portugal, the "BP Pedestrian Bridge" in Chicago (USA), the "Lunchtsingel Bridge" in Holland;
-   pedestrian bridges with water elements and over wetlands, e.g., the "Media City Footbridge' in England, the "Cirkelbroen Bridge" in Denmark, the "Iceland Bicycle Bridge" in Iceland, the "Golden Garland Bridge" and "Melkweg Bridge" in Holland and the "Merchant Square Footbridge" and "Millennium Bridge" in London.

The efforts aimed at revitalizing railway bridges and road viaducts into overpasses can be noticed both in cities and in open areas. The examples of using constructions that previously performed different functions and created by conversion of remnants are the "Promenade Plantee" in Paris and in the USA the "High Line Park", "Walkway Over the Hudson" in New York, the "Kinzua Bridge State Park" in Pennsylvania and the "Vance Creek Bridge" in Washington State.

The overpasses built in open areas are often examples of architecturally outstanding constructions, harmoniously integrated into the natural landscape. They frequently constitute impressive structures offering complex architectural solutions, e.g., the "High Trestle Trail Bridge" in Madrid, the "Moses Bridge" in Holland, the "Plitvice Lakes National Park Bridge" in Croatia, the "Zhangjijie Grand Canyon Glass Bridge" in China, the "Langkawi Sky Bridge" in Malaysia, the "Jasper National Park Glacier Skywalk" in Canada and the "Tree Top Walkway" in London.

Pedestrian bridges, especially in heavy traffic areas, were designed as a solution in the context of pedestrian safety and the continuity of the roadways. In addition to being designed engineering structures, their aesthetic aspects can be evaluated as elements of significant urban fixtures. Pedestrian overpasses are important elements that should be considered not only for their functional requirements, but also as elements that can affect the city image and aesthetics. Consequently, pedestrian bridges can be regarded as being important to the city as creative elements that can attract attention due to their larger scale [29,30]. It is worth noting two approaches in the architectural shaping of footbridges (and bridges in general), this first aiming at creating a characteristic element, recognizable and noticeable even from a distance, and the second assuming the maximum integration (concealment) of a new object in the landscape [16,18,31].

The authors of the presented research focused on the problem of safety and motivation for using an overpass by people representing different age groups and on the perception of a footbridge as a crossing over a highway.

The authors noticed that the residents of Turkish cities are reluctant to use overpasses, which results in numerous problems related to road safety (in the case of an overpass construction, traffic lights and pedestrian crossing are not located in the same place, i.e., "a zebra crossing" through the road). There is evidence that some types of facilities are generally disliked by pedestrians, which leads to a high incidence of informal road crossing behavior, away from crossing facilities [32–35].

The research addressing footbridges is focused on

- ensuring safety in crossing roads and passing over other obstacles in pedestrian traffic [10–13,20–23,32–35];
- technology, building materials and durability of the constructed footbridges [24,28],
- user comfort (vibration frequency) [36–40];
- adapting architectural form and structure to the surrounding space [16–19,29–31].

There is a gap in analyzing opinions on the multifunctionality of buildings over highways in cities in terms of assessing a viaduct in the category of an aesthetic, functional and identity type of element, which was the basis for the conducted research. However, the primary function of a footbridge (ensuring safety) cannot be neglected, which constitutes the background of the conducted research.

The following research hypotheses were put forward in the article:

Hypothesis 1 (H1). *Young people and students of different age groups Trabzon residents are more likely to use overpass than older and working people.*

Hypothesis 2 (H2). *Young people pay attention to the advantages of a footbridge other than safety.*

Hypothesis 3 (H3). *The awareness of ensuring safety and motivation for oneself and family grows with age.*

The purpose of the research was to show the preferences of Trabzon residents and visitors in terms of using the overpass built over the highway. The overpass connects the northern part of the city with the parking lot and the seaside boulevards in Trabzon. It has been constructed over the road no. D010, also known as the Black Sea Coastal Highway. This road is the main west-east highway in Turkey, serving the Black Sea Coast. In this part of the city, the road is a barrier to access the Black Sea shore. On the one hand, the city's spatial development and the density of buildings and, on the other hand, the significant traffic of motor vehicles resulted in the need to search for solutions in the form of safe access to coastal boulevards. There are seven footbridges along the section of the route running through the city of Trabzon, whereas only a small part of the road runs through an underground tunnel. The importance of this study is to reveal the design, safety, comfort, aesthetics, symbolic element and importance for transportation of an overpass in Trabzon city.

2. Materials and Methods

The Imperial Overpass, which is located within the boundaries of Ortahisar district of Trabzon city, where area field studies were carried out in the study, constitutes the main material of the study. Observations and questionnaire studies, photographs taken during field studies, SPSS 16.0 statistical program were other materials used in the research.

Imperial Overpass, which connects Gazipasa Street in the city center of Trabzon to the Black Sea Coastal Highway, has been chosen as the study area (Figure 1). The Black Sea Coastal Highway is the main west–east highway in Turkey, serving the Black Sea Coast. The reasons for choosing Imperial Overpass are as follows: it is very close to the city center of Trabzon, it is used intensively by pedestrians, there are bus stops that provide transportation to Trabzon city with other districts, it connects the city center with the coastal area and it provides access to the parking lot of the Imperial Hospital.

Figure 1. Satellite image showing the study area. Source: own study.

Steel was used in the construction of the footbridge, and the floor is of cast concrete. One abutment is located over a section of road in a residential area, while the other is connected to the hospital parking area (Figures 2 and 3).

Figure 2. The view of the overpass over the road no. D010, also known as the Black Sea Coastal Highway, after construction of the roofing. Source: own study.

Figure 3. The view of the overpass over the road no. D010, also known as the Black Sea Coastal Highway, after construction of the roofing. Source: own study.

The presented research was carried out according to the following research scheme:

1. The identification of the research problem, the selection of the research area and the formulation of the research purpose and the research questions;
2. Collecting information and developing database for the analyzed footbridge,
3. Developing a questionnaire (survey) and conducting a survey in the area of Trabzon (Turkey);
4. Analyzing findings using the descriptive method and statistical methods and seeking correlations;
5. Valuation of preferences and expectations using the Semantic Differential Scale.
6. Verification of the research hypotheses, discussion and conclusion.

Some of the issues concerning the footbridge include:

- Technical problems: These include spills on the bridge caused by the steel material, paint problems, lack of binding materials and damage caused by the users, as well as whether the bridge is not convenient for everyone to use.
- Aesthetic problems: There were no design criteria for the bridge that could add value to the space. It is simply an overpass made of a steel carriers and concrete labs.

Taking into account the overpass location along the frequently used pedestrian route and also in an important cultural and landscape place, i.e., in the center of Trabzon city, the research was carried out to identify the expectations and motives of those using the aforementioned footbridge over the road. The popular and busy communication route (highway no. D010) definitely cut off the coast and seaside boulevards from the city, and the overpass under study allows reaching these places without collisions. However, does it meet the residents' expectations? Does its form correspond to the significance of the space in which it is located? In order to obtain answers to a number of doubts and to find out the preferences of the overpass users, the following research questions were formulated:

- Are young people more likely to use the overpass than older and working people?
- Is the crossing over the highway considered safe for all users of the overpass?
- Does the awareness of ensuring safety for oneself and family grow with age?
- How do users rate the overpass in the categories of aesthetic, functional and identity-oriented elements?

In the study, the quantitative research (questionnaire) method was applied to inhabitants of Trabzon in order to determine the current usage status of the Imperial Overpass and to determine the feasibility of pedestrian and bicycle bridges in the city. For the study, a questionnaire-based field survey was administered and face-to-face interviews were carried out to determine the reasons the people of Trabzon city used the overpass and how frequently they visited it (Supplementary Material). The purpose of the quantitative research method was to qualify the numerical information. Before the questionnaire questions were prepared, on-site observations were made by going to the study area. The studies of Arslan Selcuk and Er Akan [1], Hasan and Napiah [2], Kuşkun [4], Mumcu [41] and Denli [42] were consulted during the preparation of the questionnaire. The urban population of 2019 was taken into account when determining the number of people to be surveyed. According to the data obtained from Turkish Statistical Institute, the population of Trabzon Central District in 2019 was 328,457. During the preparation of the questionnaire, care was taken to ask clear and understandable questions that could be answered, while still respecting people's privacy. The questionnaires were designed for individuals over the age of 18, with the idea that they would take a more conscientious approach. The questionnaire was carried out between March and August 2019, with 124 people living in the city center of Trabzon. The random sampling method was used to determine the sampling size for the questionnaire study. The inhabitants who participated in the questionnaire were asked a total of 12 multiple-choice questions and also provided information about their demographic characteristics. Each questionnaire took an average of 20 min.

The results of the demographic composition of the 124 people who participated in the survey were as follows:

Gender: 78 (60.5%) were female and 46 (35.7%) male.

Age: 45 (34.9%) were aged 18–23 years, 49 (38.0%) were aged 24–29 years and 30 (23.3%) were aged 30+ years. The average age of the respondents was 28.0 years of age, with a standard deviation of 9.85 years. At least half of the respondents were not older than 25. The youngest respondent was 18 years old, whereas the oldest one 63.

Educational status: 4 (3.1%) had completed primary school, 1 (0.8%) middle school, 18 (14.0%) high school, 84 (65.1%) university and 17 (13.2%) post-graduate education.

Occupational status: 27 (20.9%) were architects/engineers, 50 (38.8%) were students, and 47 (36.4%) were in other occupations.

The survey was conducted with the participation of 124 individuals, of whom 50 were randomly selected, 27 were architects/engineers and 47 were from other professions. In addition, in order to determine the value of the overpass in terms of semantic aspects, the contexts of using the overpass other than their preferences and expectations were detected in accordance with adjective pairs using the Semantic Differential Scale. In this respect, the participants were asked to evaluate the overpass using the adjectives "functional", "harmonious", "aesthetic", "attractive", "special", "memorable" and "safe" and their paired counterparts using a 7-point scale. The scale given in Table 1 below was used for scoring. The participants were people living in Trabzon who used the overpass. Therefore, assessments were facilitated by reminder photographs used during the survey (Figure 4). The data were analyzed using the SPSS 16.0 statistical program. In the study, correlation analysis was performed with descriptive statistics.

Table 1. Evaluation system of the criteria used in the survey.

Criteria	Intensity of the scale:							Criteria
	1	2	3	4	5	6	7	
Dysfunctional	1	2	3	4	5	6	7	Functional
Irregular	1	2	3	4	5	6	7	Regular
Not aesthetic	1	2	3	4	5	6	7	Aesthetic
Not attractive	1	2	3	4	5	6	7	Attractive
Ordinary	1	2	3	4	5	6	7	Private
Forgettable	1	2	3	4	5	6	7	Unforgettable
Insecure	1	2	3	4	5	6	7	Secure

1: dysfunctional; 2: somewhat dysfunctional; 3: minimally dysfunctional, 4: average; 5: minimally functional; 6: somewhat functional; 7: functional. Source: own study.

Figure 4. Photographs used in the survey.

The significance of differences in the distribution of responses between nominal variables was checked using the chi square test of independence. To calculate the chi-square statistics, the following formula was used:

$$\chi^2 = \sum_{i=1}^{n} \frac{(O_i - E_i)^2}{E_i}$$

(1)

where:

O—Observed value;

E—Expected value.

For all analyses, the maximum permissible error class I $\alpha = 0.05$ was adopted, and $p \leq 0.05$ was considered statistically significant.

3. Results and Discussion

Beyond any doubt, the Black Sea route remains in conflict with the development of the city tourist function. Urbanization processes resulted in the need to search for solutions allowing the separation of pedestrians and vehicle traffic. Unfortunately, the analyzed object differs from the contemporary design solutions, which often fit into the surrounding landscape, and their implementation uses modern engineering techniques and construction materials. Frequently, such objects make the landscape more attractive, which cannot be said about the analyzed footbridge.

When asked about the frequency of using the overpass, out of the total of 124 people, 51.2% answered "Rarely", 14.0% "1–2 times per month", 10.1% "Everyday" and "Frequently", 7.0% "1–2 times per week", 3.9% "2–3 times per week".

The overpass does not connect university buildings or any other buildings used primarily by people under 25, it is generally accessible and apart from the car park and it also provides access to the seaside boulevards. Hence, it was reasonable to analyze who the main users of the overpass are and how frequently they use it.

Correlation analysis using Spearman's rank correlation coefficient confirmed a statistically significant relationship between age and the frequency of using overpass ($p = 0.005$); however, this relationship was different from that formulated in the hypothesis. Older respondents used overpass more frequently; therefore, the discussed hypothesis should be rejected (Table 2).

Table 2. The values of Spearman's rank correlation coefficient. The relationship between age and the frequency of using overpass.

Statistical Measures	Frequency of Use
Correlation Coefficient	0.25
Sig. (2-tailed)	0.005

Source: own study.

A statistical analysis was also carried out using the chi square test of independence. To this end, the respondents were divided into two groups, where the division point was the median value for the age variable (25 years of age). Some categories related to frequencies of using overpass were also combined, so that the smallest percentage of cells in the developed cross-table had an expected number of less than 5. The respondents under 25 years of age rarely used the overpass (64%), and the percentage of such population aged over 25 was smaller and amounted to 40%. The percentage of people using the overpass often and daily in the group under or at the age of 25 was 12%, whereas among older people it was 32%. The differences are statistically significant ($p = 0.027$) (Table 3).

Table 3. The distribution of responses broken down by age.

Frequency of Using	Score	Age:	
		≤25 Years	>25 Years
Rarely	Count	43	23
	% within Age	64.2%	40.4%
1–2 times per month	Count	9	9
	% within Age	13.4%	15.8%
Few times per week	Count	7	7
	% within Age	10.4%	12.3%
Frequently or every day	Count	8	18
	% within Age	11.9%	31.6%
Pearson Chi-Square		$\chi^2 = 9.16$; *Sig.* $= 0.027$	

Source: own study.

The statistical analysis using the chi square test of independence showed a statistically significant relationship between social status and the frequency of using the overpass ($p = 0.018$). The answer often or daily was indicated by 7% of the respondents from the group of engineers, 16% by students, most often by other respondents—34% (Table 4).

Table 4. The distribution of responses broken down by social status.

Frequency of Use	Score	Social Status (Job):		
		Engineer-Architect	Student	Other
Rarely	Count	12	31	23
	% within Social status	44.4%	62.0%	48.9%
1–2 times per month	Count	8	6	4
	% within Social status	29.6%	12.0%	8.5%
Fewtimes per week	Count	5	5	4
	% within Social status	18.5%	10.0%	8.5%
Frequently or every day	Count	2	8	16
	% within Social status	7.4%	16.0%	34.0%
Pearson Chi-Square		$\chi^2 = 15.34$; *Sig.* $= 0.018$		

Source: own study.

When asked about the direction of moving through the overpass—in the first place, 76.0% of the respondents answered "To go to the coast", followed by 9.3% in second place with "To go to the city center", 7.8% in the third place with "To go home", 2.3% "To go to the hospital" and, finally, 0.8% responded "To go to the university" (Figure 5).

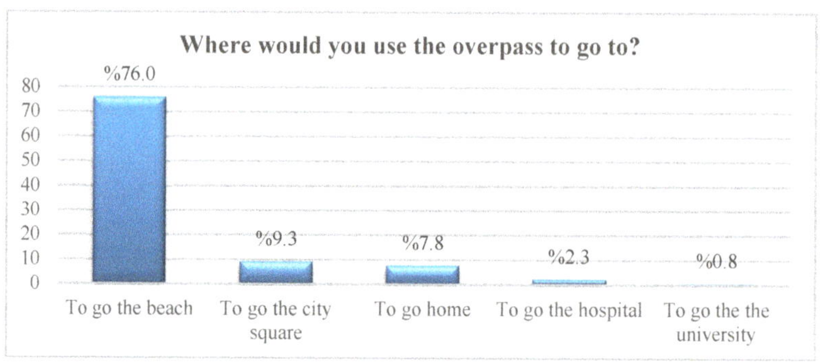

Figure 5. Reasons for overpass use.

The highway limits the possibility for taking advantage of using the coastal location of the city in many respects. First of all, by cutting off the coast and the seaside boulevards from the invested part of the city, it reduces opportunities for the development of recreation and tourism in such a large city of 240,000 residents.

As has been confirmed by the conducted research, the primary purpose of using the overpass is to ensure safety—out of the total number of 124 participants, 36.4% provided the answer "For security", 27.9% "There is no alternative", 20.2% "It's a convenient route", 10.1% "It's the way to the coast" and 1.6% "It's the way to the parking lot" (Figure 6).

Ensuring safety on the one hand and smooth traffic flow of motor vehicles on the other obviously remains the essential purpose of constructing overpasses. However, not all of the respondents indicated safety as the reason for choosing the discussed footbridge, other reasons were as follows: no alternative, convenience and the way to reach the destination (coast, parking lot). Indeed, the traffic of motor vehicles on the highway no. D010 is so intense that virtually all along its length in Trabzon, it is required to use either an overpass or an underpass to get to the other side.

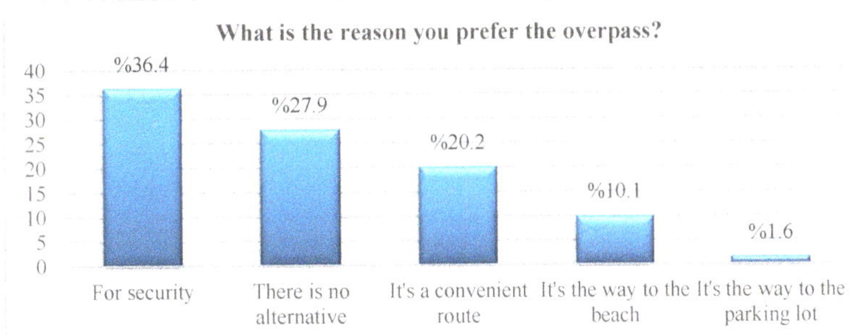

Figure 6. Reasons for preferring the overpass.

It is important that such a possibility of reaching the sea is also attractive for the pedestrians in every respect (apart from safety, the equally important factors are comfort, harmony with the surroundings and an attractive architectural form of the facility).

The respondents who chose the way through an overpass for safety reasons generally used it rarely (59%), and the percentage of such people among those selecting an overpass for another reason was smaller and amounted to 31%. Such individuals often used an overpass daily or often (39%). The analysis using the chi square test of independence showed that the discussed differences are statistically significant ($p = 0.019$) (Table 5).

Table 5. Reasons for choosing vs. frequency of using an overpass.

Frequency of Use	Score	Reasons	
		For Security	For Security
Rarely	Count	58	8
	% within Reasons	59.20%	30.80%
1-2 times per month	Count	12	6
	% within Reasons	12.20%	23.10%
Few times per week	Count	12	2
	% within Reasons	12.20%	7.70%
Frequently or every day	Count	16	10
	% within Reasons	16.30%	38.50%
Pearson Chi-Square		$\chi^2 = 9.96$; *Sig.* $= 0.019$	

Source: own study.

The statistical analysis did not show any significant correlation between the respondent's age and the choice of safety as the reason for traveling through an overpass (Table 6).

Table 6. Reasons for the choice vs. age.

Reasons	Score	Age	
		≤25 Years	>25 Years
For security	Count	54	44
	% within Age	80.6%	77.2%
Other	Count	13	13
	% within Age	19.4%	22.8%
Pearson Chi-Square		$\chi^2 = 0.22$; *Sig.* $= 0.643$	

Source: own study.

The role of pedestrian bridges is not being fully implemented, because according to Abojaradeh, in Jordan, over 60% of pedestrians choose not to use pedestrian bridges for various reasons [36]. Females use them more than males and children more than adults. In that case, such a correlation was not confirmed.

The main advantage of pedestrian footbridges is that they separate pedestrians from road traffic. As a result, footbridges (and stopping pedestrians from crossing the roadway at-grade) may reduce pedestrian accidents up to 90% [8]. It was concluded that pedestrian bridges have a positive impact and have great potential to reduce the number of pedestrian fatalities [42]. When asked the question, "Do you consider the footbridge over the road no. D010 safe?", as many as 35.5% of the respondents answered "No", however, their vast majority consider it safe (64.5%) (Figure 7).

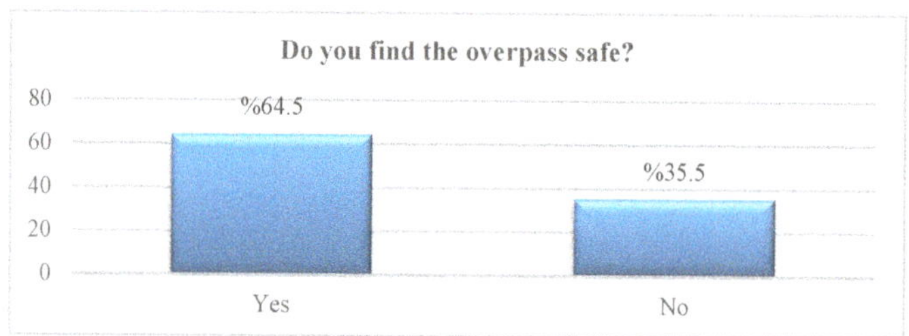

Figure 7. Reasons for preferring the overpass.

It is worth analyzing the reasons for the feeling of insecurity expressed by some of the people participating in the study. The preference for particular facilities also depends on their design and maintenance, which are associated with perceptions about crime and concerns about aesthetics and hygiene [33,35,43,44]. After performing the survey research, the overpass was covered with roofing, which definitely influenced the safety of using it (Figure 8). Unfortunately, the crossing is still not accessible, e.g., for people moving in wheelchairs, and causes many difficulties for those who have problems with mobility in general.

Figure 8. The overpass suspended over the highway no. D010 after constructing the roofing.

Taking an overpass to cross the road may be perceived as attractive or not, and the footbridge itself, apart from ensuring a safe crossing over a communication artery, can also be used as a viewing point for walking, cycling or observing the surroundings. Building a pedestrian bridge over a highway allows for other activities. When asked, "How would you like to spend time in an area without vehicle traffic?", out of 124 participants, 34.9% responded "Walking", 24.0% "I do not know", 15.5% "Riding a bicycle" and 10.9% "Sitting/Observing/Having Fun" (Figure 9).

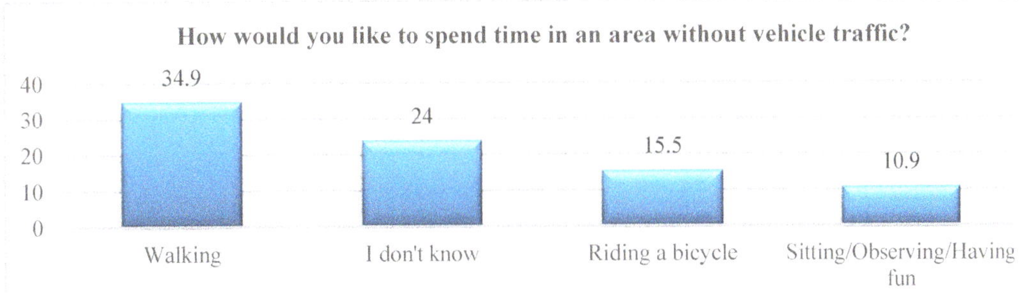

Figure 9. Activities that can be carried out on the overpass.

Taking into account the applied research methodology and the evaluation system of criteria used in the survey, presented in Table 1, the following research results were obtained. The average value of each criterion was calculated according to the results of the survey. In order to determine whether there is a relationship between the criteria, the average values given by the 124 people participating in the questionnaire for each criterion are averaged and summarized in Table 7.

Table 7. Average values of survey criteria.

Criteria	Min.	Max.	Average	Std. Deviation
Functional	1.00	7.00	5.3387	1.50277
Regular	1.00	7.00	4.1774	1.93827
Aesthetic	1.00	7.00	3.1532	1.69190
Attractive	1.00	7.00	3.2339	1.59322
Private	1.00	7.00	3.1774	1.62825
Unforgettable	1.00	7.00	3.6210	1.76503
Secure	1.00	7.00	4.3065	2.02094

Source: own study.

The existence of a relationship among the criteria of "functional", "harmonious", "aesthetic", "attractive", "special", "memorable", and "safe" was assessed by applying the Spearman correlation test. According to the Spearman correlation, a significant positive correlation was found among all the criteria ($p < 0.01$). Among the criteria, the highest correlation coefficient values were found for "functional" with "harmonious", "harmonious" with "aesthetic", "harmonious" with "safe", "aesthetic" with "attractive", "attractive" with "special" and "special" with "memorable" (Table 8). The expected result was that all the criteria were interdependent because they all supported each other.

Table 8. Relationship values among the criteria.

Relationship Criteria		Functional	Regular	Aesthetic	Attractive	Private	Unforgettable	Secure
Functional	Cor. coe.	1.000	0.569″	0.457″	0.403″	0.279″	0.303″	0.550″
	Sig. level	-	0.000	0.000	0.000	0.000	0.000	0.000
Regular	Cor.coe.	0.569″	1.000	0.613″	0.502″	0.447″	0.324″	0.574″
	Sig. level	0.000	-	0.000	0.000	0.000	0.000	0.000
Aesthetic	Cor.coe.	0.457″	0.613″	1.000	0.825″	0.633″	0.573″	0.522″
	Sig. level	0.000	0.000	-	0.000	0.000	0.000	0.000
Attractive	Cor. coe.	0.403″	0.502″	0.825″	1.000	0.670″	0.576″	0.503″
	Sig. level	0.000	0.000	0.000	-	0.000	0.000	0.000
Private	Cor. coe.	0.279″	0.447″	0.633″	0.670″	1.000	0.721″	0.424″
	Sig. level	0.000	0.000	0.000	0.000	-	0.000	0.000
Unforgettable	Cor.coe.	0.303″	0.324″	0.573″	0.576″	0.721	1.000	0.464″
	Sig. level	0.000	0.000	0.000	0.000	0.000	-	0.000
Secure	Cor.coe.	0.550″	0.574″	0.522″	0.503″	0.424″	0.464″	1.000
	Sig. level	0.000	0.000	0.000	0.000	0.000	0.000	-

Sig. level: Significance level; Cor. Coe.: Correlation coefficient; Source: own study.

The results of this study have emphasized that this overpass could carry a semantic and symbolic element in addition to the binding element. In parallel with the positive examples abroad, the overpass should be re-evaluated within a correct concept, format and scenario for its environment.

The positive adjectives that we used are a powerful semantic, syntactic and pragmatic fiction for the continuing research that needs to be done towards an overpass that can serve as an aesthetic, functional and identity element. The overpass that is the subject of this study must be recreated and, by using a correct design approach, a new pedestrian bridge with an identity must be generated. From a study of pedestrian bridge crossing conducted in Bangladesh (Dhaka), it was found that 71% of pedestrians prefer an underpass rather than using foot over bridge. Research conducted in Poland also confirmed a greater tendency to use underpasses by pedestrians compared to footbridges [9]. Always, however, one should take into account local conditions, which impose the only correct solution (in the presented case, the topography is decisive). The reasons pedestrians do not use road crossing facilities are insufficient security, the fact that they are time-consuming, poor entrances, hawker's problem, discomfort, the long walk required, etc. [10].

First of all, a good spatial composition should be prepared, and spatial components and items should be brought together in a harmonious way [41]. The composition of this architecture is formed by bringing the elements together and deliberately arranging them to create a semantically, functionally and visually satisfactory whole [45].

The space composition can be defined by the following semiotics to design a balanced pedestrian bridge. These;

Semantic → Concepts
Syntactic → Shape
Pragmatics → Scenario

In order to design a balanced pedestrian bridge, the spatial composition can be defined by following the semantics. Semantics is actually the science of meaning [30]. When we consider it in a project, this term shows the semantic value, i.e., the concept, of the resulting design. In a way, it contains the main idea of the project. A concept is determined in the context of the semantic fiction, and the project is shaped by it.

Syntactic fiction contains a sense of meaning [46]. We can think of this form of syntax in design. It can be used as a form in the concept, or it can be reflected in the appropriate activities within the concept. Pragmatics is a meaningful phenomenon [46]. We can think of pragmatic fiction in design as a set of activities and a scenario showing a sequence of these activities. A space is not created solely by its existence. The activities within the space and those who use it literally exist. For this reason, the more activities provided in a space, the more open it is to use.

The research confirmed that the presented footbridge is absolutely necessary in accessing the Black Sea coast. In the near future, it is necessary to consider the construction of a new footbridge having a modified architectural form, because apart from its basic function, i.e., safety in pedestrian traffic, it should also meet an aesthetic function and confirm the cultural identity of the city. It should be considered to construct a footbridge of the 21st century using modern technologies and taking into account the tradition of the location.

4. Conclusions

The conducted research allowed answering the formulated research questions and putting forward the following conclusions:

1. The research was carried out in the city of Trabzon (Turkey) and covered the facility built in connection with the development of urbanization and as a result of more intense traffic on the transit road running along the Black Sea. It is possible to reach the seashore in the city mainly by crossing one of the seven footbridges over the Black Sea route. The footbridge that was the research subject, like others in this city, was built to ensure pedestrian safety and maintain the flow of transit traffic; however, modern architectural solutions should also take into account other functions of such construction, i.e., those expected by its users and also the ones discussed in the presented research.

2. The research revealed that the respondents under 25 used the overpass rarely (64%); the percentage of such population over 25 was lower and amounted to 40%. The percentage of people who frequently and daily used the overpass in the group under or 25 years of age was 12%, whereas among older adults, it was 32%. The differences are statistically significant ($p = 0.027$).

3. The respondents who chose the overpass for safety reasons usually used it rarely (59%). Only 39% of those who used the overpass every day or often selected safety as the reason. This means that as a pedestrian uses the overpass more frequently, he or she begins to notice other features of the overpass. The construction should also constitute an interesting element in terms of aesthetics or architectural solutions and harmonize with the environment.

4. Unfortunately, not all the respondents consider the pedestrian bridge over D010 safe. As many as 35.5% claimed it is not safe. It was therefore necessary to find a way to increase its safety. Recently, roofing has been installed along the entire length of the overpass, which has certainly increased its safety and protection against adverse weather conditions and noise, while at the same time reducing the incidents of throwing objects onto the highway.

5. Attracting public attention to the functionality and safety of using the overpass contributed, to some extent, to its modernization (roofing). However, it still leaves a lot to be desired in terms of limitations in pedestrian traffic (e.g., no elevator), architectural form and other functions it could perform in this environment.

6. When the Semantic Differential Scale was applied, the overpass did not receive a high positive score for the seven opposite adjectives. The overpass was found to be moderately "harmonious" and "safe", while very few found it "ugly", "unattractive", "ordinary" or "memorable". The "functional" score of 5.3 was considered as only "minimally functional". This situation shows that both the survey and the Semantic Differential Scale revealed the fact that this structure held no meaning.

7. When we think about the pedestrian bridge we will design, we need to create an activity sequence. First, the main task is the binding function, which must be provided in the best and safest way. The bridge should then include seating and rest areas. In addition, an area for observation and photography can be created on the bridge to encourage people in these activities. Seating areas can also be provided in this observation area. This makes it inviting to sit and gaze at the sea view and watch the

movements of the sun. The semantic and functional aspects of Trabzon's largest and most crowded city square and Gazipaşa Avenue, an important street whose former value has been lost, are connected to the coast by this overpass.

8. The authors believe that the city authorities should aim at constructing a representative overpass, connecting the old part of the city with the Black Sea coast, which could also be a showcase of the city inhabited by 240,000 people and apart from the issue of safety could become a permanent element of its landscape as an attractive architectural object and a viewing point.

When the results of the study are evaluated in general, it is seen that the Imperial Overpass is safe, aesthetic and functional for Trabzon city and is used by different age groups to reach the coast. The results obtained through the questionnaires and observations confirm the hypotheses of the study. New activities can be created, and a new function can be given to the overpass in order to ensure that the footbridge can be used by everyone. Thus, the overpass gains a symbolic meaning for Trabzon city.

Supplementary Materials: The following are available online at https://www.mdpi.com/article/10.3390/land10040340/s1, Questionnaire S1.

Author Contributions: Conceptualization, S.S.K.K., H.G. and M.H.; methodology, H.G., S.S.K.K., M.H.; software, B.C.K.; B.P. and J.K.K.; validation, B.C.K. and B.P.; formal analysis, H.G.; investigation, M.H., S.S.K.K., B.C.K.; resources, B.P., M.H., J.K.K. and S.S.K.K.; data curation, H.G., S.S.K.K.; writing—original draft preparation, S.S.K.K., H.G., B.C.K., M.H. and B.P.; writing—review and editing, M.H., S.S.K.K., H.G., B.P. and J.K.K.; visualization, H.G., S.S.K.K. and M.H.; supervision, B.C.K.; project administration, M.H. and J.K.K.; funding acquisition, J.K.K. All authors have read and agreed to the published version of the manuscript.

Funding: This research was co-funded by the Wrocław University of Environmental and Life Sciences.

Institutional Review Board Statement: Not applicable.

Informed Consent Statement: Not applicable.

Data Availability Statement: Not applicable.

Conflicts of Interest: The authors declare no conflict of interest.

Limitations of the Study: One of the research aspects was to reveal a slightly different perception of the reality by the co-authors. The authors from Poland assess the current technical, functional and aesthetic condition of the footbridge more strictly, whereas the authors from Turkey are more focused on its reconstruction and modernization, approaching the footbridge as a permanent element of the city landscape. The research conducted in the footbridge area was accompanied by car noise, which made it difficult to interview the footbridge users.

References

1. Arslan Selçuk, S.; Er Akan, A. Bir Şehir İkonu Olma Yolunda Yaya Üst Geçitlerive Aydınlatmaları: ODTÜ Yaya ÜstGeçidiÖrneği. III. Ulusal Aydınlatma Sempozyumuve Sergisi Bildirileri Kitabı, Ankara, Turkey, 23–25 November 2005; pp. 37–43. Available online: http://www.emo.org.tr/ekler/b2b4182156b2f1f_ek.pdf (accessed on 1 August 2019). (In Turkish).
2. Hasan, R.; Napiah, M. Utilization of footbridges: Influential factors and improvement proposals. *Adv. Transp. Stud. Int. J. Sect. A* **2017**, *2017*, 43.
3. Ribbens, H. Pedestrian facilities in South Africa: Research and practice. *Transp. Res. Rec. J. Transp. Res. Board* **1996**, *1538*, 10–18. [CrossRef]
4. Kuşkun, P.A. Research on the Use of Outdoor Furniture in the Case of Erzurum. Erzurum Kent Bütününde Donatı Elemanlarının Kullanımı Üzerine Bir Araştırma. Master's Thesis, Institute of Natural and Applied Sciences, Atatürk University, Erzurum, Turkey, 2002. (In Turkish).
5. Keyvanfar, A.; Ferwati, M.S.; Shafaghat, A.; Lamit, H.A. Path Walkability Assessment Index Model for Evaluating and Facilitating Retail Walking Using Decision-Tree-Making (DTM) Method. *Sustainability* **2018**, *10*, 1035. [CrossRef]
6. Rahman, M.S.; Abdel-Aty, M.; Hasan, S.; Cai, Q. Applying Data Mining Techniques to Analyze the Pedestrian and Bicycle Crashes at the Macroscopic Level. In Proceedings of the Transportation Research Board 98th Annual Meeting, Washington, DC, USA, 13–17 October 2019; Available online: https://trid.trb.org/view/1572936 (accessed on 27 July 2019).
7. Shaaban, K. Assessing Sidewalk and Corridor Walkability in Developing Countries. *Sustainability* **2019**, *11*, 3865. [CrossRef]

8.	Miśkiewicz, M.; Pyrzowski, Ł.; Okraszewska, R. Pedestrian and Bicycle Bridges as Examples of Safe Collision-Free Road Crossings. *MATEC Web Conf.* **2017**, *122*, 01005. Available online: https://www.matec-conferences.org/articles/matecconf/pdf/2017/36/matecconf_gambit2017_01005.pdf (accessed on 13 July 2019). [CrossRef]
9.	Bartoszewski, J.; Lessear, S. *Tunele i Przejścia Podziemne w Miastach [Tunnels and Underground Passages in Cities]*; WKŁ: Warszawa, Poland, 1979. (In Polish)
10.	Nukta Ramadani, H.; Rahmani, H.; Gazali, A.A. Study of efficiency pedestrian bridge crossing in the road of Pangerang Antasari, Banjarmasin. In Proceedings of the MATEC Web of Conferences 181, Makassar, Indonesia, 4 November 2017; Available online: https://www.matec-conferences.org/articles/matecconf/pdf/2018/40/matecconf_istsdc2017_06009.pdf (accessed on 13 July 2019). [CrossRef]
11.	Binti Kadzim, H.N. Study on Effectiveness of Pedestrian Bridge Utilization. Bachelor's Thesis, Faculty of Civil Engineering & Earth Resources University Malaysia Pahang, Pahang, Malaysia, 2012. Available online: http://umpir.ump.edu.my/id/eprint/14470/1/FKASA%20-%20NURUL%20HIDAYAH%20KADZIM.PDF (accessed on 4 August 2019).
12.	Cohn, J.; Sperling, E. Improving Pedestrian and bicycle Connectivity During Rehabilitation of Existing Bridges. In *Transportation Research Board*; Pedestrian and Bicycle Information Center: Chapel Hill, NC, USA, 2016; pp. 1–13. Available online: http://www.pedbikeinfo.org/cms/downloads/PBIC_WhitePaper_Bridges.pdf (accessed on 27 July 2019).
13.	Biliszczuk, J.; Onysyk, J.; Prabucki, P.; Toczkiewicz, R. Kładki jako elementy infrastruktury zwiększające bezpieczeństwo pieszych [Footbridges as infrastructure elements increasing pedestrian safety]. *Inżynieria I Bud.* **2020**, *6*, 269–273. (In Polish)
14.	North Seattle College. The Seattle Department of Transportation Northgate Pedestrian and Bicycle Bridge Alternative Development and Selection. Available online: https://www.seattle.gov/Documents/Departments/SDOT/BridgeStairsProgram/AlternativeDevelopmentandSelectionFINAL.pdf (accessed on 5 August 2019).
15.	Altınçekiç, H.; Koç, H. Peyzaj Tasarımında Kent Mobilyalarıve Kalite Beklentileri. II. In Proceedings of the Uluslararası Kent Mobilyaları Sempozyumu, İstanbul, Turkey, 24–27 April 2003. (In Turkish).
16.	Biliszczuk, J.; Machelski, C.; Onysyk, J.; Prabucki, P.; Węgrzyniak, M. Kładki dla pieszych jako punkty orientacyjne na autostradzie [Pedestrian bridges as landmarks on the highway]. *Inżynieria I Bud.* **2001**, *11*, 643–648. (In Polish)
17.	Eyre, J. Aesthetics of Footbridge Design. In Proceedings of the International Conference on the Design and Dynamic Behaviour of Footbridges, Paris, France, 20–22 November 2002; pp. 96–103.
18.	Idelberger, K. *The World of Footbridges: From the Utilitarian to the Spectacular*; Ernst & Sohn: Berlin, Germany, 2011.
19.	Flaga, K.; Januszkiewicz, K. On the aesthetic and technical efficiency of current arched footbridges. In Proceedings of the International Conference on the Design and Dynamic Behaviour of Footbridges, Wrocław, Poland, 6–8 July 2011; pp. 126–137.
20.	Jamroz, K.; Gaca, S.; Michalski, L.; Kieć, M.; Budzyński, M.; Gumińska, L.; Kustra, W.; Mackun, T.; Oskarbska, L.; Rychlewska, J.; et al. Ochrona Pieszych. Podręcznik dla organizatorów ruchu pieszego. In *Protection of Pedestrians. Guidelines for Pedestrian Traffic Organizer*; National Road Safety Council: Gdansk, Poland; Warsaw, Poland; Cracow, Poland, 2014. (In Polish)
21.	Ivan, K.; Benedek, J.; Ciobanu, S.M. School-Aged Pedestrian-Vehicle Crash Vulnerability. *Sustainability* **2019**, *11*, 1214. [CrossRef]
22.	Congiu, T.; Sotgiu, G.; Castiglia, P.; Azara, A.; Piana, A.; Saderi, L.; Dettori, M. Built Environment Features and Pedestrian Accidents: An Italian Retrospective Study. *Sustainability* **2019**, *11*, 1064. [CrossRef]
23.	Arroya, R.; Mars, L.; Ruiz, T. Perception of Pedestrian and Cyclist Environments, Travel Behaviors, and Social Networks. *Sustainability* **2018**, *10*, 3241. [CrossRef]
24.	Gimsing, N. *Cable Supported Bridges—Concept and Design*; John Wiley and Sons: New York, NY, USA, 1997.
25.	Flaga, A. *Mosty Dla Pieszych [Pedestrian Bridges]*; Wydawnictwa Komunikacji i Łączności: Warszawa, Poland, 2011. (In Polish)
26.	Keil, A. *Pedestrian Bridges: Ramps, Walkways, Structures*; Walter de Gruyter: Berlin, Germany, 2013.
27.	Kasuga, A. Footbridge construction—A challenge for innovation and sustainability. In Proceedings of the International Conference on the Design and Dynamic Behaviour of Footbridges, London, UK, 16–18 July 2014; pp. 86–95.
28.	Chen, W.; Duan, L. *Handbook of International Bridge Engineering*; Tylor & Francis Group: New York, NY, USA, 2014.
29.	Aksu, Ö. Yaya Üst Geçitlerinde Tasarım Ölçütlerinin İrdelenmesi. *İstanbul Üniversitesi Orman Fakültesi Derg.* **2014**, *64*, 12–28. [CrossRef]
30.	Salamak, M.; Fross, K. Bridges in Urban Planning and Architectural Culture. *Procedia Eng.* **2016**, *161*, 207–212. [CrossRef]
31.	Dietrich, R. *Faszination Brücken*; Callwey: München, Germany, 1998.
32.	Demiroz, Y.I.; Onelcin, P.; Alver, Y. Illegal road crossing behavior of pedestrians at overpass locations: Factors affecting gap acceptance, crossing times and overpass use. *Accid. Anal. Prev.* **2015**, *80*, 220–228. [CrossRef]
33.	Sinclair, M.; Zuidgeest, M. Investigations into pedestrian crossing choices on Cape Town freeways. *Transp. Res. F* **2016**, *42*, 479–494. [CrossRef]
34.	Obeng-Atuah, D.; Poku-Boansi, M.; Cobbinah, P.B. Pedestrian crossing in urban Ghana: Safety implications. *J. Transp. Health* **2017**, *5*, 55–69. [CrossRef]
35.	Anciaes, P.A.; Jones, P. Estimating preferences for different types of pedestrian crossing facilities. *Transp. Res. Part F: Traffic Psychol. Behav.* **2018**, *52*, 222–237. [CrossRef]
36.	Banaś, A. Ocena komfortu pieszego na kładkach. Case study [Assessment of comfort on the footbridges. Case study]. *Builder* **2020**, *8*, 277. [CrossRef]
37.	Cross, E.J.; Koo, K.Y.; Brownjohn, J.M.W.; Worden, K. Long-term monitoring and data analysis of the Tamar Bridge. *Mech. Syst. Signal Process.* **2013**, *35*, 16–34. [CrossRef]

38. Lievens, K.; Lombaert, G.; De Roeck, G.; Van den Broeck, P. Robust design of a TMD for the vibration serviceability of a footbridge. *Eng. Struct.* **2016**, *123*, 408–418. [CrossRef]
39. Wang, D.; Wu, C.; Zhang, Y.; Li, S. Study on vertical vibration control of long-span steel footbridge with tuned mass dampers under pedestrian excitation. *J. Constr. Steel Res.* **2019**, *154*, 84–98. [CrossRef]
40. Matsumoto, Y.; Maeda, S.; Iwane, Y.; Iwata, Y. Factors affecting perception thresholds of vertical whole-body vibration in recumbent subjects: Gender and age of subjects, and vibration duration. *J. Sound Vib.* **2011**, *330*, 1810–1828. [CrossRef]
41. Mumcu, S. Investigation of Place Preferences in Open Spaces in Terms of Spatial Properties: The Case of Trabzon Atapark. AçıkMekânlardakiYerTercihlerinin, MekânsalÖzelliklerAçısındanİncelenmesi: Trabzon Atapark Örneği. Master's Thesis, Institute of Natural and Applied Sciences, Karadeniz Technical University, Trabzon, Turkey, 2002. (In Turkish).
42. Abojaradeh, M. Evaluation of Pedestrian Bridges and Pedestrian Safety in Jordan. *Civil Environ. Res.* **2013**, *3*, 66–78. Available online: https://www.iiste.org/Journals/index.php/CER/article/view/3868/3927 (accessed on 3 August 2019).
43. Sharples, J.M.; Fletcher, J.P. Pedestrian Perceptions of Road Crossing Facilities Scottish Executive Central Research Unit, Edinburgh, United Kingdom. 2001. Available online: http://www.gov.scot/resource/doc/156816/0042173.pdf (accessed on 11 July 2019).
44. James, E.; Millington, A.; Tomlinson, P.; Understanding community severance—Part 1: Views of practitioners and communities. Report to the UK Department for Transport, United Kingdom. 2005. Available online: http://webarchive.nationalarchives.gov.uk/+/http:/www.dft.gov.uk/adobepdf/163944/Understanding_Community_Sev1.pdf (accessed on 11 July 2019).
45. Gür, Ş.Ö. *Mekan Örgütlenmesi*; Trabzon, Turkey, 1996. (In Turkish)
46. Denli, S. The Analysis of Graphical Indication Meanings Regarding Semiology. Göstergebilim Açısından Grafik Gösterge Anlamlarının İncelenmesi. Master's Thesis, Institute of Social Sciences, Atatürk University, Erzurum, Turkey, 1997. (In Turkish).

MDPI
St. Alban-Anlage 66
4052 Basel
Switzerland
Tel. +41 61 683 77 34
Fax +41 61 302 89 18
www.mdpi.com

Land Editorial Office
E-mail: land@mdpi.com
www.mdpi.com/journal/land